Edited by
Uwe Bovensiepen, Hrvoje Petek, and Martin Wolf

Dynamics at Solid State Surfaces and Interfaces

Related Titles

Kolasinski, K. W.

Surface Science

Foundations of Catalysis and Nanoscience

2008

ISBN: 978-0-470-03304-3

Breme, J., Kirkpatrick, C. J., Thull, R. (eds.)

Metallic Biomaterial Interfaces

2008

ISBN: 978-3-527-31860-5

Wetzig, K., Schneider, C. M. (eds.)

Metal Based Thin Films for Electronics

2006

ISBN: 978-3-527-40650-0

Bordo, V. G., Rubahn, H.-G

Optics and Spectroscopy at Surfaces and Interfaces

2005

ISBN: 978-3-527-40560-2

Butt, H.-J., Graf, K., Kappl, M.

Physics and Chemistry of Interfaces

Second, Revised and Enlarged Edition

2006

ISBN: 978-3-527-40629-6

Edited by
Uwe Bovensiepen, Hrvoje Petek,
and Martin Wolf

Dynamics at Solid State Surfaces and Interfaces

Volume 2: Fundamentals

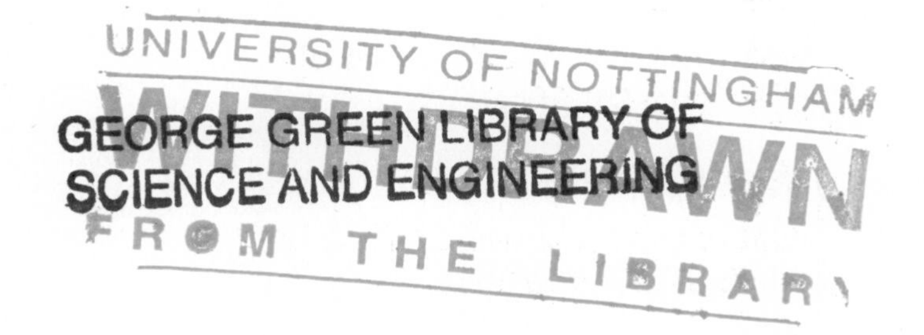

WILEY-
VCH

WILEY-VCH Verlag GmbH & Co. KGaA

The Editors

Prof. Dr. Uwe Bovensiepen
University of Duisburg-Essen
Faculty of Physics
Germany
uwe.bovensiepen@uni-due.de

Prof. Hrvoje Petek
University of Pittsburgh
Department of Physics and Astronomy
USA
petek@pitt.edu

Prof. Dr. Martin Wolf
Fritz-Haber Institute of the Max-Planck Society
Department of Physical Chemistry
Berlin
Germany
wolf@fhi-berlin.mpg.de

Cover
The cover figure depicts (i) a time-resolved experiment at an interface using time-delayed pump and probe femtosecond laser pulses (left) and the detected response (right) being either reflected light or a photoemitted electron. In addition (ii) charge transfer from an excited resonance of an alkali atom to single crystal metal substrate is shown. The false color scale represents the wave packet propagation which was calculated by A. G. Borisov including the many-body response of the metal. The figure was designed and created by A. Winkelmann.

Library of Congress Card No.: applied for

British Library Cataloguing-in-Publication Data
A catalogue record for this book is available from the British Library.

Bibliographic information published by the Deutsche Nationalbibliothek
The Deutsche Nationalbibliothek lists this publication in the Deutsche Nationalbibliografie; detailed bibliographic data are available on the Internet at http://dnb.d-nb.de.

Composition Thomson Digital, Noida, India
Printing and Binding Strauss GmbH, Mörlenbach
Cover Design Adam Design, Weinheim

Printed in the Federal Republic of Germany
Printed on acid-free paper

Print ISBN: 978-3-527-40924-2
ePDF ISBN: 978-3-527-64649-4
oBook ISBN: 978-3-527-64646-3
ePub ISBN: 978-3-527-64648-7

Contents

Preface

The dynamics of elementary processes in solids are decisive for various physical properties of solid state materials and their application in devices. This book intends to provide an introductory and comprehensive overview of the fundamental concepts, techniques and underlying elementary processes in the field of ultrafast dynamics of solid state surfaces and interfaces. While the first volume addresses recent research on quasiparticle dynamics, collective excitations, electron transfer and photoinduced dynamics, the focus of this second volume lies on fundamentals and provides introductory information on elementary processes and light-matter interaction. Our goal is to make these concepts accessible also to non-experts and, in particular, to newcomers and younger researchers in the field of ultrafast dynamics of solids, their interfaces and nanostructured materials. We hope that both volumes will help to further new research directions and developments in this field.

We acknowledge support from our funding agencies, important contributions from our co-workers, stimulating discussions with colleagues and the understanding from our families that were essential to realize this book.

Duisburg, Pittsburgh and Berlin, January 2012

Uwe Bovensiepen,
Hrvoje Petek,
and *Martin Wolf*

List of Contributors

Silke Biermann
École Polytechnique
Centre de Physique Theorique
91128 Palaiseau Cedex
France

Uwe Bovensiepen
Universität Duisburg-Essen
Fakultät für Physik
Lotharstr. 1
47048 Duisburg
Germany

Paweł Buczek
Max-Planck-Institut für
Mikrostrukturphysik
Weinberg 2
06120 Halle
Germany

Evgueni V. Chulkov
Donostia International Physics Center
(DIPC)
Paseo de Manuel Lardizabal 4
20018 San Sebastián/Donostia
Basque Country
Spain

and

Centro de Física de Materiales (CFM)
Departamento de Física de Materiales
Paseo de Manuel Lardizabal 4
20018 San Sebastian/Donostia
Basque Country
Spain

Thomas Fauster
Universität Erlangen-Nürnberg
Lehrstuhl für Festkörperphysik
Staudtstr. 7
91058 Erlangen
Germany

Kunie Ishioka
University of Tsukuba
National Institute for Materials Science
Graduate School of Pure and Applied
Sciences
Advanced Nano Characterization Center
1-2-1 Sengen
Tsukuba 305-0047
Japan

Mackillo Kira
Philipps-Universität
Fachbereich Physik
Renthof 5
35032 Marburg
Germany

Stephan W. Koch
Philipps-Universität
Fachbereich Physik
Renthof 5
35032 Marburg
Germany

Eugen E. Krasovskii
Universität Kiel
Institut für Theoretische Physik und Astrophysik
Leibnizstraße 15
24098 Kiel
Germany

Christoph Lienau
Carl von Ossietzky Universität
Institut für Physik
26129 Oldenburg
Germany

Ricardo Díez Muiño
Donostia International Physics Center (DIPC)
Paseo de Manuel Lardizabal 4
20018 San Sebastián/Donostia
Spain

Luca Perfetti
École Polytechnique
91128 Palaiseau Cedex
France

Hrvoje Petek
University of Pittsburgh
Physics and Astronomy Department
100 Allen Hall
3941 O'Hara Street
Pittsburgh, PA 15260
USA

Jose M. Pitarke
Euskal Herriko Unibertsitatea
Materia Kondentsatuaren Fisika Saila
Zientzi Fakultatea
644 Posta kutxatila
48080 Bilbo
Basque Country
Spain

Daniel Sánchez-Portal
Donostia International Physics Centre (DIPC)
Paseo Manuel Lardizábal 4
20018 San Sebastián
Spain

and

Dep. Física de Materiales (UPV/EHU)
Facultad de Química
Apartado 1072
20080 San Sebastián
Spain

Leonid M. Sandratskii
Max-Planck-Institut für Mikrostrukturphysik
Weinberg 2
06120 Halle
Germany

Jörg Schäfer
Universität Würzburg
Physikalisches Institut
Am Hubland
97074 Würzburg
Germany

Wolfgang Schattke
Universität Kiel
Institut für Theoretische Physik und Astrophysik
Leibnizstraße 15
24098 Kiel
Germany

Irina Sklyadneva
Donostia International Physics Center (DIPC)
Paseo de Manuel Lardizabal 4
20018 San Sebastián/Donostia
Basque Country
Spain

and

Tomsk State University
pr. Lenina 36
634050 Tomsk
Russian Federation

Julia Stähler
Fritz-Haber-Institut der Max-Planck-Gesellschaft
Abteilung für Physikalische Chemie
Faradayweg 4-6
14195 Berlin
Germany

Martin Weinelt
Freie Universität Berlin
Fachbereich Physik
Arnimallee 14
14195 Berlin
Germany

Martin Wolf
Fritz-Haber-Institut der Max-Planck-Gesellschaft
Abteilung für Physikalische Chemie
Faradayweg 4-6
14195 Berlin
Germany

Xiaoyang Zhu
University of Texas at Austin
Department of Chemistry & Biochemistry
1 University Station A5300
Austin, TX 78712
USA

1
The Electronic Structure of Solids

Uwe Bovensiepen, Silke Biermann, and Luca Perfetti

The discussion of dynamics at interfaces is based on the motion of ion cores and electronic excitations that are mostly optically driven. Hence, the electronic structure is of fundamental importance here. In solids such as molecular or ionic crystals, the valence electron distribution is not considerably distorted from the respective isolated atoms, ions, or molecules. Hence, their cohesion is entirely given by the classical potential energy of negligibly deformed electron distributions of bare particles, and van der Waals or Coulomb interactions are responsible for the formation of solid materials. This ceases to be so in metals and covalent crystals because the valence electron distribution plays the decisive role in bonding the constituents to a solid. In turn, the valence electron distribution can be considerably modified from the isolated atom or ion. A general description of solids must, therefore, consider the electronic structure in the first place. Furthermore, the dynamical processes discussed in this book are mostly optically excited or electron mediated.

This chapter introduces the basic concepts widely used in the description of the electronic structure in solid materials. In Section 1.1, we present the description of the nearly free electron approximation that is motivated by optical excitations of a solid following the Drude model. We introduce the Fermi sphere and the dispersion of electronic bands in momentum space. In Section 1.2, the influence of the periodic potential in a crystal is considered, which leads to the description of the electronic band structure by Bloch's theory for delocalized states. There is a considerable variety of materials that is not described by band theory, which originates from electron–electron interaction. In Section 1.3, we introduce Mott insulators that manifest deviations from the band picture. In Section 1.4, we introduce established concepts to describe the electronic structure of materials with strong electron correlations and give examples.

Dynamics at Solid State Surfaces and Interfaces: Volume 2: Fundamentals, First Edition.
Edited by Uwe Bovensiepen, Hrvoje Petek, and Martin Wolf.

1.1 Single-Electron Approximation

Although a solid contains about 10^{23} electrons/cm^3, for a number of materials (but not for all) it is sufficient to neglect the explicit interaction among these particles. In this one-electron approximation, the energy of individual electrons is renormalized to account for the electron–electron interaction, which simplifies the description enormously. These electrons in the material are then termed quasi-particles.

1.1.1 The Drude Model of the Free Electron Gas

We start by considering propagation of electrons in a metal. Such dynamical processes have been essential for Drude's theory of electrical conductivity in metals [1] and will be discussed in detail in Chapter 5. Here, we introduce the concept briefly in order to motivate the description of the electronic structure in solids.

Drude applied the kinetic theory of gases to a metal that is represented by a gas of electrons occupying the interstitial region between the ion cores. Without an electric field $\mathbf{E}$, the current density $\mathbf{j} = -ne\mathbf{v}$ averages to zero because the electrons have no preferential direction to move at velocity $\mathbf{v}$ in between two collisions with scattering centers (which Drude imagined to be the ion cores); here, n represents the electron density and e the elementary charge. In the presence of an electric field, a net current density $\mathbf{j} = -ne\mathbf{v}$ develops because during the time interval $\tau = 1/\Gamma$ between two collisions the electron is accelerated in a preferential direction to $\mathbf{v} = -e\mathbf{E}\tau/m$. The electrical DC conductivity σ_0 is hence proportional to the time τ between two scattering events.

$$\mathbf{j} = \sigma_0 \mathbf{E}; \quad \sigma_0 = \frac{ne^2\tau}{m}. \tag{1.1}$$

Here, m is the electron mass. To estimate the order of magnitude of τ, the measured DC conductivity is taken, for example, for Cu at a temperature $T = 77$ K and one finds $\tau \approx 2.1 \times 10^{-13}$ s or 210 fs being well in the femtosecond regime.

If the electric field is time dependent, the result can be generalized for a frequency ω to

$$\mathbf{j}(\omega) = \sigma(\omega)\mathbf{E}(\omega); \quad \sigma(\omega) = \frac{\sigma_0}{1 - i\omega\tau}. \tag{1.2}$$

Using the wave equation for the electric field $-\nabla^2\mathbf{E} = \omega^2/c^2\varepsilon(\omega)\mathbf{E}$, the complex dielectric function $\varepsilon(\omega)$ is introduced.

$$\varepsilon(\omega) = 1 + \frac{4\pi i\sigma(\omega)}{\omega}. \tag{1.3}$$

With ω_p being the plasma frequency, the dielectric constant according to Drude's theory reads after linearization in $1/\tau$

$$\varepsilon(\omega) = 1 - \frac{\omega_p^2}{\omega^2} + i\frac{\omega_p^2}{\omega^3}\frac{1}{\tau}; \quad \omega_p^2 = \frac{4\pi n e^2}{m}. \tag{1.4}$$

Considering the reflectivity $R = ((1-n)^2 + k^2)/((1+n)^2 + k^2)$ with $n = \mathrm{Re}\sqrt{\varepsilon}$ and $k = \mathrm{Im}\sqrt{\varepsilon}$, an electromagnetic wave cannot penetrate into the bulk of a metal for $\omega < \omega_p$ because it is reflected. For large frequencies $\omega > \omega_p$, it does propagate in a metal. The absorption is proportional to the scattering rate $\Gamma = 1/\tau$ and $1/\omega^3$. The latter is usually referred to as the free carrier response, which will become clear further below. Please note that Drude's considerations are very fundamental throughout this book and will be used with emphasis on different aspects in Chapter 5.

The absorption is determined by $\mathrm{Im}\ \varepsilon = (4\pi/\omega)\mathrm{Re}\ \sigma$ and can be deduced from a measurement of the reflectivity using, for example, the Kramers–Kronig relations. Figure 1.1 shows the experimental results for alkali metals. At low frequency, the pronounced increase in absorption according to the free carrier response (Eq. (1.4)) is readily visible. At higher frequency, the behavior exhibits particular signatures that are characteristic for the respective material. For an understanding of the absorption spectrum in the visible and the ultraviolet spectral range, we consider next the electronic structure in the single (or independent) electron approximation.

A considerable improvement of Drude's theory was achieved by including the Pauli principle and Sommerfeld implemented the Fermi–Dirac velocity distribution

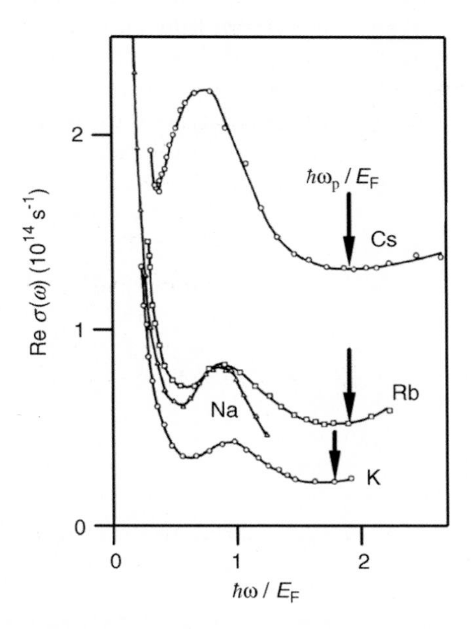

Figure 1.1 Optical absorption spectra of alkali metals (adapted from Ref. [2]). At low frequency the Drude-like increase in absorption originating from the free carrier response is clearly seen. At higher frequency, specific absorption peaks due to interband excitation are found. Arrows indicate the plasma frequency ω_p in units of $E_F/\hbar$ for Cs, Rb, and K.

of electrons leading to the free electron gas. In this description, the electrons do not interact with each other, which is usually termed the independent electron approximation. Electrons populate single-particle states beginning from the smallest energy. The states with different wave vectors $k = (k_x, k_y, k_z)$ in a particular three-dimensional volume $V = L_x \cdot L_y \cdot L_z$ are defined by $k_i = 2\pi n_i / L_i$ with $i = x, y, z$ and n_i being integer numbers. Hence, the electrons populate the states up to a maximum energy E_F, which defines the Fermi sphere in k-space. The momentum $\hbar k_F$ is termed the Fermi momentum and $E_F = \hbar^2 k_F^2 / 2m$ is the Fermi energy.

1.1.2
The Electronic Band Structure: Metals, Insulators, and Semiconductors

Before the interband transitions observed in Figure 1.1 can be understood, the periodic arrangement of ion cores in a crystal lattice must be considered.

Indeed, both the ionic potential and the electronic density follow the periodicity of the solid. It is convenient to consider this periodicity in the description of the electronic structure. For free electrons, it is reasonable to consider that the electron wave vector components k_x, k_y, k_z attain absolute values between zero and infinity in a continuous and uniquely defined manner. For electrons in a periodic potential, however, an electronic wave can experience Bragg reflection. This adds the wave vector of the respective periodicity in the crystal, which is described by a reciprocal lattice vector with an absolute value $G = 2\pi / a$ with a being the lattice constant responsible for the periodic potential, to the electron wave vector. Therefore, the electron wave vector in a periodic crystal is defined by $2\pi/\lambda \pm n \cdot 2\pi/a$, with n being an integer number. Consequently, the electron wave vector in a periodic potential is no longer uniquely defined. It is helpful to transfer a wave vector by the appropriate number of Bragg reflections to its smallest absolute value within the interval $-\pi/a < k < \pi/a$. Taking the lattice symmetries and the respective variation of the lattice constant a along different high-symmetry directions into account, the smallest volume in the three-dimensional wave vector space that contains all available back-folded wave vectors is defined as the first Brillouin zone. In Figure 1.2, the electron band is plotted in the first Brillouin zone along the direction from Γ to N in the bcc lattice of sodium.

In the independent electron approximation, the effects of electron–electron interactions are entirely accounted for by an effective potential $U(\mathbf{r})$, which also respects this periodicity. Within this framework, the electronic wave functions are determined by the time-independent Schrödinger equation with the periodic potential $U(\mathbf{r})$. Vibrations of the ion cores and deviations from a perfect periodicity due to defects, for example, are neglected at this stage. In the case of the nearly free electron approximation, the periodic potential is included as a weak perturbation of the free electron gas.

It turns out that the nearly free electron approximation is quite reasonable for simple metals with predominant *s,p*-like valence electrons. Prominent examples are the alkali metals and aluminum. To assign the peak in the optical absorption spectrum of alkalis observed in Figure 1.1 at $\hbar\omega / E_F \approx 1$, the occurrence of band gaps at the boundary of the Brillouin zone is important. Free electron wave functions, which resemble traveling waves $\psi_k \propto \exp i\mathbf{kr}$, are found in the center of the Brillouin zone in the vicinity of the band bottom. The crystal periodicity offers the possibility for Bragg reflection of electron waves in the vicinity of the Brillouin zone boundary. The respective wave functions are a superposition of waves traveling in opposite directions. The different phase of the resulting standing waves leads to accumulation of electron probability density either in the vicinity of the ion cores, where the potential is deep, or in the interstitial region where the potential is shallow and rather flat. The different energy of these two situations results in an energy gap close to the Brillouin zone boundary. The gap size is determined by the periodic perturbation of the free electron gas potential.

It is a well-known exercise of introductory textbooks to calculate the interband transition threshold for alkali metals in the nearly free electron approximation [3]. The resulting threshold is $0.64\,\hbar\omega/E_\mathrm{F}$, which agrees well with the noticeable rise in absorption at energies above $0.64\,\hbar\omega/E_\mathrm{F}$ reported in Figure 1.1. This energy represents a threshold because for higher optical energies a combination of an occupied and unoccupied state can be found that can be excited by a direct optical transition. Such a direct transition proceeds vertically in the electron dispersion diagram as illustrated in Figure 1.2 and zero momentum change occurs concomitantly with the transition. Energy and momentum are also conserved for indirect excitations as depicted also in Figure 1.2 for an excitation where the initial and final

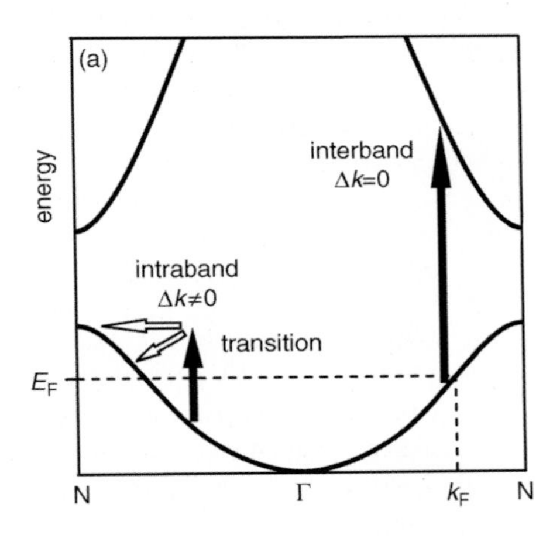

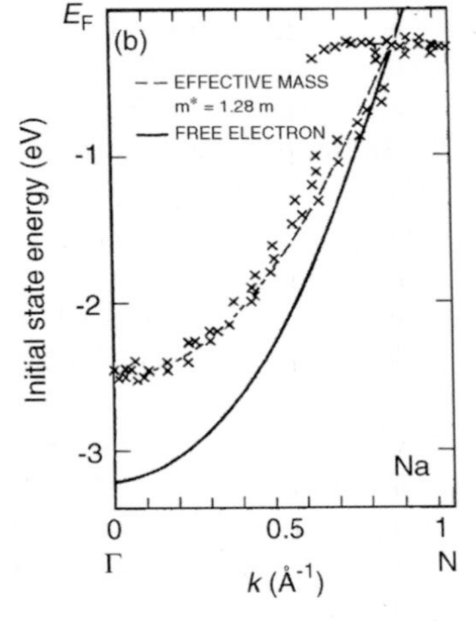

Figure 1.2 Typical dispersion of electron wave functions with wave vector in the nearly free electron approximation. In (b) experimental results from angle-resolved photoelectron spectroscopy as a function of photon energy from Ref. [4] are shown by crosses. For an introduction to this technique, we refer to Chapter 3. The solid line represents a calculation on the nearly free electron approximations; the dashed line fits a modified effective electron mass to the data. In (a) direct interband and indirect intraband transitions are indicated. The latter require a compensation of the momentum change to maintain energy and momentum conservation.

states belong to the same band. Here, a vertical transition is not possible in case of perfect crystal periodicity. However, a defect or a lattice distortion can absorb the required momentum change Δk and facilitate the transition in an elastic or inelastic manner. Further details are beyond the present introduction and can be found in Chapter 5.

By including the d-orbitals into electronic structure calculations, optical interband transitions from d to s, p states add significant excitation pathways and are responsible for the characteristic optical reflection spectrum of noble metals and hence their color.

To describe the electric conductivity, the population of the bands is essential. A material for which the valence electrons fill the uppermost populated band completely will be an insulator because an applied electric field will not cause electrical current to flow. If a band is partially filled, the electric field can move the carriers and the material is a metal. Considering the Pauli principle, a crystal can be an insulator only if the number of valence electrons in the primitive unit cell is an even integer. Otherwise, it is expected to be metallic. As will be discussed in the forthcoming section, this expectation is in general not always met, albeit it holds for many simple materials. Insulators and semiconductors are distinguished although both have a completely occupied valence band because thermal excitations in semiconductors facilitate a considerable electrical conductivity, which increases further with increasing temperature.[1)] In insulators, the gap is in general large enough to suppress thermally induced conductivity. However, under particular conditions, insulators can conduct an electric current, for example, after optical excitation or by ion propagation.

1.2
From Bloch Theory to Band Structure Calculations

1.2.1
Bloch Theory

Bloch's theorem teaches us that the eigenstates of a quantum particle in a periodic potential are of the form $\Psi_{n\mathbf{k}}(r) = u_{n\mathbf{k}}(\mathbf{r}) \exp(i\mathbf{k}\mathbf{r})$, where $u_{n\mathbf{k}}(\mathbf{r}+\mathbf{R}) = u_{n\mathbf{k}}(\mathbf{r})$ for all $\mathbf{R}$ of the Bravais lattice. The wave vector $\mathbf{k}$ lies in the first Brillouin zone and the band index n is an additional quantum number. The corresponding probability of finding the particle at a given point in the solid is thus a periodic function with the periodicity of the lattice: the electron is delocalized over the whole solid.

The eigenstates of a system of N noninteracting electrons are Slater determinants of Bloch states. The same remains true if interactions are taken into consideration but approximated by a static mean field, such as the electrostatic Hartree potential or a more refined effective potential resulting, for example, from a description by density

1) The temperature dependence of electrical conductivity facilitates a clear distinction of metals and semiconductors. While in semiconductors the conductivity can increase with temperature, it generally decreases in metals because of the increased electron–phonon scattering.

functional theory. Indeed, in the absence of two-particle terms in the Hamiltonian, the latter can be expressed as a sum of N (identical) one-particle Hamiltonians. This "independent electron approximation" thus leads necessarily to a band picture of the solid. In the ground state of the N-electron system, the resulting one-particle Bloch states corresponding to the N lowest lying one-particle (band) energies are filled. If the occupied state at the highest energy lies within a band, the system is metallic in this picture. Conversely, if a fully filled conduction band and an empty valence band are separated by a gap, we are dealing with an insulator.

1.2.2
The Tight Binding Approach to the Solid

An elementary approach to the band structure of a solid, complementary to the nearly free electron approach, is the tight-binding approximation. It describes the solid as a collection of atoms, weakly perturbed by their neighbors. The deviation from atomic behavior originates from the so-called overlap integrals, matrix elements of electronic wave functions centered on neighboring atoms: $\gamma_{ij}(\mathbf{R}) = -\int_i d\mathbf{r}\psi_i^*(\mathbf{r})\Delta U(\mathbf{r})\psi_j^*(\mathbf{r})$, where the Hamiltonian $H = H_{at} + \Delta U(\mathbf{r})$ is separated into an atomic part and a correction. The latter is responsible for the coupling, that is, the delocalization of electrons over the solid. Mathematically speaking, the tight binding approach consists in searching for the Bloch functions as solutions of Schrödinger's equation in the periodic potential of the solid in the form of linear

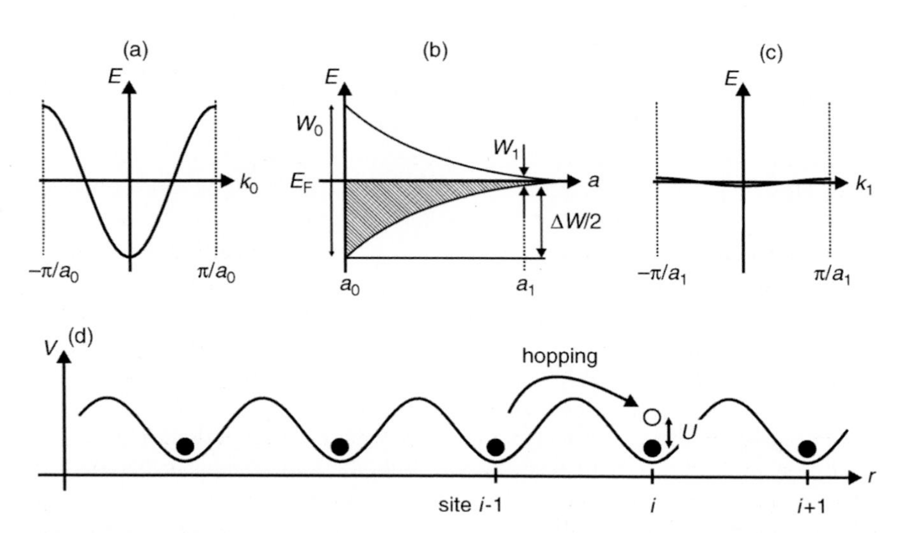

Figure 1.3 Schematic representation of the competition between kinetic energy or band width W and Coulomb repulsion U. (b) depicts the reduction in band width with increasing lattice constant a. In (c) and (a) the effect of band width on the dispersion of the single-electron band structure is depicted. In (d) the energy U required to populate two electronic states at one localized site is illustrated.

combinations of atomic orbitals. It is clear from this picture that quite extended orbitals lead to large overlap integrals and thus large band width, whereas core or semicore states give rise to dispersionless bands of negligible band width; see Figure 1.3a–c for illustration.

1.2.3
Band Structure Calculations

As seen above, once the independent electron approximation has been chosen, the electronic states are of Bloch form, and the corresponding eigenvalues are known as the band structure of the solid. To calculate it quantitatively, a choice for the effective interaction potential has to be made. The most natural choice that one may think of is the electrostatic potential created by a charge distribution corresponding to the electronic density, the so-called Hartree potential. This is, however, not the approximation chosen for most quantitative calculations nowadays. Indeed, the "Kohn–Sham potential" of density functional theory often yields band structures in better agreement with experimental observations. Density functional theory is an exact theoretical framework that associates with the interacting electron system an auxiliary noninteracting system of the same density in an effective potential: these two systems have the same ground-state properties. Strictly speaking, the theoretical framework does not attribute any physical meaning to the corresponding eigenvalues; nevertheless, they often represent a not too bad approximation to the electronic band structure observed in experiments. We do not enter here in the vast literature on the subject, but the reader interested in the relation of the Kohn–Sham gap and an electronic excitation gap of an insulator may be directed to the work of Perdew [5]. An additional difficulty consists in the fact that while the existence of the effective Kohn–Sham potential is established by the fundamental theorems of density functional theory, its explicit form is unknown, and in practical calculations further approximations have to be made. The most popular among them is the so-called local density approximation (LDA), which adds to the Hartree term an exchange–correlation contribution based on the homogeneous electron gas.

1.3
Beyond the Band Picture

The importance of band theory for the development of modern solid-state physics can hardly be overestimated. In combination with theories proposing appropriate effective potentials approximating the electronic Coulomb interactions by one-particle potentials, band theory gave a framework to the first attempts of *quantitative* descriptions and predictions of electronic properties of solids. Nevertheless, one did not have to wait for the more and more complex materials synthesized in modern materials science to find examples where band theory fails.

1.3.1 Mott's Hydrogen Solid

A simple argument by Mott [6] already shows the limits of the band description. Consider a collection of atoms with a single valence electron and a single orbital, for example, hydrogen, and let us forget for the moment their tendency to molecule formation. Arranging the atoms in a periodic structure with one atom per unit cell and a lattice constant a of the order of Å results – according to Bloch's theorem – in a half-filled band and thus in a metallic state; see Figure 1.3. The physical mechanism behind the band formation is the kinetic energy gain associated with the delocalization of the electrons. Now consider the following *Gedankenexperiment* (see Figure 1.3a–c): let us increase the lattice constant to artificially large values a_1, so as to decrease the kinetic energy gain and in particular to make it smaller than the price of the Coulomb interaction U that we have to pay when two electrons occupy the same atom, as illustrated in Figure 1.3d. In this situation, hopping becomes unfavorable, and the ground state consists of electrons being localized on their respective atoms. The system is just reduced to a collection of independent atoms. In other words, in this "atomic limit" the ground state is *not* the Slater determinant of Bloch states postulated by Bloch's theorem. It is the competition with the Coulomb interaction associated with double occupancies that destroys the Bloch state. In fact, in this situation it is more appropriate to think in terms of real space rather than in k-space. To first approximation, we can entirely neglect the kinetic energy and write the effective Hamiltonian for the Coulomb interactions:

$$H_{\text{int}} = \sum_i U n_{i\uparrow} n_{i\downarrow} \tag{1.5}$$

Here, $n_{i\sigma}$ denotes the occupation number operators corresponding to an electron in the s-orbital on site i with spin σ; they commute with the Hamiltonian and are good quantum numbers. The Hamiltonian is thus diagonal in the eigenbasis of these operators, and the eigenstates can be labeled simply by these atomic occupation numbers. We are, therefore, naturally led to a real space picture. This is in contrast to the kinetic energy part of the Hamiltonian

$$H_{\text{kin}} = \sum_{ij\sigma} t c^{\dagger}_{i\sigma} c_{j\sigma}, \tag{1.6}$$

which is diagonal in momentum space. Indeed, Fourier transformation of the creation and annihilation operators for an electron at site i and with spin σ, $c^{\dagger}_{i\sigma}$, and $c_{j\sigma}$ according to

$$c^{\dagger}_{k\sigma} = \sum_{j\sigma} \exp(ikR_j) c^{\dagger}_{j\sigma} \tag{1.7}$$

$$c_{k\sigma} = \sum_{j\sigma} \exp(ikR_j) c_{j\sigma} \tag{1.8}$$

diagonalizes the Hamiltonian

$$H_{\text{kin}} = \sum_{k\sigma} \varepsilon_k c_{k\sigma}^{\dagger} c_{k\sigma} \tag{1.9}$$

Here, the R_n denotes the lattice sites and the k-vectors are the crystal momenta as in Bloch's theorem.

Despite its simplicity, the above example illustrates the physical mechanisms at work in systems that require a description beyond one-particle approaches. Indeed, in between the two limits – the band picture of noninteracting Bloch electrons and the atomic limit of localized electrons – the situation becomes much more complicated. The physics is determined by the competition of kinetic energy W and Coulomb interaction U, and the general N-particle state is not just a Slater determinant of one-particle states. The Hamiltonian is diagonal neither in real nor in Fourier space.

Before entering into a more detailed theoretical modeling of the resulting N-particle physics, we will discuss some examples of real materials where correlated electron physics enters into play.

1.3.2
Mott Insulators in Nature

At the early stage of quantum mechanics, Verwey reported that many transition metal oxides with partially filled d-electron bands were poor conductors or even insulators [7]. This finding is in sharp contrast with the prediction of band theory, thus questioning the basic hypothesis of the independent electron approximation. Indeed, in these compounds the Coulomb repulsion U between two electrons occupying a d-orbital could become larger than the energy gain due to the electronic delocalization. When this is the case, a half-filled electronic system gains energy by a complete localization of the charges into a Mott insulating phase. In reality, the orbital degeneracy of the d-electron systems can oversimplify the description in terms of a single-conduction band [8]. This orbital degeneracy is an unavoidable source of complicated behavior, as the colossal magnetoresistance observed in the manganites. In many materials, the physics of low-energy excitation is also determined by the overlap of the d-orbitals of the transition metals with the p-band of ligand atoms. If the p-band is situated close to the Fermi energy, the character of the minimum charge excitation gap corresponds to the transition from a state with fully occupied p-band to a state with an extra electron in the d-band and a p-like hole. The existence of such charge transfer insulators has been experimentally demonstrated by resonant photoemission spectra of NiO [9]. Zaanen *et al.* classified the transition metal oxides on the basis of their charge gap energy and on-site Coulomb repulsion U [10]. It follows that lighter and heavier transition metals tend to generate a charge transfer and Mott insulator, respectively. However, it should be noticed that a clear separation between charge transfer insulators and Mott insulators is not possible in the presence of significant hybridization of the p–d orbitals. As an example, the oxygen p-orbitals of high-temperature superconductors are strongly hybridized with $d_{x^2-y^2}$ orbitals, leading to a single band crossing the Fermi level. As a consequence, an effective

Hubbard model with a single band is often considered to account for the physics of the high-temperature superconductors.

The phase diagram of Mott insulators can be experimentally determined by varying the temperature, by electronic filling, or by applying external perturbations. For example, the hydrostatic pressure reduces the interatomic distances and changes the crystal field. These effects increase the conducting band width and affect the orbital degeneracy. Alternatively, the same process can be obtained by varying the original stoichiometry through chemical substitution of isoelectronic elements. The phase diagram for $V_{1-x}Cr_xO_3$ shows that pressure and chemical substitution induce comparable effects on the physical properties [11]. As shown by Figure 1.4, a first-order transition connects the paramagnetic insulating phase to the paramagnetic metallic phase. At low temperature, the material enters an antiferromagnetic phase that extends to relatively high pressure. The transition from the paramagnetic insulator to the metal has for a long time been attributed to a gradual increase in the

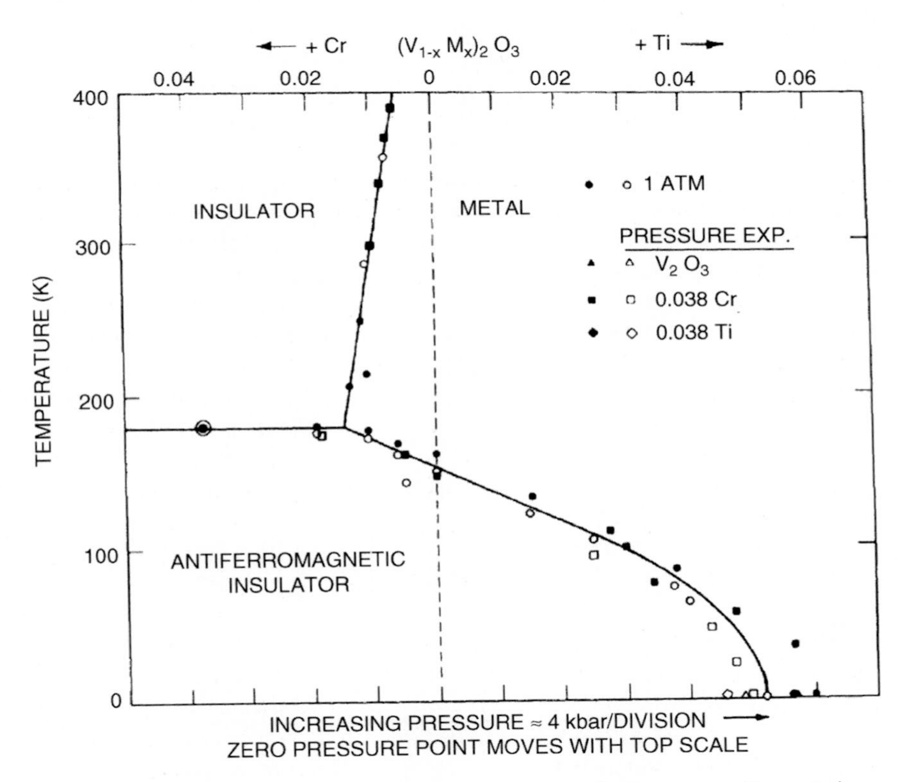

Figure 1.4 Phase diagram for the metal insulator transition in $(V_{1-x}M_x)_2O_3$ as a function of temperature (left axis) and pressure (bottom axis). Square symbols are for M = Cr and $x = 0.038$. The electronic substitution of V with Cr or Ti impurities induces a lattice distortion acting as an effective pressure. The equivalence is shown by the circular symbols, which have been obtained at ambient pressure and varying the concentration of Cr or Ti (top axis); reprinted with permission from Ref. [11]. Copyright (1971) by the American Physical Society.

electronic band width of an occupied d-orbital with respect to the electronic correlation. This process is known as a band width-controlled transition and can be qualitatively explained by the Hubbard-like models of the electronic Hamiltonian. However, it is currently understood that the metal–insulator transition of $V_{1-x}Cr_xO_3$ is possible only because of a larger – and correlation-enhanced – crystal field of the insulating phase [12]. The crystal field removes some of the orbital degeneracy of the d-shell. In the case of a cubic symmetry, the crystal field splits the d-states in a threefold degenerate t_{2g} level and a doubly degenerate e_g level. The stretching of the oxygen octahedra around the transition metal atom may further reduce the point group symmetry of the crystal field. It appears that an external pressure induces a structural distortion that may modify the orbital energies even more than the electronic band width. Indeed, more examples have been recently analyzed, where the orbital degrees of freedom play a crucial role in metal–insulator transitions. We do not enter here the involved description of the multiorbital Mott transition. Instead, we discuss the metal–insulator transition in the model compound $1T$-$TaSe_2$ that has been investigated in the framework of a one-band Hubbard model. Theory predicts a large transfer of spectral weight from the Fermi level to a localized feature that is termed a lower Hubbard band. An unavoidable coupling to the lattice degrees of freedom results in a first-order characterization of this transition. As shown in Figure 1.5, the evolution of electronic states across a band width-controlled transition of $1T$-$TaSe_2$ is in reasonable agreement with theoretical expectations [13].

Another, qualitatively different method of inducing the transition from a Mott insulator to a metallic phase is obtained by changing the band filling. This is experimentally realized by doping the material with elements having different valence. Even small deviations from an integer occupation of the orbitals generates dramatic effects on the electronic properties. Figure 1.6 shows the optical conductivity spectra of $La_{1-x}Sr_xTiO_{3+y}$ and $Y_{1-x}Ca_xTiO_3$ [15]. The hole doping of such tinanates is achieved by varying the concentration x of the alkaline earth element. It

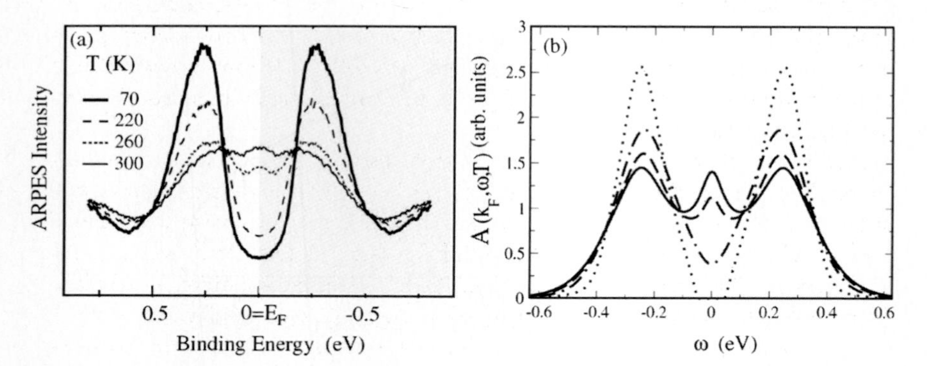

Figure 1.5 (a) Temperature-dependent ARPES data for $1T$-$TaSe_2$ that represent the spectral function of the system. The curves have been obtained by symmetrization of the raw spectra around the Fermi level. (b) Temperature-dependent spectral function of the half-filled Hubbard model calculated within DMFT for the temperatures corresponding to the experiment; reprinted with permission from Ref. [13]. Copyright (2003) by the American Physical Society.

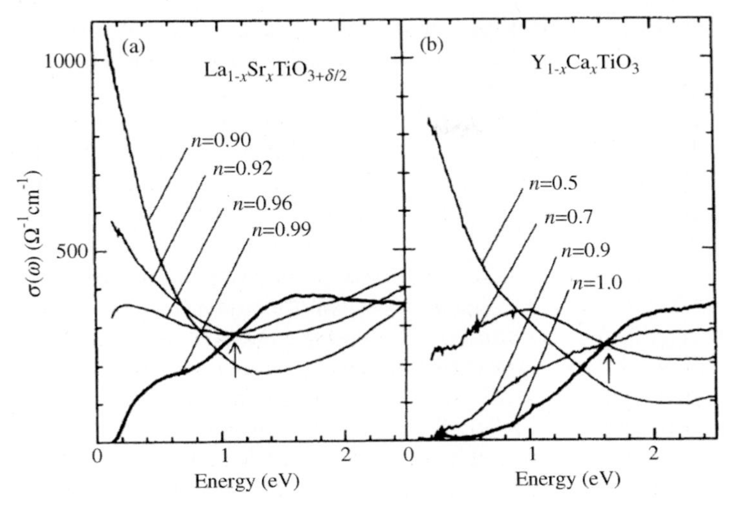

Figure 1.6 Real part of the optical conductivity determined from optical spectra for two types of carrier-doped perovskite titanates with change of the band filling n; reprinted with permission from Ref. [15]. Copyright (1995) by the American Physical Society.

follows a filling-controlled metal–insulator transition that causes an abrupt increase for a photon energy of 0.1 eV and above, as can be seen in Figure 1.6 with decreasing n. This finding is consistent with the occurrence of an insulating ground state in such a material. Upon doping $La_{1-x}Sr_xTiO_{3+y}$ with few percentages of electrons per unit cell, the transfer of spectral weight takes place on an energy scale 10 times larger than the energy of the gap. This redistribution of the spectral properties over a wide interval of energy is a hallmark of strong electronic interactions. Furthermore, the low-energy excitations appearing below the gap value are largely "incoherent." In the DC limit, the electronic mean free path predicted by the Drude model becomes comparable or even smaller than the lattice constant. Such anomalous scattering rate makes the electron dynamics of such materials very different from the usual Drude behavior, discussed in Section 1.1, and precludes any description in terms of nearly independent particles.

Finally, the photoexcitation of electron–hole pairs is a novel and interesting approach to perturb the Mott insulating phase. Notice that photoexcitation is very different from filling or band width control. In most cases, the charge fluctuations generated by the laser field thermalize within few femtoseconds, leading to a quasi-stationary distribution of the electronic excitations. Under this condition, the average filling of the electronic states is not changed with respect to the equilibrium state. A first idea of the photoexcitation can be obtained by considering the ratio U/W as constant. At sufficiently large excited carrier density, the Mott insulator becomes unstable and the correlation gap collapses. This process can be modeled by a large increase in the electronic temperature while leaving the lattice degrees of freedom unchanged. The outcome of photoexcitation is a pseudogao-like phase corresponding to an incoherent metal with reduced density of the electronic states close to the

Fermi level. The presence of a quasi-particle peak in such a pseudogap-like state has been detected in $1T$-TaS_2 and requires further experimental and theoretical investigations [16].

1.4 Electronic Structure of Correlated Materials

In the following sections, we provide an introduction to electronic states of materials in which effects of electronic correlation are essential. Their description is involved and real materials are further complicated by dimensionality effects and crystal fields, and the emergence of ordered states, which makes these systems very interesting. We concentrate on essential concepts and provide illustrative examples that we consider helpful for an introduction.

1.4.1 The Hubbard Model

The above theoretical considerations and the given experimental results indicate the importance of the interplay of delocalization due to the kinetic energy gain and the localization due to Coulomb interactions. The most simple model that takes this competition into account was formulated by John Hubbard, Martin Gutzwiller, and Junjiro Kanamori:

$$H = \sum_{ij\sigma} t c_{i\sigma}^{\dagger} c_{j\sigma} + \sum_{i} U n_{i\uparrow} n_{i\downarrow} \tag{1.10}$$

It describes the competition between the energy gain due to delocalization of the electrons and the Coulomb interaction. Indeed, the expectation value of the double occupation $\langle n_{i\uparrow} n_{i\downarrow} \rangle$ is finite in the delocalized state, leading to an interaction energy that can – for narrow band systems – compete with the kinetic energy contribution.

The microscopic justification of the Hubbard model relies on the introduction of a localized basis set, spanning the low-energy Hilbert space of the one-particle part of the Hamiltonian, and on the calculation of the effective Coulomb interaction acting on those degrees of freedom. To simplify the discussion, let us assume for the moment that we are dealing with a system where the low-energy band states can be derived from a single atomic orbital, replicated to all atomic sites R, $\chi_R(r)$ is then a Wannier function of the system, and in this case, the Fourier transform of the Bloch eigenstate $\chi_k(r)$. When writing the full crystal Hamiltonian on the basis of the atomic orbitals, among the matrix elements of the Coulomb interaction

$$U(R, R') = \int \mathrm{d}r\mathrm{d}r' |\chi_R(r)|^2 |\chi_{R'}(r')|^2 v_{\text{screened}}(r, r') \tag{1.11}$$

the on-site ($R = R'$) term dominates because it describes the high Coulomb energy required to create doubly occupied atomic sites.

For an account of the microscopic derivation, we refer the reader to the original papers [17]. In one dimension, the model (1.10) is exactly solvable and exhibits the effects special to this case, which are in particular Tomonaga–Luttinger behavior and spin–charge separation. Here, we focus on the basic physical effects that this model displays in higher dimensions. Apart from the one-dimensional case, this model allows for yet another limit where the exact solution can be investigated, which is the case of infinite lattice coordination. Indeed, in this case the dynamical mean field theory described below gives the exact solution.

The phase diagram for this case is given in Figure 1.7. At sufficiently low temperature, the phase diagram of the Hubbard model displays a metal–insulator transition as a function of the interaction, as expected from the qualitative discussion of the atomic and band limits above. The transition – except at zero temperature – is of first order, displaying a pronounced coexistence region, where both an insulating and a metallic phase can be stabilized. The sharp metal–insulator transition ends at a high-temperature critical point, above which a crossover between a phase that is referred to as "bad metal" and insulating behavior can be found.

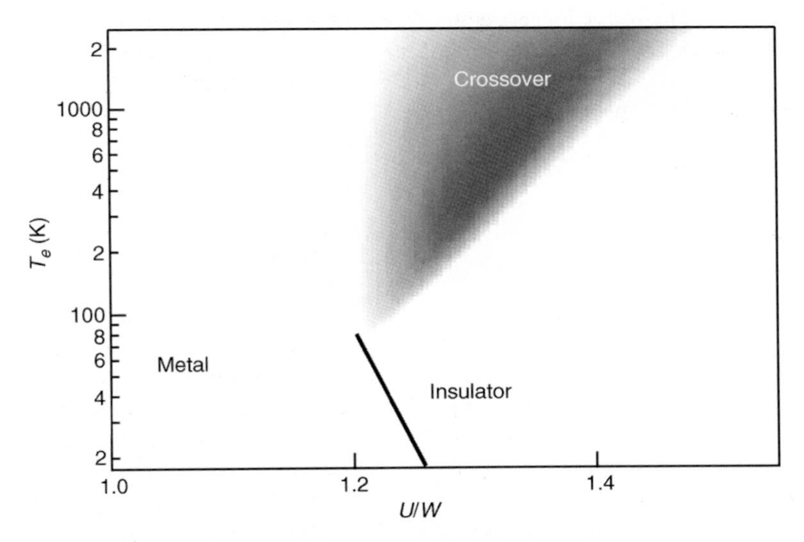

Figure 1.7 Phase diagram of the half-filled one-band Hubbard model within dynamical mean field theory. The physics of the model depends on two parameters: the ratio of the Coulomb interaction U over the band width W in eV and the temperature T. A paramagnetic metal-to-insulator transition from a Fermi liquid phase at weak interactions (small U) to a Mott insulator at large interactions (large U) is observed at intermediate temperatures. The transition in this framework is first order with a coexistence region plotted in gray scale where both the metallic and the insulating phases can be stabilized. At low temperatures, depending on the lattice geometry, magnetically ordered phases are more favorable. At high temperatures, the metal–insulator transition is replaced by a crossover from a bad metal to a semiconducting or insulating regime.

Formally, the exact solution holds for infinite dimensions, meaning for a lattice model where each site has an infinity of nearest neighbors. The above scenario is believed to be a reasonable approximation also in the three dimensions in the regime not too close to the metal–insulator transition.

The spectral properties of the Hubbard model have been calculated by means of dynamical mean field theory (DMFT). As shown in Figure 1.5, the electronic spectrum of the insulating phase displays two pronounced peaks separated by the on-site repulsion U. In the atomic limit, this term corresponds to the expected amount of energy that is necessary to add an electron to the system. Figure 1.5 shows that the Hubbard peaks at finite binding energy are present even in the metallic phase. Therefore, the electronic excitations still have considerable projection on localized states. On the other hand, a third peak emerges at the Fermi level. This feature has the character of a quasi-particle band with weak energy dispersion. Therefore, the electronic excitations also have a finite projection on delocalized states, but with an effective mass that is largely renormalized by the Coulomb interaction. This duality between the localized and itinerant character of the electronic excitations is probably the most surprising and interesting outcome of the model.

It is, therefore, more than an academic exercise to describe how the above solution can be obtained. This is the subject of the following section, devoted to DMFT.

1.4.2
Dynamical Mean Field Theory

The basic idea of DMFT is to replace a lattice problem (or in the realistic case, the solid) by an effective local system, coupled to a bath and subject to a self-consistency condition, in analogy to conventional Weiss mean field theory in statistical mechanics. Contrary to the latter, however, the intervening mean field is energy dependent, hence the notion of a dynamical MFT.

Let us illustrate the method with the example of the single-band Hubbard model, as defined by the Hamiltonian

$$H = \sum_{ij\sigma} t_{ij} \left(c_{i\sigma}^{\dagger} c_{j\sigma} + c_{j\sigma}^{\dagger} c_{i\sigma} \right) + \sum_{i} U n_{i\uparrow} n_{i\downarrow}. \tag{1.12}$$

Dynamical mean field theory associates with this Hamiltonian an auxiliary *local* problem, defined by the *Anderson impurity* Hamiltonian

$$H = \sum_{l} \left(\varepsilon_l d_{l\sigma}^{\dagger} d_{l\sigma} + V_l c_{0\sigma}^{\dagger} d_{l\sigma} + V_l^{*} d_{l\sigma}^{\dagger} c_{0\sigma} \right) + U n_{0\uparrow} n_{0\downarrow} \tag{1.13}$$

The latter describes the physics of one of the original lattice sites (labeled here as 0), coupled to a bath (operators d, $d^{\dagger}$) through a hybridization V_l. The bath has an infinity of degrees of freedom labeled by l, at energies ε_l. This model was originally introduced by P. W. Anderson for describing an impurity in a host material [18]. Within dynamical mean field theory, it is reinterpreted in the following way. The impurity site 0 represents an arbitrary site of the original lattice, and the bath plays the role of the mean field representation of the lattice. In contrast to the original ideas of Anderson who wrote an impurity model in order to describe a physical impurity in a (given) host material, the impurity within the DMFT context is representative of a correlated orbital at a given site of a translationally invariant solid. The bath is thus a quantity that has to be determined self-consistently in order to maintain the translational invariance of the lattice.

In practice, the computational task consists in calculating the local Green's function $G_\sigma^{\mathrm{imp}} = -\left\langle \hat{T} c_{L\sigma} c_0^\dagger \right\rangle$ of the above local impurity model. Here, the brackets denote the thermodynamical average taken with the density matrix of the Anderson impurity model Hamiltonian (1.13), and $\hat{T}$ is the time-ordering operator. For the explicit calculation of the Green's function, a variety of techniques, ranging from Monte Carlo simulations to approximate schemes, are available. As in standard mean field theory, there is a second step related to restoration of the translational invariance of the original model. Imposing that all equivalent sites behave in this same way yields a self-consistency condition that eventually determines the parameters of the Anderson impurity model. Mathematically speaking, one imposes the local Green's function G of the solid to equal the impurity Green's function G^{imp}. To this effect, the self-energy of the impurity model $\Sigma_{\mathrm{imp}} = \mathcal{G}_0^{-1} - G^{-1}$ is calculated and used as an approximation to the full self-energy of the lattice. Thus,

$$G(i\omega_n) = \sum_{\mathbf{k}} \left[i\omega_n + \mu - H_0(\mathbf{k}) - \Sigma^{\mathrm{imp}}(i\omega_n) \right]^{-1}. \tag{1.14}$$

In practice, this set of equations is solved iteratively, starting from a guess for the bath parameters ε_l, V_l, solving the impurity model, inserting the corresponding self-energy into the self-consistency equation (1.14), recalculating the bath from Dyson's equation, using the result to update the impurity model, and so on.

1.4.3
Electronic Structure Calculations

We have discussed in the preceding sections two strategies of theoretical modeling: the first one, a band structure approach, is appealing in that it treats the solid from *first principles* (i.e., without any adjustable parameters). The electronic Coulomb interactions, however, are included within static mean field theory. We have mentioned above why this strategy fails for solids with localized electrons. The second strategy, shown in the example of the Hubbard model, consists of studying the interplay of localized and delocalized behavior within the most simple models that incorporate the competition between kinetic energy and Coulomb interactions. While very successful for assessing the basic mechanisms, for example, of correlation–driven metal–insulator transitions, this approach obviously takes a very simplistic view on the chemistry of a compound, namely, by including only one band and an effective hopping between sites. It provides no information about the effect of multiorbital Coulomb correlations, hybridization between different bands, the influence of crystal field splittings, and so on. To address a given compound, an attractive way to arrive at a material-specific description is to include hopping, hybridization, and crystal field in the band structure calculations, and then to use the resulting one-particle Hamiltonian as a starting point for a *multiorbital* Hubbard-type Hamiltonian that can be treated within many-body techniques. This strategy is realized in the recent combination of density functional theory with dynamical mean field theory, the so-called "LDA + DMFT" scheme.

The basic idea of constructing a local model for the purpose of calculating a local self-energy as an approximation to the full many-body self-energy of the system carries directly over from the model context to the case of a real solid. The most basic version of the combined "LDA + DMFT" scheme [19, 20] can be viewed as a DMFT solution of a multiorbital Hubbard model, where the parameters are calculated from DFT-LDA. The impurity represents the correlated orbitals of a type of atoms in the solid, and the self-consistency condition attributes the same self-energy to all equivalent correlated atoms – up to rotations in orbital space.

From a conceptual point of view, LDA + DMFT in its general framework can be viewed as an approximation to a functional of both the density and the spectral density of correlated orbitals, as formulated within spectral density functional theory [25].

As an illustration of DMFT, we discuss calculations of correlated metals and Mott insulators for the series of d^1 perovskite compounds $SrVO_3$, $CaVO_3$, $LaTiO_3$, and $YTiO_3$. These compounds, while isoelectronic and (nearly) isostructural, display a radically different behavior. The vanadates are moderately correlated metals with a mass enhancement of about a factor of 2, while the titanates are insulating. LDA + DMFT calculations for these compounds were able to unravel the underlying mechanism [26]: a tiny difference in the orbital occupations, which are induced by $GdFeO_3$-type distortions already at the band structure level, is amplified by the electronic Coulomb interactions and leads to an orbitally polarized insulating behavior. For $YTiO_3$, for example, the orbital polarization results in an effective single-band model; this reduction of the degeneracy suppresses the kinetic energy, and even moderate Coulomb interactions fully localize the system.

1.4.4
Ordered States

We briefly mentioned in the previous section that metal insulator transitions can also arise from the symmetry breaking of spin, orbital, or lattice degrees of freedom. In the case of symmetry breaking, the insulating phase can be reached through the vanishing carrier number with no need for mass enhancement. It was pointed out by Landau that it is not possible to analytically deform a state of one phase into the state of a phase having different symmetry [28, 29]. Typically, the more symmetric phase is on the high-temperature side of the accessible states and the less symmetric phase on the low-temperature side. This happens because the Hamiltonian of a system usually exhibits all the possible symmetries of the system, whereas the low-energy states lack some of these symmetries. At low temperatures, the system tends to be confined to the low-energy states. At higher temperatures, thermal fluctuations increase the probability to find states with higher energy. The system explores a larger portion of phase space, thus favoring a more symmetric solution. Symmetry-breaking transitions are usually classified into first- or second-order. During a first-order phase transition, the system either absorbs or releases a fixed amount of energy that is called latent heat. Because energy cannot be instantaneously transferred between the system and its environment, first-order transitions are associated with mixed phase regimes with coexistence of the low- and high-temperature states. Conversely, a second-order phase transition does not have an associated latent heat. In this case, the order parameter defining the low-temperature phase changes continuously when the critical temperature T_c is crossed. Some examples of second-order phase transitions are the transition between the ferromagnetic and paramagnetic, superconducting and normal conducting, and charge density wave (CDW) formation; see e.g. Refs. [4, 27] for further reading.

In the following paragraphs, we will consider the case of a CDW transition and elucidate the role of electron–phonon coupling in such an instability. These transitions break the symmetry of the lattice and generally occur in low dimensional materials. The reduction of phase space from three dimensions to two or even one dimension has several important consequences. In the vicinity of the respective critical temperature, where the transition occurs, interaction effects and fluctuations are essential. As depicted in Figure 1.8a,b the Fermi surface of a one-dimensional electron gas consists of two parallel sheets that are at a distance of twice the Fermi wavevector apart (namely, $2k_F$). This peculiar topology of the Fermi surface leads to a response to an external perturbation that is dramatically different from that in higher dimensions. In particular, the charge susceptibility at zero temperature diverges at

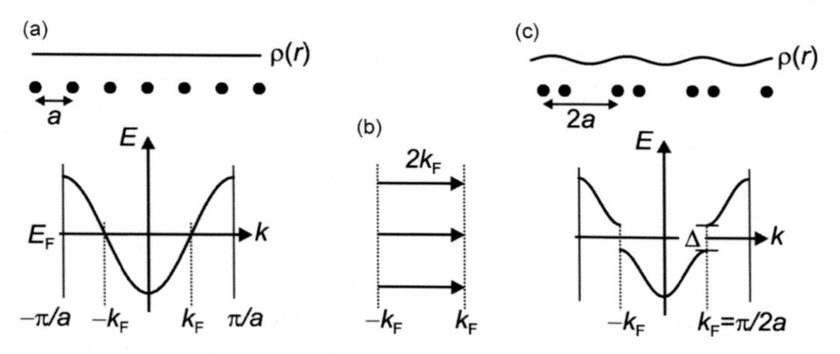

Figure 1.8 Illustration of charge density wave formation: (a) shows a half-filled band expected for a linear chain of atoms with lattice constant a for the situation of considerable wave function overlap of neighboring electrons. The first Brillouin zone extends up to $\pm\pi/a$ and the charge density $\varrho(r)$ is constant. The Fermi surface of a linear chain consist of lines at $\pm k_F$ perpendicular to the line extension in real space (b). This state is highly susceptible to excitations with momenta $2k_F$, which nest the two parts of the Fermi surface. Corresponding rearrangement of atoms leading to a lattice constant $2a$ reduces the first Brillouin zone to $\pi/2a$, which is identical to k_F. The band gap Δ that opens as a consequence of rearrangement of ion cores turns the linear chain to an insulator. Simultaneously, $\varrho(r)$ is corrugated with a periodicity of $2a$. Such a situation in low dimensional materials is referred to as a charge density wave. See Ref. [27] for more details.

wave vector $2k_F$ [27], thus implying that any infinitesimal external perturbation leads to a macroscopic charge redistribution. This phase transition leads to a modulation of the charge density with $2k_F$ periodicity in the direction perpendicular to the Fermi surface sheets. Since the CDW formation comes along with the distortion of the lattice, the transition costs elastic energy. However, the development of the $2k_F$ periodicity creates an energy gap that removes the Fermi surface, which is illustrated in Figure 1.8c. As a consequence, the energy lowering of electronic states compensates the elastic energy that is necessary for the distortion. In this case, the materials become insulating because of the vanishing carrier density associated with formation of the CDW periodicity. If the Fermi surface is not completely degenerate with respect to a $2k_F$ translation, the so-called "nesting condition" is not perfect. In this case, the Fermi surface is only partially removed and the compound remains metallic down to low temperature. A typical family of CDW materials with imperfect nesting comprehends the rare earth tellurides RTe_3 (with R = La, Ce, Y, . . .). Figure 1.9 shows the Fermi surface of $CeTe_3$ in the low-temperature CDW phase [30]. Notice that part of the spectral intensity on the Fermi surface is removed by the coherent scattering associated with the CDW periodicity. Even if the long-range order of the CDW vanishes above a temperature T_c, the low dimensional fluctuations of the density wave potential persist over a much broader temperature range. As a consequence, the charge gap of the CDW phase evolves in a pseudogap phase above T_c. An approximate theory accounting for the strong effects of fluctuations on the electronic excitation has been developed by Lee *et al.* [31].

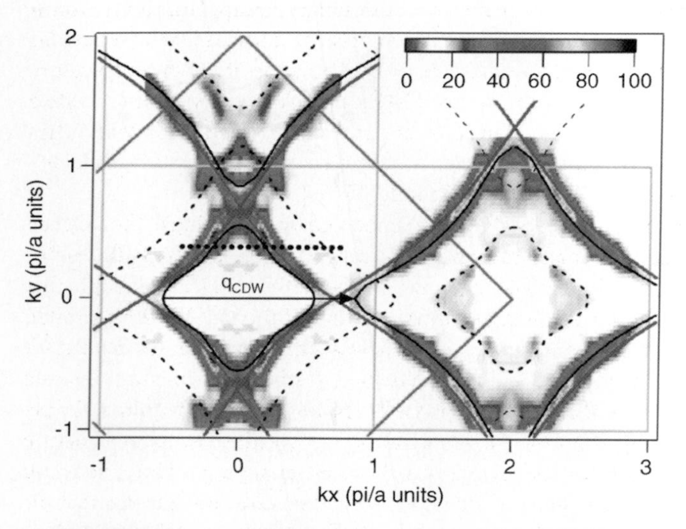

Figure 1.9 Fermi surface of $CeTe_3$ obtained by ARPES at the temperature of 25 K. In the horizontal direction, the Fermi surface is gapped due to charge density wave formation along this direction due to Fermi surface nesting with a momentum q_{CDW}; reprinted with permission from Ref. [30]. Copyright (2004) by the American Physical Society. (Please find a color version of this figure on the color plates.)

Like the electrons, the lattice excitations also display anomalies that are related to the charge density wave. Notably, the quantum energy of the phonon mode that is responsible for the Fermi surface nesting exhibits a strong temperature dependence. The lowering of quantum energy in the CDW phase is also known as a Kohn anomaly and explains the occurrence of a static lattice distortion.

In the context of the present book, we note that the excitation of such a "soft phonon" by a femtosecond optical perturbation can directly modify the amplitude of the charge density wave. Due to the impulsive character of the photoexcitation, the coherent oscillations of CDW amplitude modulate the electronic states in the gapped region of the Fermi surface. Recent time-dependent angle-resolved photoelectron spectroscopy (ARPES) experiments showed that the perturbation of the CDW turns out to be highly enhanced in the region of reciprocal space where the nesting is optimal [32]. It follows that optical perturbation is a promising tool for the exploration of charge ordering in highly correlated materials. In particular, short-range ordered patterns with checkerboard structure have been detected in several high-temperature superconductors. It is still an open question whether such charge-ordered phases are related to the electronic pseudogap of the cuprates.

1.4.5 Cooperation Between Lattice Instabilities and Electronic Correlations: The Example of Vanadium Dioxide

We end this chapter with a discussion of another exemplary material that is vanadium dioxide (VO_2). It has triggered much interest over decades due to its metal–insulator transition as a function of temperature [33]. It is metallic in its high-temperature phase of rutile structure and cooling beneath 340 K provokes a phase transition to a monoclinic insulating phase. The nature of this transition – Peierls or Mott – has been the subject of a long-standing debate. For a review, see, for example, Ref. [34]. Recent work can be found, for example, in Ref. [35–37].

The metal–insulator transition in VO_2 is accompanied by a doubling of the unit cell due to the dimerization of vanadium atoms and tilting of VO octahedra with respect to the crystallographic *c*-axis. According to early considerations by Goodenough, this dimerization leads to a bonding–antibonding splitting of the half-filled a_{1g} orbital, responsible for hopping along that axis, and a subsequent opening of the insulating gap between the bonding a_{1g} band and the conduction e_g^{π} states. This picture suggests that the gap opening in the monoclinic phase could be described within a simple band scenario, resulting from the Peierls distortion. Nevertheless, band structure calculations within density functional theory in the local density approximation result in metallic Kohn–Sham densities of states for both the rutile and the monoclinic phase. It is only when the dimerization is artificially exaggerated [34] that the Kohn–Sham band structure exhibits a gap.

Vanadium dioxide has served as an example for a compound where correlation effects are important for inducing the insulating state in the early work by Sir Neville Mott. This view was supported by experimental findings by Pouget *et al.*, who investigated mixed phases of VO_2 under uniaxial pressure or Cr doping. These phases could, in fact, be classified as Mott insulators on the basis of their magnetic response. In the pure monoclinic phase, the magnetic susceptibility is constant and of low value, indicating the formation of a singlet state in the bonding a_{1g} band of the V-dimers.

Dynamical mean field calculations allowed to resolve the apparent contradictions outlined above shown results for the local spectral function [38–40] agree well with experimental findings as provided by (angle-integrated) photoemission and X-ray absorption spectra. In particular, the characteristic features of the metal, with a pronounced peak at the Fermi level, and a lower Hubbard satellite have been experimentally confirmed [41]. Besides one-particle spectra, the optical conductivity of experiments and LDA + DMFT calculations are in agreement [42].

Particularly interesting is the debate on the band versus Mott mechanism [37] where LDA + DMFT obtains the insulating nature of the monoclinic phase as a consequence of subtle changes in the electronic structure due to the dimerization (giving rise to pronounced bonding–antibonding splitting in the a_{1g} states) and correlations enhancing both the orbital polarization and the bonding–antibonding splitting. The resulting state was thus described as a "correlation-assisted Peierls insulator." It was moreover shown (see Figure 1.10) that starting from the full LDA + DMFT solution, an effective static, albeit orbital-dependent, potential can be constructed that reproduces the full k-dependent spectral function. It would be too simple, however, to conclude from the existence of such an effective band structure for the insulating phase of VO_2 on *weak* correlation effects in this compound in general. This is already clear from an analysis of the *metallic* phase, which displays clear correlated features. From transport experiments, it was inferred that rutile VO_2

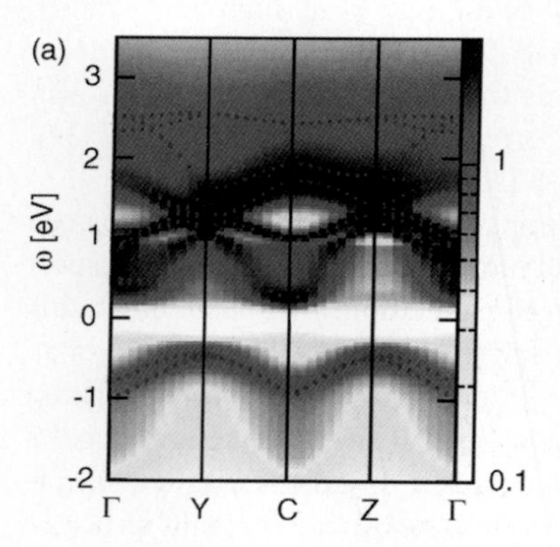

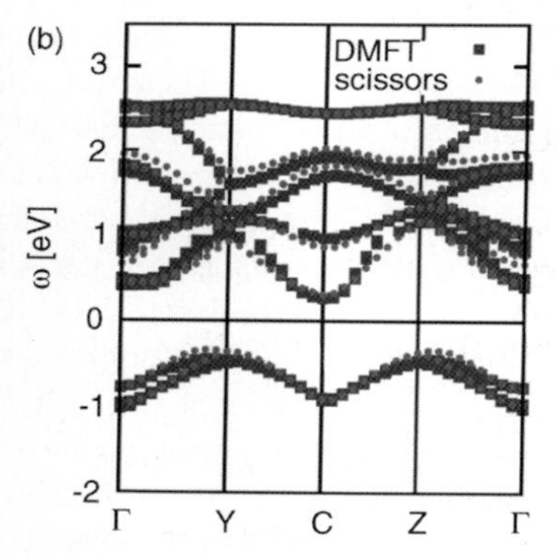

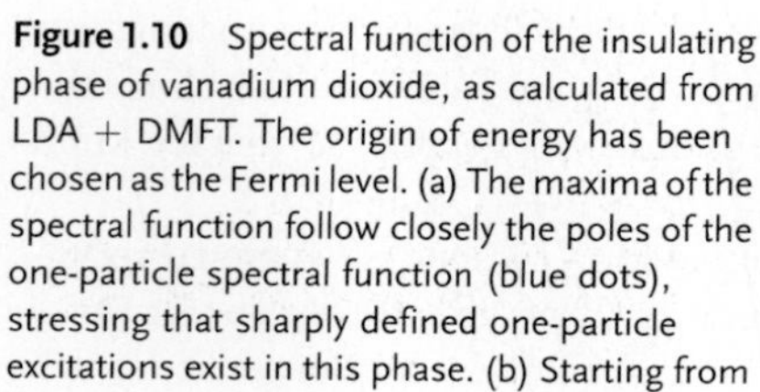

Figure 1.10 Spectral function of the insulating phase of vanadium dioxide, as calculated from LDA + DMFT. The origin of energy has been chosen as the Fermi level. (a) The maxima of the spectral function follow closely the poles of the one-particle spectral function (blue dots), stressing that sharply defined one-particle excitations exist in this phase. (b) Starting from the full LDA + DMFT solution, a static orbital-dependent potential has been constructed. The band structure corresponding to this potential is plotted in red dots and compared with the DMFT result of (a); reprinted with permission from Ref. [37]. Copyright (2008) by the American Physical Society. (Please find a color version of this figure on the color plates.)

might be a "bad metal" in the sense that the Ioffe–Regel–Mott resistivity limit is surpassed [43]. Photoemission spectroscopy identifies a lower Hubbard band, and the DMFT calculations interpret the broad features as manifestations of very poorly defined band states, due to lifetime effects. This is also supported by the proximity to the Mott insulating mixed phases obtained under Cr doping. From a fundamental physics point of view, this coexistence of an insulating phase, whose spectral function is well described by an effective band structure (even if beyond LDA and – due to the orbital-dependence of the effective potential – even beyond DFT), and strongly correlated metallic and Mott insulating phases in the same compound is particularly intriguing.

Several time-resolved experiments provided insight into the structural deformation and the optical response of VO_2 upon a photoinduced transition. The absorption of a near-IR photon removes an electron from the bonding orbital, destabilizing the dimer and launching a coherent structural deformation that oscillates with frequency of 6 THz around the new potential minimum. The atomic motion has been imaged by X-ray diffraction [44], by X-ray absorption [45], or by means of electron diffraction [46]. The authors observed two types of dynamics, indicating a stepwise atomic motion along different directions. It has been concluded that the initial fs motion is along *a*-axis, which is the direction of the V–V bond in the monoclinic structure. On the other hand, a subsequent and slower structural transformation projects along the *c*- and *b*-axes. The optical response does follow the fast dynamics of the Vanadium dimerization, namely, on a timescale of 75 fs. The authors conclude that the coherently initiated structural motion, brought about by optical phonons, is the most likely explanation for the collapse of the band gap [47]. Further experimental evidence for such behavior has been brought about by time-resolved THz experiments on thin films [48].

References

1 Drude, P. (1900) *Ann. Phys.*, **1**, 535.
2 Smith, N.V. (1970) *Phys. Rev. B*, **2**, 2840.
3 Ashcroft, N.W. and Mermin, N.D. (1976) *Solid State Physics*, Hartcourt College Publishers.
4 Jensen, E. and Plummer, E.W. (1985) *Phys. Rev. Lett.*, **55**, 1912.
5 Perdew, J.P. and Levy, M. (1983) *Phys. Rev. Lett*, **51** 1884.
6 Mott, N.F. (1949) *Proc. Phys. Soc. Lond.*, **62**, 416.
7 Verwey, E.J.W. (1939) *Nature*, **144**, 327.
8 Imada, M., Fujimori, A., and Tokura, Y. (1998) *Rev. Mod. Phys.*, **70**, 1039.
9 Davis, L.C. (1982) *Phys. Rev. B*, **25**, 2912.
10 Zaanen, J., Sawatzky, G.A., and Allen, J.W. (1985) *Phys. Rev. Lett.*, **55**, 418.
11 McWhan, D.B., Remeika, J.P., Rice, T.M., Brinkman, W.F., Maita, J.P., and Menth, A. (1971) *Phys. Rev. Lett.*, **27**, 941.
12 Poteryaev, A.I., Tomczak, J.M., Biermann, S., Georges, A., Lichtenstein, A.I., Rubtsov, A.N., Saha-Dasgupta, T., and Andersen, O.K. (2007) *Phys. Rev. B*, **76**, 085127.
13 Perfetti, L., Georges, A., Florens, S., Biermann, S., Mitrovic, S., Berger, H., Tomm, Y., Höchst, H., and Grioni, M. (2003) *Phys. Rev. Lett.*, **90**, 166401.

14 Carter, S.A., Rosenbaum, T.F., Metcalf, P., Honig, J.M., and Spalek, J. (1993) *Phys. Rev. B*, **48**, 16841.
15 Okimoto, Y., Katsufuji, T., Okada, Y., Arima, T., and Tokura, Y. (1995) *Phys. Rev. B*, **51**, 9581.
16 Perfetti, L., Loukakos, P.A., Lisowski, M., Bovensiepen, U., Wolf, M., Berger, H., Biermann, S., and Georges, A. (2008) *New J. Phys.*, **10**, 053019.
17 Hubbard, J. (1963) *Proc. Roy. Soc. A*, **276**, 238.
18 Anderson, P.W. (1961) *Phys. Rev.*, **124**, 41.
19 Lichtenstein, A.I. and Katsnelson, M.I. (1998) *Phys. Rev. B*, **57**, 6884.
20 Anisimov, V.I., Poteryaev, A.I., Korotin, M.A., Anokhin, A.O., and Kotliar, G. (1997) *J. Phys. Condens. Matter*, **9**, 7359.
21 Lechermann, F., Georges, A., Poteryaev, A., Biermann, S., Posternak, M., Yamasaki, A., and Andersen, O.K. (2006) *Phys. Rev. B*, **74**, 125120.
22 Savrasov, S.Y., Kotliar, G., and Abrahams, E. (2001) *Nature*, **410**, 793.
23 Minár, J., Chioncel, L., Perlov, A., Ebert, H., Katsnelson, M.I., and Lichtenstein, A.I. (2005) *Phys. Rev. B*, **72**, 045125.
24 Pourovskii, L., Amadon, B., Biermann, S., and Georges, A. (2007) *Phys. Rev. B*, **76**, 235101.
25 Savrasov, S.Y. and Kotliar, G. (2004) *Phys. Rev. B*, **69**, 245101.
26 Pavarini, E., Biermann, S., Poteryaev, A., Lichtenstein, A.I., Georges, A., and Andersen, O.K. (2004) *Phys. Rev. Lett.*, **92**, 176403.
27 Grüner, G. (2000) *Density Waves in Solids, Frontiers in Physics*, Vol. 89, Perseus Publishing, Cambridge, MA, USA.
28 Landau, L.D. (1957) *Sov. Phys. JETP*, **3**, 920.
29 Landau, L.D. (1957) *Sov. Phys. JETP*, **5**, 101.
30 Brouet, V., Yang, W.L., Zhou, X.J., Hussain, Z., Ru, N., Shin, K.Y., Fisher, I.R., and Shen, Z.-X. (2004) *Phys. Rev. Lett.*, **93**, 126405.
31 Lee, P.A., Rice, T.M., and Anderson, P.W. (1973) *Phys. Rev. Lett.*, **31**, 462.
32 Schmitt, F., Kirchmann, P.S., Bovensiepen, U., Moore, R.G., Rettig, L., Krenz, M., Chu, J.-H., Ru, N., Perfetti, L., Lu, D.H., Wolf, M., Fisher, I.R., and Shen, Z.-X. (2008) *Science*, **321**, 1649.
33 Morin, F.J. (1959) *Phys. Rev. Lett.*, **3**, 34.
34 Eyert, V. (2002) *Ann. Phys.*, **11** (9), 650.
35 Tanaka, A. (2006) *Phys. Rev. B*, **378–80**, 269–270.
36 Tomczak, J.M. and Biermann, S. (2007) *J. Phys. Condens. Matter*, **19**, 365206.
37 Tomczak, J.M., Aryasetiawan, F., and Biermann, S. (2008) *Phys. Rev. B*, **78**, 115103.
38 Laad, M.S., Craco, L., and Müller-Hartmann, E. (2006) *Phys. Rev. B*, **73**, 195120.
39 Liebsch, A., Ishida, H., and Bihlmayer, G. (2005) *Phys. Rev. B*, **71**, 085109.
40 Biermann, S., Poteryaev, A., Lichtenstein, A.I., and Georges, A. (2005) *Phys. Rev. Lett.*, **94**, 026404.
41 Koethe, T.C., Hu, Z., Haverkort, M.W., Schüß ler-Langeheine, C., Venturini, F., Brookes, N.B., Tjernberg, O., Reichelt, W., Hsieh, H.H., Lin, H.-J., Chen, C.T., and Tjeng, L.H. (2006) *Phys. Rev. Lett.*, **97**, 116402.
42 Tomczak, J.M. and Biermann, S. (2009) *Phys. Rev. B*, **80**, 085117; Tomczak, J.M. and Biermann, S. (2009) *Europhys. Lett.*, **86**, 37004; Tomczak, J.M. and Biermann, S. (2009) *Phys. Stat. Sol. B*, **246**, 9.
43 Qazilbash, M.M., Brehm, M., Andreev, G.O., Frenzel, A., Ho, P.-C., Chae, B.-G., Kim, B.-J., Yun, S.J., Kim, H.-T., Balatsky, A.V., Shpyrko, O.G., Maple, M.B., Keilmann, F., and Basov, D.N. (2009). *Phys. Rev. B*, **79**, 075107.
44 Cavalleri, A., Tóth, Cs., Siders, C.W., Squier, J.A., Ráksi, F., Forget, P., and Kieffer, J.C. (2001) *Phys. Rev. Lett.*, **87**, 237401.
45 Cavalleri, A., Rini, M., Chong, H.H.W., Fourmaux, S., Glover, T.E., Heimann, P.A., Kieffer, J.C., and

Schoenlein, R.W. (2005) *Phys. Rev. Lett.*, **95**, 067405.

46 Baum, P., Yang, D.-S., and Zewail, A.H. (2007) *Science*, **318**, 788.

47 Cavalleri, A., Rini, M., Chong, H.H.W., Fourmaux, S., Glover, T.E., Heimann, P.A., Kieffer, J.C., and Schoenlein, R.W. (2004) *Phys. Rev. B*, **70**, 161102.

48 Kübler, C., Ehrke, H., Huber, R., Lopez, R., Halabica, A., Haglund, R.F., Jr., and Leitenstorfer, A. (2007) *Phys. Rev. Lett.*, **99**, 116401.

2
Quasi-Particles and Collective Excitations

Evgueni V. Chulkov, Irina Sklyadneva, Mackillo Kira, Stephan W. Koch, Jose M. Pitarke, Leonid M. Sandratskii, Paweł Buczek, Kunie Ishioka, Jörg Schäfer, and Martin Weinelt

2.1
Introduction

A solid contains about 10^{22} electrons and vibrating ions per cubic centimeter all of which interact with each other via the long-range Coulomb interaction. In this respect, it is astounding that we can describe the electronic band structure of a solid in a single-particle picture and obtain in most cases reasonable agreement with experimental data. In such a single-particle approach, we hide the electron–electron interaction in some average potential and neglect all the other elementary excitations of the solid. Essentially, we consider a bare electron moving through the modified periodic potential of the crystal. The outcome of such a simplified model are Bloch states characterized by the wave vector $\mathbf{k}$ and the dispersion relation $E(\mathbf{k})$. The electron's velocity and mass are modified by the crystal potential. They can be expressed by the first and second derivatives of $E(\mathbf{k})$ with respect to $\mathbf{k}$ (see, for example, Appendix E in Ref. [1]). The interaction of the electron with the crystal potential is now included in the dispersion relation. The electron is no longer free but can be considered as a quasi-electron with a $\mathbf{k}$-dependent velocity and effective mass.

We can extend this concept of the quasi-electron and argue that all kinds of elementary excitations in a solid that modify the properties of the electron contribute to its quasi-particle nature. For example, each electron repels all the other electrons of the crystal by Coulomb interaction. This must lead to a positive cloud around each electron relative to the average electron density of the solid. The charge of the electron becomes screened and the long-range Coulomb interaction is replaced by the screened Coulomb interaction. This screened Coulomb interaction can be described by individual electron–electron scattering processes in which two electrons

Dynamics at Solid State Surfaces and Interfaces: Volume 2: Fundamentals, First Edition.
Edited by Uwe Bovensiepen, Hrvoje Petek, and Martin Wolf.

transfer among themselves a particular momentum Δk and energy ΔE. In contrast to the steady state of Bloch electrons, the screened Coulomb interaction among the electrons leads to a finite lifetime τ of all interacting electronic states. Furthermore, the dynamics of the screening process may lead to collective excitations. At the birth of a quasi-particle, such as a photoexcited electron or hole, the long-range Coulomb interaction is barely suppressed by screening. This can lead to collective oscillations of the electron gas, the so-called plasmons. At lower electron densities, we retrieve some of the long-range character of the Coulomb interaction. This becomes particularly important for systems with an energy gap between ground and first excited electronic state, where the mutual interaction between excited electrons and holes leads to the formation of excitons. Besides carrier–carrier scattering, the Coulomb interaction of the moving electron with the ion cores leads to deformation of the lattice, which can be described by emission and reabsorption of phonons, the quanta of the lattice vibrations. The electron becomes dressed by a virtual phonon cloud that it carries through the crystal. This interaction, where the polarization of the lattice acts back on the electron, can be transformed into a quasi-particle called a polaron. The electron may become heavier and slower and the band dispersion $E(\mathbf{k})$ will have a smaller slope near the Fermi level, where such low-energy excitations are important. The electron with its phonon cloud behaves like a noninteracting particle with a **k**-dependent effective mass.

So far we have neglected the spin of electrons. It is the spins, in combination with the Pauli principle, that underlie the exchange interaction. The latter can lead to alignment of the electron spins and support a magnetically ordered ground state. The low-energy excitations of this ground state are spin waves and the corresponding quanta are called magnons. These collective magnetic excitations will couple to electrons and phonons and enrich the response function of magnetic materials.

The coarse energy- and timescales of quasi-particle and collective excitations are illustrated in Figure 2.1. The decay rate $\hbar/\tau$, and thus lifetime τ of a quasi-particle, relates to the natural linewidth Γ of the corresponding photoemission line, and thus to the imaginary part of the self energy Σ of the corresponding state via

$$\hbar/\tau = \Gamma = -2\,\mathrm{Im}\,\Sigma \tag{2.1}$$

This relationship allows us to experimentally detect and analyze the signatures of quasi-particle and collective excitations in both energy- and time-resolved spectroscopy. However, one must be careful when evaluating lifetimes from energy-resolved spectroscopy via Eq. (2.1) since besides the decay rate, the band dispersion, the inhomogeneous broadening, and the dephasing processes may add to the linewidth in a solid-state system. Though often used as a rule of thumb, the lifetime of a quasi-particle should not be mistaken for the time it requires to excite the quasi-particle [2, 3].

This chapter is intended as a short introduction to the physics of elementary excitations in solids. Each section is written by experts in the field who are listed below. To structure the theoretical description, we make a distinction between quasi-particle and collective excitations that should not be taken axiomatic.

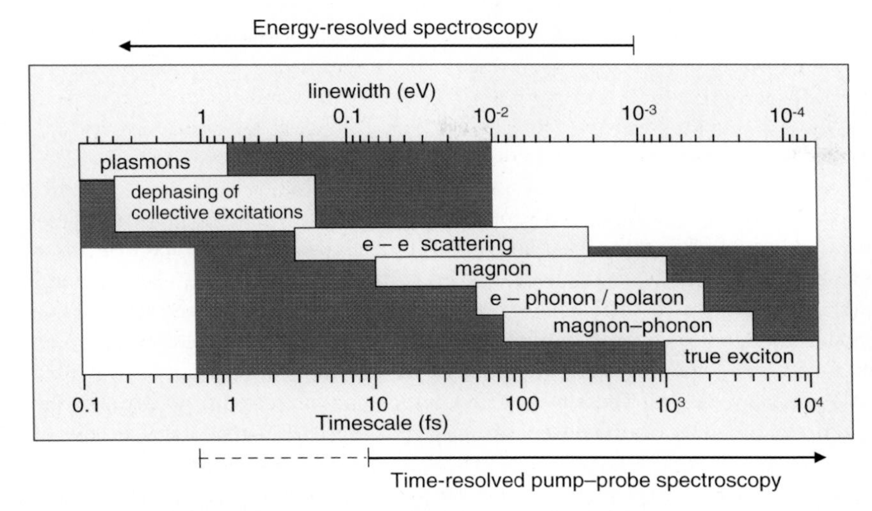

Figure 2.1 Quasi-particle and collective excitations. The lifetime of an excitation τ and its approximate energy Γ are related via $\Gamma \cdot \tau = \hbar = 658$ meV fs. After Ref. [4].

In Section 2.2, electrons and phonons are introduced by Evgueni Chulkov and Irina Sklyadneva with an emphasis on electron–phonon coupling in metals. This is followed by a discussion of the physics of coherent and incoherent excitons in semiconductor quantum wells and their signatures in terahertz and photoluminescence spectroscopy by Mackillo Kira and Stephan W. Koch. A contribution by Jose M. Pitarke on polarons, that is, electron–phonon coupling in polar and ionic materials, completes the description of quasi-particle excitations.

Section 2.3 deals with collective excitations. Jose M. Pitarke gives an introduction to plasmons, the collective oscillations of the electron gas in metals. Magnons are the low-energy collective excitations in magnetically ordered systems. Their physics is discussed by Leonid M. Sandratskii and Paweł Buczek.

Many experimental techniques allow us to study quasi-particle formation and decay, most of which are discussed in Volume 1 of this series. Section 2.4 restricts itself to an overview of three techniques quite prominent in the field. First, the all-optical ultrafast spectroscopy of coherent phonons, as an example of collective excitations, is introduced by Kunie Ishioka. Jörg Schäfer summarizes recent developments in high-resolution photoemission spectroscopy, which gives experimental access to elementary excitations in the energy domain. Finally, the signatures of quasi-particle and collective exciations in time-resolved photoelectron spectroscopy are outlined by Martin Weinelt.

There is a wealth of books and review articles on quasi-particle and collective excitations. For a comprehensive introduction, we refer the reader to the textbooks by Kittel, Madelung, and Czycholl [5–7]. Furthermore, there are a number of excellent reviews, for example, [4, 8–12], which may be useful.

2.2 Quasi-Particles

2.2.1 Electrons and Holes

As outlined in the introduction, various experiments, in particular photoemission measurements, show that excitation spectra of metals can be understood in terms of excited single particles, electrons and holes, with certain momentum **k** and energy $E(\mathbf{k})$ [13]. This picture is analogous to that of a free (noninteracting) electron gas (FEG) model where an electron excitation is formed by transition of an electron from an occupied state (below the Fermi level, E_F) to an unoccupied state (above E_F) with a hole left below E_F [14]. These two excited particles are protected from decay by the absence of any interaction between the electrons in the FEG model irrespective of the excitation energy.

In real systems despite that electrons strongly interact with each other, a single-particle excitation picture is still valid for relatively low excitation energies since electron–electron interaction results in the decay of excited electrons and holes, that is, final lifetimes of these particles. The final lifetime or lifetime broadening is also influenced by the interaction of the excited electron with other (quasi)particles [15], for example, with single-particle excitations, electrons and holes [9], and with collective excitations, phonons [16], magnons [17], and plasmons [18].

Electron excitation energies and the excited particle decay being closely related to the electron band structure of a metal are also related to the phonon, magnon, and plasmon spectra of a metal via the interaction with these quasi-particles and depend on the dimensionality of a system. For instance, on metal surfaces the decay of electrons excited in surface electron states will be modified by the interaction with surface phonons, magnons, and plasmons [9, 16–18]. Therefore, decay mechanisms on surfaces and a single-particle interpretation of electron excitations may be very different from those in bulk. Bulk metals exibit three-dimensional translational symmetry that results in discrete electron spectra at any selected momentum in the Brillouin zone. With such an electronic structure, the decay of excited electrons in the absence of defects can occur only via inelastic scattering of the excited particle with other electrons and other quasi-particles (phonons, magnons, and plasmons). The formation of a metal surface leads to the loss of translational symmetry in the direction perpendicular to the surface. As a result, such a semiinfinite metal exibits only two-dimensional (2D) translational symmetry and the bulk states form a continuum of electronic states with energy gaps in the projection of the bulk band structure onto the 2D surface Brillouin zone. As a consequence, at metal surfaces there exist not only gap surface states and gap image potential states but also surface and image potential resonances can be found lying outside the energy gaps and, therefore, are degenerate in energy with bulk electronic states. The change in the character of the electronic states from gap state to resonance results in a change in the dominating decay mechanism. Gap states decay via many-body electron–electron (inelastic) scattering, while resonance decay via one-electron transition, that is, energy conserving resonant electron transfer into the bulk metal states. The latter mechanism is significatly more efficient than inelastic electron–electron scattering for electrons excited into resonance states. This one-electron decay mechanism does not exist in bulk metals. However, it is ubiquitous on clean metal surfaces, on surfaces with adatoms/clusters/islands, and on surfaces covered with adlayers making a single particle interpretation of electron excitations more complicated.

2.2.2 Phonons

One of the most important physical phenomena in solids is the vibration of atoms about their equilibrium positions. In general, a description of such a motion is a formidable problem. In solids, it can be simplified (i) by taking into account the lattice translational symmetry and (ii) making approximations such as adiabatic and harmonic. The first one allows us to reduce the problem to the atomic motion in the primitive unit cell containing a reduced number of atoms and electrons, the second simplifies the equation of motion.

2.2.2.1 Adiabatic Approximation

To exclude from our consideration the motion of electrons around the ion cores, the well-known adiabatic approximation is used. This approximation means that the nuclei and electron degrees of freedom can be decoupled due to the considerable difference in their masses. In this case, (i) the values concerning the motion of electrons are calculated at the fixed equilibrium positions of the ion cores and (ii) the dynamics of the ion cores is considered as a motion in an averaged potential field generated by the electrons.

2.2.2.2 Harmonic Approximation

Let us consider the potential energy of a crystal as a function of atomic positions $\mathbf{R}_{m\alpha} = \mathbf{R}^{\circ}_{m\alpha} + \mathbf{S}_{m\alpha}$. Here, $\mathbf{R}^{\circ}_{m\alpha} = \mathbf{R}^{\circ}_{m} + \tau^{\circ}_{\alpha}$, superscript $\circ$ indicates the equilibrium position expressed in terms of the position $\mathbf{R}^{\circ}_{m}$ of unit cell m, and the position τ°_{α} of atom α in the unit cell, and $\mathbf{S}_{m\alpha}$ is a displacement of atom α in unit cell m from the equilibrium position. In most of the physically interesting cases, the relative displacements of the atoms are small compared to interatomic distances. Therefore, one can expand the potential energy in powers of the atomic displacements from the equilibrium position:

$$V(\ldots \mathbf{R}_{m\alpha} \ldots) = V(\ldots \mathbf{R}^{\circ}_{m\alpha} \ldots) + \sum_{m\alpha i} \left. \frac{\partial V}{\partial R_{m\alpha i}} \right|_{\mathbf{R}_{m\alpha} = \mathbf{R}^{\circ}_{m\alpha}} S_{m\alpha i} + \frac{1}{2} \sum_{\substack{m\alpha i \\ n\beta j}} \left. \frac{\partial^2 V}{\partial R_{m\alpha i} \partial R_{n\beta j}} \right|_{\substack{\mathbf{R}_{m\alpha} = \mathbf{R}^{\circ}_{m\alpha} \\ \mathbf{R}_{n\beta} = \mathbf{R}^{\circ}_{n\beta}}} S_{m\alpha i} S_{n\beta j} + \cdots \tag{2.2}$$

Here, $R_{m\alpha i}$, $R^{\circ}_{m\alpha i}$, and $S_{m\alpha i}$ are the components of vectors $\mathbf{R}_{m\alpha}$, $\mathbf{R}^{\circ}_{m\alpha}$, and $\mathbf{S}_{m\alpha}$, respectively. Since we are interested in the atomic motion, the first term can be omitted: it does not depend on $\mathbf{S}_{m\alpha}$. The linear term vanishes as the potential energy is expanded at the equilibrium position of atoms where the potential energy has to have an absolute or relative minimum to stabilize the crystal lattice. The first term that determines the dynamical properties of a crystal is the quadratic one. If we take the potential energy of a crystal in this quadratic form and neglect the higher order terms in the Expansion 2.2, we express the potential energy in the so-called harmonic approximation.

2.2.2.3 **Lattice Dynamics**

Let us define

$$\Phi_{\substack{m\alpha i\\ n\beta j}} = \left.\frac{\partial^2 V}{\partial X_{m\alpha i}\partial X_{n\beta j}}\right|_{\substack{\mathbf{R}_{m\alpha} = \mathbf{R}^{\circ}_{m\alpha}\\ \mathbf{R}_{n\beta} = \mathbf{R}^{\circ}_{n\beta}}} \tag{2.3}$$

Now one can write the equation of motion where each atom can be considered as a simple harmonic oscillator:

$$M_\alpha \ddot{S}_{m\alpha i} + \sum_{n\beta j} \Phi_{\substack{m\alpha i\\ n\beta j}} S_{n\beta j} = 0 \tag{2.4}$$

Here, M_α is the mass of atom α and $\Phi_{\substack{m\alpha i\\ n\beta j}}$ are interatomic force constants that determine the force acting on atom α in unit cell m along direction i when atom β in unit cell n shifts in j-direction over distance $S_{n\beta j}$, while all other atoms keep their equilibrium positions.

The simplest solutions of the equation of motion are normal crystal vibrations when all atoms vibrate with the same frequency ω. The number of such normal vibrations and frequencies is equal to the number of atoms N multiplied by 3. Such vibrations are equivalent to a system of independent harmonic oscillators with frequencies equal to the normal frequencies of the lattice. A normal vibration mode can be taken in a general form of $S_{m\alpha i} = A_{\alpha i} \cdot \exp\left[i(\mathbf{q}\cdot\mathbf{R}^{\circ}_{m\alpha} - \omega t)\right]$, where $\mathbf{q}$ is a wave vector showing the direction of wave propagation, ω is the circular frequency of the wave, and $A_{\alpha i}$ is the amplitude of vibration. The energies of the normal modes are quantized and the quantum of energy, $\omega_{\mathbf{q}}$, is associated with an elementary excitation, the phonon. In the harmonic approximation, the lattice dynamics can be expressed in terms of a system of noninteracting phonons that can be considered as quasi-particles since, for convenience, a momentum $\hbar\mathbf{q}$ is assigned to each phonon in mode $\mathbf{q}$. Thus, the problem is to find the energies (or frequencies) of the independent phonons as a function of their wave vector, $\omega = \omega(\mathbf{q})$, the so-called phonon dispersion.

By substituting $S_{m\alpha i}$ into the equation of motion (Eq. (2.4)) and multiplying by $\exp\left[-i(\mathbf{q}\mathbf{R}^{\circ}_{m\alpha} - \omega t)\right]$, we obtain

$$\sum_{\beta j}\left[\left(\sum_{\mathbf{h}} \Phi^{\mathbf{h}}_{\substack{\alpha i\\ \beta j}} \exp\left[i\mathbf{q}(\mathbf{h} + \tau_\beta - \tau_\alpha)\right]\right) - \sqrt{M_\alpha M_\beta}\,\omega^2\,\delta_{\alpha\beta}\,\delta_{ij}\right] A_{\beta j} = 0 \tag{2.5}$$

where $\mathbf{h} = \mathbf{R}_n^\circ - \mathbf{R}_m^\circ$. Let us define the dynamical matrix of a crystal as

$$D_{\substack{\alpha i \\ \beta j}}(\mathbf{q}) = \frac{1}{\sqrt{M_\alpha M_\beta}} \sum_{\mathbf{h}} \Phi^{\mathbf{h}}_{\substack{\alpha i \\ \beta j}} \exp\left[i\mathbf{q}(\mathbf{h} + \tau_\beta - \tau_\alpha)\right] \tag{2.6}$$

Now, one can rewrite Eq. (2.5) as a system of linear equations:

$$\sum_{\beta j} \left[D_{\substack{\alpha i \\ \beta j}}(\mathbf{q}) - \omega^2 \delta_{\alpha\beta} \delta_{ij} \right] \varepsilon_{\beta j}(\mathbf{q}) = 0, \tag{2.7}$$

which is equivalent to

$$\det \left| D_{\substack{\alpha i \\ \beta j}}(\mathbf{q}) - \omega^2 \delta_{\alpha\beta} \delta_{ij} \right| = 0, \tag{2.8}$$

where $\varepsilon_{\beta j}(\mathbf{q}) = \sqrt{M_\beta} \cdot A_{\beta j}$. Equations (2.7) and (2.8) are general and can be used for 1D, 2D, and 3D translationally invariant systems. For each $\mathbf{q}$, there are $3n$ vibrational modes or phonon branches (here n is the number of atoms in the unit cell) with frequencies $\omega_{\mathbf{q},\nu} (\nu = 1, \ldots, 3n)$ and $3n$ eigenvectors $\hat{\varepsilon}_{\mathbf{q},\nu}$ describing the polarization of the wave of displacements.

To evaluate the phonon dispersion of normal modes, various approximations are used. For instance, if we consider only acoustic modes (phonons) that at low frequencies have a linear relationship between their frequency and wave vector $\mathbf{q}$ and extend the low-frequency limit to the maximum phonon frequency $\omega_{\max}$, we obtain the Debye model for phonons where $\omega_{\max}$ is replaced by the Debye frequency ω_{D}. In this model, optical branches of the phonon spectrum correspond to large values of $\mathbf{q}$. The optical phonon modes can also be taken into account in the Einstein model where the phonon density of states (DOS) is simply $F(\omega) = \delta(\omega - \omega_{\mathrm{E}})$. In this approximation, the frequencies of optical phonons do not depend on $\mathbf{q}$ and are equal to the Einstein frequency ω_{E}.

2.2.2.4 **Phonons at Surfaces**

At surfaces the number of nearest neighbors changes when compared to the bulk. As a result, vibrational modes appear that are not allowed within the solid. These modes propagate along the surface with wave vector $\mathbf{q}$ ($\mathbf{q}$ is now two dimensional) and decay exponentially in amplitude into the bulk. Such modes are called surface phonons and include both atomic vibrations localized at the surface and manifestations of bulk vibrations at the surface, the so-called surface resonances. The surface-localized modes exist in the gaps of the bulk phonon spectrum where particular values of $\mathbf{q}$ are not allowed for bulk phonons. The surface atoms can also acquire relatively large (or small) displacement amplitudes compared to atoms in the bulk due to the change in the force constants near the surfaces. The corresponding decrease (or increase) in coupling constants tends to lower (or elevate) the vibrational mode frequencies such that they lie beyond the limits of the bulk phonon frequencies.

As far as the surface resonances are concerned, they exist inside the bulk continuum and have both surface-localized and bulk-like components. To illustrate

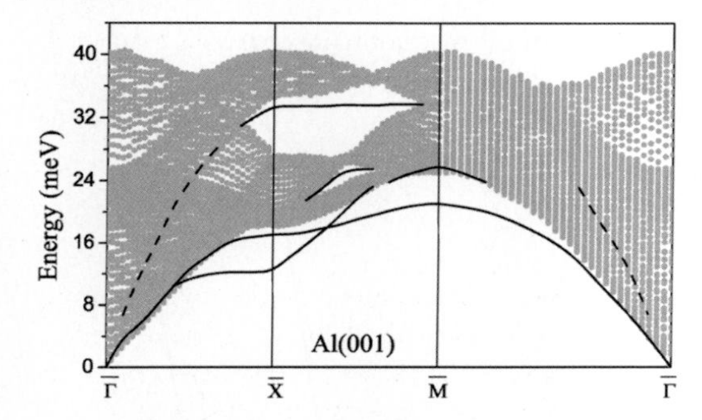

Figure 2.2 Example of a calculated dispersion of surface vibrational modes: the surface phonons (solid lines) and surface resonances (dashed lines) on Al(001) are shown together with the bulk vibrational modes projected on the two-dimensional Brillouin zone of the Al(001) surface (gray dots).

the above, Figure 2.2 shows the dispersion of the surface phonons and resonances on the Al(001) surface.

2.2.3
Electron–Phonon Coupling in Metals

The effect produced by the moving lattice on the conducting electrons can be considered as the scattering of electrons by phonons or as the electron–phonon (e–ph) coupling. As a result of this interaction, an electron can change its momentum and (i) emit a phonon to be released from its excess energy or (ii) a phonon can be absorbed by an electron.

The e–ph interaction is a fundamental many-body process that can be tested both experimentally and theoretically. For example, this interaction contributes to the finite lifetime of excited electrons (cf. Section 2.4.3.3). In addition, at the Fermi level, within a typical phonon energy $\hbar\omega_{max}$, the electronic dispersion is renormalized due to the scattering of electrons by phonons such that the dispersion becomes flatter (cf. Section 2.4.2.2). As a result, the electron's effective mass at the Fermi energy and the density of states is enhanced [19]. The increase in the effective mass can be described by the electron–phonon mass enhancement parameter λ: $m^* = m_0(1+\lambda)$, where m^* and m_0 are the effective masses with and without taking into account the e–ph interaction, respectively.

The effect of the e–ph coupling on the dispersion and lifetime of electronic states can be expressed in terms of the complex e–ph self-energy Σ_{e-ph}. The real part of the self-energy, $\mathrm{Re}\Sigma_{e-ph}$, determines the renormalization of an electronic energy band i close to the Fermi level.

$$m^*_{(i)}(\mathbf{k}) = m_0\left(1 - \frac{\partial \mathrm{Re}\Sigma_{e-ph}(\mathbf{k};\omega)}{\partial\omega}\right) \equiv m_0(1+\lambda_{(i)}(\mathbf{k})). \tag{2.9}$$

The phonon-induced decay rate (lifetime broadening) Γ_{e-ph} of an electronic state can be obtained from the imaginary part $\mathrm{Im}\Sigma_{e-ph}$ of the self-energy:

$\Gamma_{e-ph} = 2\text{Im}\,\Sigma_{e-ph}$. All these quantities are related to the Eliashberg spectral function [19, 20] $\alpha^2 F(\omega)$. For an initial electron state (i) with momentum $\mathbf{k}$ and energy ε_i

$$\alpha^2 F^{E(A)}_{(i)}(\mathbf{k};\omega) = \sum_{\nu,f} \int d\mathbf{q}\delta\left(\varepsilon_i - \varepsilon_f \mp \omega_{\mathbf{q},\nu}\right) \times |g^{i,f}(\mathbf{k},\mathbf{q},\nu)|^2 \delta(\omega - \omega_{\mathbf{q},\nu}). \tag{2.10}$$

It is nothing else than the phonon density of states weighted by the e–ph coupling due to phonon emission (E, sign " $-$ " in the delta function) and phonon absorption (A, sign " $+$ " in the delta function) processes. During phonon emission, an electron is scattered from $\mathbf{k}$ to $\rightarrow \mathbf{k}' = \mathbf{k} - \mathbf{q}$ creating a phonon with momentum $\mathbf{q}$. In the phonon absorption process, an electron is scattered from $\mathbf{k}$ to $\rightarrow \mathbf{k}' = \mathbf{k} + \mathbf{q}$ absorbing a phonon with momentum $\mathbf{q}$. The summation in Eq. (2.10) is carried out over all possible final states (f) and phonon modes ν. The probability of electron scattering from an initial state (i) with momentum $\mathbf{k}$ to a final electronic state (f) with momentum $\mathbf{k}' = \mathbf{k} \mp \mathbf{q}$ by emission (absorption) of a phonon ($\mathbf{q}, \nu$) is described by the e–ph matrix element $g^{i,f}(\mathbf{k},\mathbf{q},\nu)$:

$$g^{i,f}(\mathbf{k},\mathbf{q},\nu) = \sqrt{\frac{1}{2M\omega_{\mathbf{q},\nu}}} \langle \Psi_{\mathbf{k}i} | \hat{\varepsilon}_{\mathbf{q},\nu}\, \delta V_{\mathbf{q},\nu} | \Psi_{\mathbf{k}\mp\mathbf{q}f} \rangle. \tag{2.11}$$

Here, M is the atomic mass, $\Psi_{\mathbf{k}i}$ and $\Psi_{\mathbf{k}\mp\mathbf{q}f}$ are the electronic wave functions for the initial and final states, respectively. $\delta V_{\mathbf{q},\nu}$ gives the gradient of the crystal potential with respect to atomic displacements induced by the phonon mode ($\mathbf{q}, \nu$) with frequency $\omega_{\mathbf{q},\nu}$ and phonon polarization vector $\hat{\varepsilon}_{\mathbf{q},\nu}$. As an example, Figure 2.3 shows

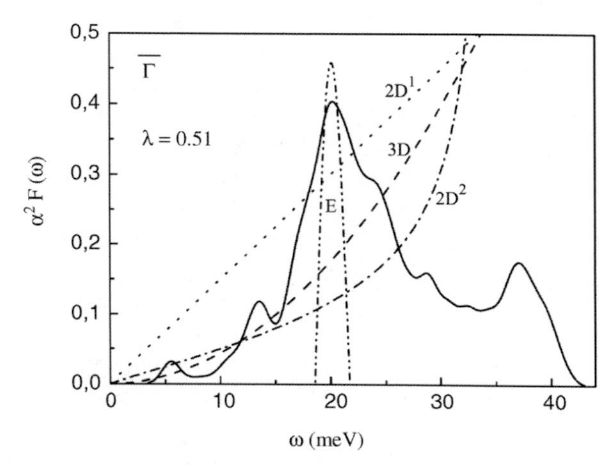

Figure 2.3 Example of a calculated Eliashberg spectral function, $\alpha^2 F(\omega)$, for the surface electronic state at the $\bar{\Gamma}$ point of the surface Brillouin zone of Al(001) (solid line). Also shown are the spectral functions obtained in the 3D Debye model (dashed line), in the 2D Debye model when only the first-order diagram is included (2D^1, dotted line), in the 2D Debye model with the first- and second-order diagrams included (2D^2, dashed-dotted line), and in the Einstein approximation (E).

the Eliashberg spectral function for the surface electronic state at the $\bar{\Gamma}$ point of the Al (001) surface (solid line) [21].

The inverse lifetime or lifetime broadening $\Gamma_{e-ph} = \hbar/\tau$ is the integral over all the scattering events that conserve energy and momentum [19].

$$\begin{aligned}\Gamma^{(i)}_{e-ph}(\mathbf{k};T) &= -2\,\mathrm{Im}\,\Sigma_{e-ph}\\ &= 2\pi\hbar\int_0^{\infty}\{\alpha^2F^{E}_{(i)}(\mathbf{k};\omega)[1+n(T;\omega)-f(T;\varepsilon_i-\omega)]\\ &\quad+\alpha^2F^{A}_{(i)}(\mathbf{k};\omega)[n(T;\omega)+f(T;\varepsilon_i+\omega)]\mathrm{d}\omega.\end{aligned} \tag{2.12}$$

Here, f and n are the Fermi and Bose distribution functions, which introduce the temperature dependence of Γ_{e-ph}. Since the phonon energies are small compared to the electronic energy scale, one can neglect the change in energy in the e–ph scattering processes and assume that $\delta(\varepsilon_i-\varepsilon_f \mp \omega_{q,\nu}) \approx \delta(\varepsilon_i-\varepsilon_f)$. This so-called quasi-elastic assumption allows us to use the same Eliashberg function for both emission and absorption processes:

$$\begin{aligned}\Gamma^{(i)}_{e-ph}(\mathbf{k};T) &= 2\pi\hbar\int_0^{\infty}\alpha^2F_{(i)}(\mathbf{k};\omega)[1-f(T;\varepsilon_i-\omega)\\ &\quad+f(T;\varepsilon_i+\omega)+2n(T;\omega)]\mathrm{d}\omega.\end{aligned} \tag{2.13}$$

Let us obtain the behavior of the e–ph induced linewidth in the limiting cases. When $T\to 0$ and the electronic energy ε_i exceeds the maximum phonon energy $\hbar\omega_{max}$

$$T\to 0:\Gamma^{(i)}_{e-ph}(\mathbf{k}) = 2\pi\hbar\int_0^{\omega_{max}}\alpha^2F(k;\omega)\mathrm{d}\omega. \tag{2.14}$$

Since at $T = 0$ all the electronic states above the Fermi energy, E_F, are empty, electrons cannot scatter into a hole at the Fermi level. In the same way, the hole cannot decay into a lower energy state as only phonon emission is allowed. Therefore, the linewidth for holes at E_F is equal to zero and they have infinite lifetime. As the temperature is increased, some electronic states above E_F become occupied and can now fill the hole by emitting phonons or the holes can decay to a lower energy state by phonon absorption. Γ_{e-ph} increases monotonically with temperature up to a maximum value at $\omega = \omega_{max}$. An important feature of the e–ph interaction is the linear temperature dependence of Γ_{e-ph} at elevated temperatures, when k_BT is higher than the maximum phonon energy. In this limit, Eq. (2.13) can be written as

$$\Gamma^{(i)}_{e-ph}(\mathbf{k};T) = 2\pi\lambda_{(i)}(\mathbf{k})\,\mathrm{k_B T} \tag{2.15}$$

with a slope determined by the electron–phonon coupling parameter λ [19]

$$\lambda_{(i)}(\mathbf{k}) = 2\int_0^{\omega_{max}}\frac{\alpha^2F_{(i)}(\mathbf{k};\omega)}{\omega}\mathrm{d}\omega. \tag{2.16}$$

As an example, the temperature dependence of Γ_{e-ph} is shown in Figure 2.4 (solid line) for the surface electronic state at the $\bar{\Gamma}$ point of the Al(001) surface [21].

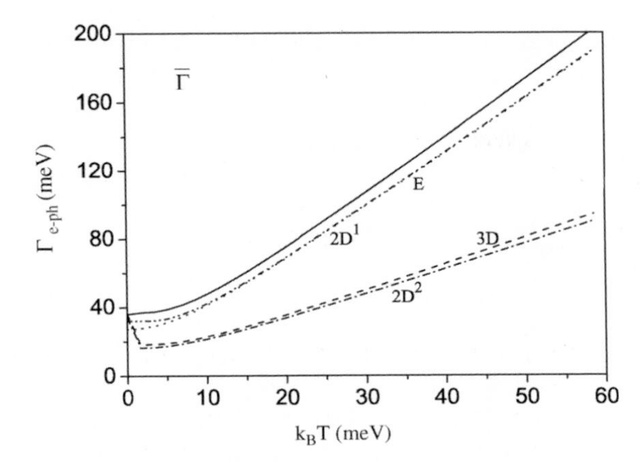

Figure 2.4 Example of the temperature dependence of the phonon-induced inverse lifetime, $\Gamma_{e-ph}(T)$, for the surface electronic state at the $\bar{\Gamma}$ point of the surface Brillouin zone of Al(001) (solid line). Also shown are $\Gamma_{e-ph}(T)$ in the Debye and Einstein models, the notations are the same as in Figure 2.3.

The linear dependence has often been used to extract the e–ph coupling parameter λ for electronic states with energies much larger than the maximum phonon energy. The method uses the temperature dependence of the linewidths of quasi-particle peaks measured in angle-resolved photoemission spectroscopy (ARPES). At temperatures much higher than the Debye temperature ($T \gg T_D$), this relation becomes linear and independent of the details of the phonon spectrum. Thus, if the Debye temperature T_D is not too high, λ can be deduced experimentally from the slope $d\Gamma/dT$ of the linewidth Γ plotted versus the temperature T [19]. In most other cases, the e–ph parameter can be obtained by fitting the experimental data with a simple model for the phonon density of states. This method with the assumption that the predominant contribution to the linewidth's temperature dependence can be attributed to e–ph coupling has widely been used to obtain λ for many surface electronic states except for those located close to E_F [16].

For a first estimate of the e–ph coupling and its contribution to the lifetime of excited electrons, the Debye and Einstein models are widely used to describe the phonon spectrum of a system. With the drastic simplification of a constant e–ph matrix element $g^{i,f}(\mathbf{k}, \mathbf{q}, \nu)$, one can obtain analytical expressions for both Eliashberg function and Γ_{e-ph} in terms of the e–ph coupling parameter and the characteristic Debye (Einstein) frequency ω_D (ω_E):

i) In the 3D Debye model for the bulk [19], $\alpha^2 F(\omega) = \lambda(\omega/\omega_D)^2$ for $\omega < \omega_D$ and zero otherwise. The lifetime broadening for $\omega < \omega_D$ is then given by

$$\Gamma_{e-ph} = \frac{2\pi\hbar\lambda\omega_D}{3}\left(\frac{\omega}{\omega_D}\right)^3 \tag{2.17}$$

and $\Gamma_{e-ph} = 2\pi\hbar\lambda\omega_D/3$ for $\omega \geq \omega_D$.

ii) In the 2D Debye model of a surface when only the first-order diagram is included, the Eliashberg spectral function $\alpha^2 F(\omega) = \lambda(\omega/\omega_D)$ and

$$\Gamma_{e-ph} = \frac{\pi\hbar\lambda\omega_D}{2}\left(\frac{\omega}{\omega_D}\right)^2 \tag{2.18}$$

for $\omega < \omega_D$ and $\Gamma_{e-ph} = \pi\hbar\lambda\omega_D/2$ otherwise [19].

If both the first- and the second-order diagrams are included, the 2D Debye model yields

$$\alpha^2 F(\omega) = (\lambda/\pi)\omega(\omega_D^2 - \omega^2)^{-1/2}, \ \omega < \omega_D, \tag{2.19}$$

$$\Gamma_{e-ph} = 2\hbar\lambda\omega_D\left(1 - \sqrt{1 - \left(\frac{\omega}{\omega_D}\right)^2}\right), \quad \omega < \omega_D \tag{2.20}$$

and $\alpha^2 F(\omega) = 0$ and $\Gamma_{e-ph} = 2\hbar\lambda\omega_D$ for $\omega \geq \omega_D$.

Obviously, for $\omega < \omega_D$ both $\alpha^2 F(\omega)$ and Γ_{e-ph} are linear functions of λ and scale with ω/ω_D.

iii) In the Einstein model, electrons interact only with a single-phonon mode with frequency ω_E. In this case,

$$\alpha^2 F(\omega) = \frac{\lambda\omega_E}{2}\delta(\omega - \omega_E). \tag{2.21}$$

And one obtains $\Gamma_{e-ph} = \pi\hbar\lambda\omega_E$.

The corresponding Eliashberg spectral functions and the temperature behavior of Γ_{e-ph} in the Debye ($\omega_D = 34$ meV) and Einstein ($\omega_E = 20$ meV) models are shown for the surface electronic state at the $\bar{\Gamma}$ point of the Al(001) surface in Figures 2.3 and 2.4 by dashed and dotted lines.

2.2.4 Excitons: Electron–Hole Pairs in Semiconductor Quantum Wells

The discussion in this subsection deals with three kinds of quasi-particles: electrons, holes, and excitons. As our model, we take a semiconductor quantum well (QW), that is, a quasi-two-dimensional system where the charge carriers are free to move in a planar structure. Therefore, most of the concepts can also be used for the description of surface excitations.

The unexcited system is an insulator where all low-lying bands are fully occupied by electrons and the conduction-band states are free. The highest occupied band – the valence band – is separated by an energy gap from the lowest conduction band. Here, we focus on systems where this band gap is "direct", that is, the valence-band maximum and the conduction-band minimum (CBM) correspond to the same electron-momentum state, usually zero momentum, which is the center of the Brillouin zone (see Figure 2.5a).

Optical excitation of semiconductors in the spectral vicinity of the direct band gap causes electron transitions from the valence into the conduction band. Referring to

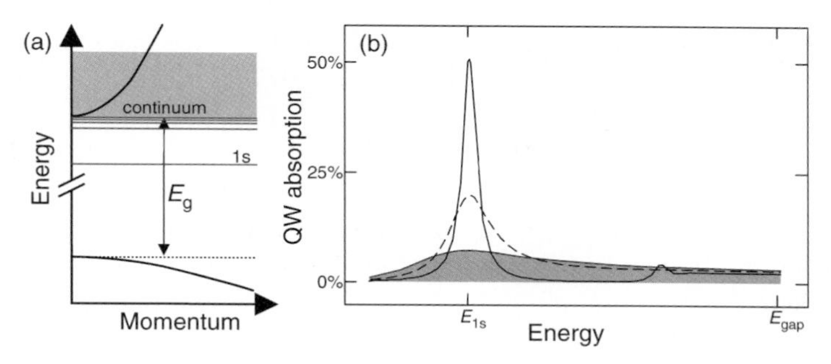

Figure 2.5 (a) Schematic conduction-band and valence-band structure showing energy versus in-plane carrier momentum. The horizontal lines mark the energies of the energetically lowest excitonic states. (b) Quantum well absorption spectra for increasing carrier densities, 10^9 cm^{-2} (solid line), 4×10^{10} cm^{-2} (dashed line), and 9×10^{10} cm^{-2} (gray solid line) assuming a temperature of 40 K.

the missing valence band electrons as "holes," we speak of optically-induced electron–hole pair transitions. The properties of the valence-band holes are those of the missing electrons, that is, they are Fermions and have a spin, charge, effective mass, and so on, which are opposite to that of the valence-band electrons. Since the electron charge is $-|e|$, where e is the elementary charge, the holes are positively charged and we have an attractive Coulomb interaction between the valence-band holes and the conduction-band electrons. (For more details and background information, see, for example, Refs [22–25] and work cited therein.)

Mathematically, the problem of a single electron–hole pair in parabolic bands is identical to that of the hydrogen atom, where the hole plays the role of the proton. In semiconductor physics, the corresponding Schrödinger equation is known as the Wannier equation, and the bound pair-state solutions are the (Wannier) excitons [26–30]. However, instead of the roughly 13.1 eV binding energy (Rydberg energy) in atomic hydrogen, the excitonic Rydberg is typically in the range of a few to 100 meV. This reduction in the pair-state binding energy is observed because the effective electron and hole masses are substantially lighter than the free electron and proton masses. Furthermore, the background dielectric constant reduces the Coulomb interaction strength roughly by one order of magnitude compared to hydrogen.

2.2.4.1 **Microscopic Theory**

For the model of a simple two-band system with a parabolic band structure, see Figure 2.5a, we have the standard many-body Hamiltonian

$$H = H_0 + H_\mathrm{C} + H_\mathrm{I}. \tag{2.22}$$

Here,

$$H_0 = \sum_{\mathbf{k}_{||}} \varepsilon^c_{\mathbf{k}_{||}} a^\dagger_{c,\mathbf{k}_{||}} a_{c,\mathbf{k}_{||}} + \sum_{\mathbf{k}_{||}} \varepsilon^v_{\mathbf{k}_{||}} a^\dagger_{v,\mathbf{k}_{||}} a_{v,\mathbf{k}_{||}} \tag{2.23}$$

is the single-particle part,

$$\begin{aligned}
H_C &= \frac{1}{2}\sum_{\mathbf{k}_{||},\mathbf{k}'_{||},\mathbf{q}_{||}\neq 0} V_{\mathbf{q}_{||}} a^{\dagger}_{c,\mathbf{k}_{||}+\mathbf{q}_{||}} a^{\dagger}_{c,\mathbf{k}'_{||}-\mathbf{q}_{||}} a_{c,\mathbf{k}'_{||}} a_{c,\mathbf{k}_{||}} \\
&+ \frac{1}{2}\sum_{\mathbf{k}_{||},\mathbf{k}'_{||},\mathbf{q}_{||}\neq 0} V_{\mathbf{q}_{||}} a^{\dagger}_{v,\mathbf{k}_{||}+\mathbf{q}_{||}} a^{\dagger}_{v,\mathbf{k}'_{||}-\mathbf{q}_{||}} a_{v,\mathbf{k}'_{||}} a_{v,\mathbf{k}_{||}} \\
&- \sum_{\mathbf{k}_{||},\mathbf{k}'_{||},\mathbf{q}_{||}\neq 0} V_{\mathbf{q}_{||}} a^{\dagger}_{c,\mathbf{k}_{||}+\mathbf{q}_{||}} a_{v,\mathbf{k}'_{||}} a^{\dagger}_{v,\mathbf{k}'_{||}-\mathbf{q}_{||}} a_{c,\mathbf{k}_{||}}
\end{aligned} \tag{2.24}$$

is the Coulomb interaction Hamiltonian, and

$$H_I = -E(t)\sum_{\mathbf{k}_{||}} (d^{*}_{cv} a^{\dagger}_{c,\mathbf{k}_{||}} a_{v,\mathbf{k}_{||}} + d_{cv} a^{\dagger}_{c,\mathbf{k}_{||}} a_{v,\mathbf{k}_{||}}) \tag{2.25}$$

describes the interaction with a classical electromagnetic field. To account for the QW geometry, we have split all three-dimensional momentum vectors $\mathbf{k}$ into their components in z-direction and parallel to the plane of the QW,

$$\mathbf{k} = (\mathbf{k}_{||}, k_z). \tag{2.26}$$

Furthermore, we introduced the creation and annihilation operators $(a^{\dagger}_{\lambda,\mathbf{k}_{||}}, a_{\lambda,\mathbf{k}_{||}})$ for the conduction ($\lambda = c$) and valence ($\lambda = v$) band electrons.

In Eq. (2.23), $\varepsilon^{c}_{\mathbf{k}_{||}}$ ($\varepsilon^{v}_{\mathbf{k}_{||}}$) denotes the single-particle energy of an electron in the conduction (valence) band with in-plane momentum $\mathbf{k}_{||}$. In Eq. (2.24), the first two sums describe the repulsion among electrons in the same bands and the last term includes the interband attraction. Here, the Fourier transform of the Coulomb interaction potential is denoted by $V_{\mathbf{q}_{||}}$ that incorporates the QW confinement. The system is coupled to the light field via the interband dipole matrix-element d_{cv} in Eq. (2.25), showing that the light field either creates or destroys pairs of electrons and holes (missing valence-band electrons).

To compute the semiclassical optical properties of a material system, we have to solve Maxwell's equations. Assuming a light field that propagates normal to the QW plane, we can write the wave equation for the one-dimensional propagation as

$$\left[\frac{\partial^2}{\partial z^2} + \frac{n^2(z)}{c^2}\frac{\partial^2}{\partial t^2}\right] E = +\mu_0 \frac{\partial^2}{\partial t^2} P. \tag{2.27}$$

Here, E is the electromagnetic field, z is the space coordinate normal to the surface of the QW, t denotes the time, and μ_0 is a constant prefactor depending on the system of units. The response of the material has been divided into a resonant part, treated dynamically in terms of the macroscopic optical polarization P, and the nonresonant part lumped into the (background) refractive index n.

We expand the optical polarization P into a Bloch basis [22]

$$P = \sum_{\mathbf{k}_{||}} d_{cv} P_{\mathbf{k}_{||}} + \text{c.c.}, \tag{2.28}$$

where $P_{\mathbf{k}_{||}}$ is the microscopic polarization. The carrier occupation probabilities $f^{e,h}_{\mathbf{k}_{||}}$ and the microscopic polarization $P_{\mathbf{k}_{||}}$ constitute the diagonal and off-diagonal elements of the reduced single-particle density matrix ϱ,

$$\varrho = \begin{pmatrix} \langle a^\dagger_{c,\mathbf{k}_{||}} a_{c,\mathbf{k}_{||}} \rangle & \langle a^\dagger_{v,\mathbf{k}_{||}} a_{c,\mathbf{k}_{||}} \rangle \\ \langle a^\dagger_{c,\mathbf{k}_{||}} a_{v,\mathbf{k}_{||}} \rangle & \langle a_{v,\mathbf{k}_{||}} a^\dagger_{v,\mathbf{k}_{||}} \rangle \end{pmatrix} = \begin{pmatrix} f^{\mathrm{e}}_{\mathbf{k}_{||}} & P_{\mathbf{k}_{||}} \\ P^*_{\mathbf{k}_{||}} & f^{\mathrm{h}}_{\mathbf{k}_{||}} \end{pmatrix}. \tag{2.29}$$

We calculate the equations of motion using the Heisenberg equation

$$i\hbar \frac{\partial}{\partial t} \varrho = [\varrho, H]. \tag{2.30}$$

Working out the commutators in Eq. (2.30), we obtain the semiconductor Bloch equations (SBE) [10, 22, 31]

$$\begin{aligned} \left[i\hbar \frac{\partial}{\partial t} - \varepsilon_{\mathbf{k}_{||}}(t) \right] P_{\mathbf{k}_{||}}(t) &= -\left[1 - f^{\mathrm{e}}_{\mathbf{k}_{||}}(t) - f^{\mathrm{h}}_{\mathbf{k}_{||}}(t) \right] \Omega_{\mathbf{k}_{||}}(t) + \Gamma^{v,c}_{\mathbf{k}_{||}} \\ \frac{\partial}{\partial t} f^{a}_{\mathbf{k}_{||}}(t) &= -\frac{2}{\hbar} \mathrm{Im} \left[\Omega_{\mathbf{k}_{||}}(t) P^*_{k}(t) \right] + \Gamma^{c,c}_{\mathbf{k}_{||}} \end{aligned} \tag{2.31}$$

where the Hartree–Fock terms are shown explicitly and the contributions beyond Hartree–Fock are denoted by $\Gamma^{\lambda,\lambda'}_{\mathbf{k}_{||}}$. In Eq. (2.31),

$$\Omega_{\mathbf{k}_{||}}(t) = d_{cv} E(t) + \sum_{\mathbf{k}'_{||} \neq \mathbf{k}_{||}} V_{|\mathbf{k}_{||} - \mathbf{k}'_{||}|} P_{\mathbf{k}'_{||}}(t) \tag{2.32}$$

is the renormalized field (Rabi energy) and

$$\varepsilon_k(t) = \varepsilon^c_k - \varepsilon^v_k - \sum_{k'_{||} \neq k_{||}} V_{|\mathbf{k}_{||} - \mathbf{k}'_{||}|} \left[f^{\mathrm{e}}_{k'_{||}}(t) + f^{\mathrm{h}}_{k'_{||}}(t) \right] \tag{2.33}$$

is the renormalized transition energy.

At the Hartree–Fock level, the SBE contain neither dephasing of the polarization nor screening of the interaction potential nor relaxation of the carrier distributions. To include these effects, we have to go beyond the Hartree–Fock approximation and specify the correlation contributions $\Gamma^{\lambda,\lambda'}_{\mathbf{k}_{||}}$.

2.2.4.2 **Excitonic Resonances and Populations**

The linear optical response can be computed analytically. For an unexcited semiconductor system, that is, all polarizations and occupations vanish before the system is excited, we can linearize the polarization equation in the interaction with the external field. After a Fourier transform to real space, we obtain

$$\left[i\hbar \frac{\partial}{\partial t} + \frac{\hbar^2 \nabla^2_{\mathbf{r}_{||}}}{2m_r} - V(\mathbf{r}_{||}) \right] P(\mathbf{r}_{||}, t) = -d_{cv}\, E(t)\, \delta(\mathbf{r}_{||}), \tag{2.34}$$

where m_r is the reduced electron–hole mass and $\mathbf{r}_{||}$ is a position within the QW plane.

The homogeneous part of Eq. (2.34) leads to the Wannier equation

$$\left[\frac{\hbar^2\nabla^2_{\mathbf{r}_{||}}}{2m_r} - V(\mathbf{r}_{||})\right]\phi_\nu(\mathbf{r}_{||}) = E_\nu\phi_\nu(\mathbf{r}_{||}). \tag{2.35}$$

It is mathematically identical to the Schrödinger equation for the relative motion in the hydrogen problem. The solutions of the Wannier equation determine both the bound and the inonized exciton states. From the solution of the inhomogeneous Eq. (2.34), we obtain the electron–hole pair susceptibility, which yields the Elliott formula [28] for the linear semiconductor susceptibility

$$\chi(\omega) = 2|d_{cv}|^2 \sum_\lambda \frac{|\phi_\lambda(r=0)|^2}{E_\lambda - \hbar\omega - i\gamma}. \tag{2.36}$$

Here, we wrote only the resonant contribution and introduced γ as phenomenological background dephasing of the polarization. The Elliott formula shows that absorption resonances occur at the frequencies $\omega = E_\lambda/\hbar$. An example of such an excitonic spectrum is plotted in Figure 2.5b.

It is interesting to note that $\chi(\omega)$ displays excitonic resonances even though it describes the linear response of the system. In this limit, the analysis at the single-particle level, that is, the Hartree–Fock approximation, is exact. In fact, one can even find the many-body wave function [10] in the form of a Slater determinant. Hence, the appearance of excitonic resonances in the polarization and therefore in $\chi(\omega)$ cannot imply the presence of bound pair populations, that is, true excitons, since no populations are generated in the linear regime. One therefore refers to the polarization resonances either as "excitonic polarization" or as "coherent excitons." Here, each electronic excitation is in a superposition state between the valence and the conduction bands.

To describe true excitons as well as carrier screening, dephasing, or relaxation contributions, we must extend the theory beyond the Hartree–Fock level and derive equations for the correlations. Systematic and microscopically consistent approximations can be obtained by using the cluster expansion approach [10]. In general, the correlation terms $\Gamma^{\lambda,\lambda'}$ introduce microscopic couplings to the two-particle Coulomb and carrier–phonon correlations,

$$\Gamma^{\lambda,\lambda'}_{\mathbf{k}_{||}} \equiv \sum_{\nu,\mathbf{k}'_{||},\mathbf{q}_{||}\neq 0} V_{\mathbf{q}_{||}}\left[c^{\mathbf{q}_{||},\mathbf{k}'_{||},\mathbf{k}_{||}}_{\lambda,\nu;\nu,\lambda'} - \left(c^{\mathbf{q}_{||},\mathbf{k}'_{||},\mathbf{k}_{||}}_{\lambda',\nu;\nu,\lambda}\right)^*\right] + \Gamma^{\lambda,\lambda'}_{\mathbf{k}_{||},\text{phonon}}, \tag{2.37}$$

where we used the abbreviation

$$c^{\mathbf{q}_{||},\mathbf{k}'_{||},\mathbf{k}_{||}}_{\lambda,\nu;\nu',\lambda'} \equiv \Delta\left\langle a^\dagger_{\lambda,\mathbf{k}_{||}} a^\dagger_{\nu,\mathbf{k}'_{||}} a_{\nu',\mathbf{k}'_{||}+\mathbf{q}_{||}} a_{\lambda',\mathbf{k}_{||}-\mathbf{q}_{||}}\right\rangle. \tag{2.38}$$

True excitons follow from

$$c_X^{\mathbf{q}_{||},\mathbf{k}'_{||},\mathbf{k}_{||}} \equiv c^{\mathbf{q}_{||},\mathbf{k}'_{||},\mathbf{k}_{||}}_{c,\nu;c,\nu} = \Delta\left\langle a^\dagger_{c,\mathbf{k}_{||}} a^\dagger_{\nu,\mathbf{k}'_{||}} a_{c,\mathbf{k}'_{||}+\mathbf{q}_{||}} a_{\nu,\mathbf{k}_{||}-\mathbf{q}_{||}}\right\rangle \tag{2.39}$$

because it describes pair-wise correlations among different electrons and holes.

When the nonlinear absorption is studied, the presence of both coherent and true excitons as well as $f^{\mathrm{e}}, f^{\mathrm{h}}$, and $c_{\lambda,\nu;\nu',\lambda'}$ can significantly modify the probe response. All these contributions appear in the sources for the $\Gamma^{\lambda,\lambda'}_{\mathbf{k}_{||}}$ and $\Gamma^{\lambda,\lambda'}_{\mathbf{k}_{||},\text{phonon}}$ contributions in Eq. (2.37). As one effect, they introduce excitation-induced dephasing that yields spectral broadening of the absorption resonances due to many-body scattering among the excited charge carriers (see Figure 2.5b and Ref. [10] for more details).

To identify coherent and/or true excitions in an excited semiconductor QW, quantitative optical measurements under resonant excitation conditions have recently been performed [32]. For the analysis, the optical response was calculated for more than a 100 000 different electronic many-body states, including two-particle correlations. A detailed comparison between the quantitative experiments and the theory shows that the absorptive nonlinearities depend so sensitively on the many-body configuration that the role of coherent polarization, exciton, and electron–hole plasma contributions can be identified with great confidence.

For a fixed overall carrier density, Figure 2.6a demonstrates how the percentage of carriers bound into excitonic pairs modifies the absorption resonance of a weak probe beam. We see that the presence of true excitons is signified by a resonance narrowing because these excitons – as charge neutral particles – yield a reduced Coulomb scattering relative to the scattering from separate electrons and holes. In a case where the probed system already contains coherent excitons, there is the possibility to transfer energy between the excitonic and the probe polarization. This mechanism can yield significant excitonic gain, as shown in Figure 2.6b. The experimental verification of these effects is presented in Ref. [32].

2.2.4.3 **Terahertz Spectroscopy of Exciton Populations**

A well-established method to detect small concentrations of a particular species of atoms or molecules is to use an optical probe that is sensitive to transitions between the eigenstates of the respective species. If the characteristic absorption resonances are observed in the probe spectrum, the atoms or molecules must be present, and

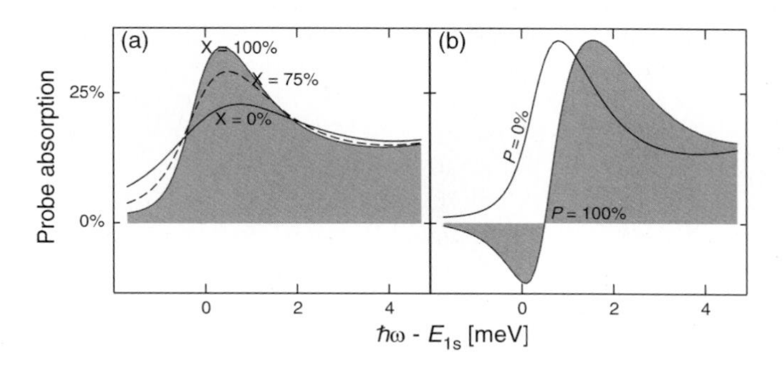

Figure 2.6 (a) Probe absorption spectra for a GaAs/AlGaAs semiconductor quantum well for the fixed carrier density (4×10^{10} cm^{-2}) and different exciton population fractions X. (b) Probe absorption spectra without (solid line) and with (shaded area) coherent excitons, assuming a constant carrier density of (3×10^{10} cm^{-2}).

through proper normalization of the respective oscillator strength one can deduce their relative concentration.

In order to apply this concept to the identification of excitonic populations in typical III–V or II–VI semiconductors, one needs terahertz (THz) fields since these can be resonant with transitions from the exciton's 1s to 2p state [33, 34]. The energy corresponding to the THz regime is way below the fundamental band gap of most direct-gap semiconductors. Hence, THz fields do not generate electron–hole interband excitations in those systems. Instead, they lead to intraband transitions, that is, to transitions between different many-body states.

If the system is in the completely incoherent, quasi-stationary regime, we can find an analytic expression for the linear THz susceptibility [10]. In that case, we obtain the THz susceptibility as

$$\chi_{\mathrm{THz}}(\omega) \propto \sum_{\lambda} (S_{\lambda}(\omega) - [S_{\lambda}(-\omega)]^{*}) N_{\lambda}^{ex}, \tag{2.40}$$

$$S_{\lambda}(\omega) = \sum_{\beta} \frac{|D_{\lambda,\beta}|^2}{E_{\beta} - E_{\lambda} - \hbar\omega - i\kappa_{\lambda}}, \tag{2.41}$$

where the number of excitons in the state λ is defined as

$$N_{\lambda}^{\mathrm{ex}} = \sum_{\mathbf{q}_{||}} N_{\lambda}^{\mathrm{ex}}(\mathbf{q}_{||}) = \sum_{\mathbf{k}_{||},\mathbf{k}'_{||},\mathbf{q}_{||}} \phi_{\lambda}^{*}(k)\phi_{\lambda}(k') c_{X}^{\mathbf{q}_{||},\mathbf{k}'_{||},\mathbf{k}_{||}} \tag{2.42}$$

and $D_{\lambda,\beta}$ is the excitonic dipole matrix-element between the states λ and β. The form of Eq. (2.41) is typical of an atomic absorption spectrum when different atomic levels are populated according to N_{λ}^{ex}.

Our theory shows that the THz response follows from the entire distributions since Eq. (2.42) contains a sum over all center-of-mass momenta $\mathbf{q}_{||}$. The THz absorption can therefore directly identify truly incoherent exciton populations in all center-of-mass momentum states, that is, both bright and dark excitons.

To illustrate the emergence of excitonic population signatures, we present in Figure 2.7 a series of computed THz absorption spectra for a situation where a short-pulse continuum excitation generates initially unbound electron–hole pairs. At the early times after the carrier generation, the THz spectrum in Figure 2.7a is very broad and has no resonances. This shape is characteristic of an electron–hole plasma. Figure 2.7b shows the development of a small peak around the excitonic $1s-2p$ transition energy. This peak grows into a pronounced resonance approximately 1 ns after the excitation, signifying the buildup of an incoherent exciton population. The asymetric shape of this resonance is a consequence of transitions between the lowest and all the energetically higher exciton states.

2.2.4.4 **Excitonic Signatures in the Photoluminescence**

In many situations, excitonic features can also be observed in the spontaneous emission, that is, the photoluminescence (PL) of a pre-excited system. Under incoherent conditions, the presence of an emission signal is a clear indication of radiatively decaying electron–hole pair populations. However, without a detailed microscopic

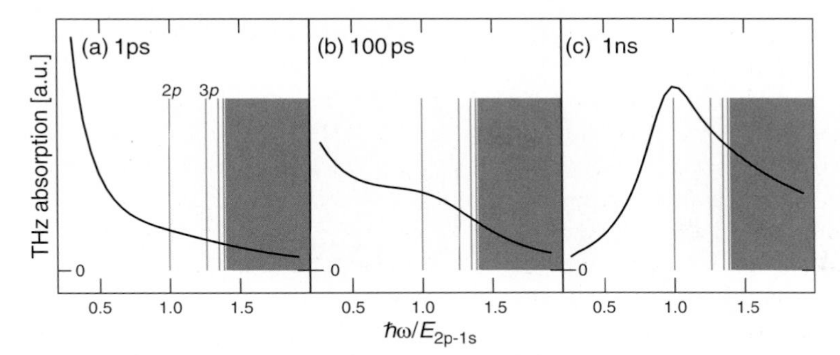

Figure 2.7 Computed THz absorption spectra for differnt delay times after the short-pulse interband excitation of a semiconductor quantum well. (a) The plasma-like spectrum after 1 ps, (b) a resonance develops around the energy corresponding to the intraexcitonic 1s-2p transition, and (c) pronounced excitonic resonance. The lines indicate the energies of the energetically higher excitonic states and the shaded area shows the ionization continuum.

analysis it is not obvious how much one can deduce about the state of the population, that is, about the possible presence of incoherent excitons and their distribution.

Since the process of light emission via spontaneous electron–hole recombination is an intrinsically quantum mechanical effect, we have to quantize the light field. For this purpose, we introduce bosonic creation and annihilation operators $B^{\dagger}_{\mathbf{p}}$ and $B_{\mathbf{p}}$ where the index $\mathbf{p}$ refers to a specific light mode with momentum $\mathbf{p}$ and energy $\hbar\omega_{\mathbf{p}} = \hbar c|\mathbf{p}|$.

Extending the semiclassical light–matter coupling Hamiltonian (2.25) into the full quantum optical regime, we can again derive an equation hierarchy that contains the quantum optical in addition to the Coulomb coupling effects already present in the SBE. The resulting equations yield the semiconductor luminescence equations (SLE) [10].

Under quasi-stationary weak excitation conditions, one can simplify the SLE to obtain an expression for the photoluminescence spectrum that is given as the steady-state photon flux,

$$I_{\mathrm{PL}}(\omega) \propto \mathrm{Im}\left[\sum_{\lambda} \frac{|\phi_{\lambda}(r=0)|^2 N_{\lambda}}{E_{\lambda}-\hbar\omega-i\delta_{\lambda}}\right], \tag{2.43}$$

where

$$N_{\lambda} = \sum_{\mathbf{k}_{||}} |\phi_{\lambda}(\mathbf{k}_{||})|^2 f^{\mathrm{e}}_{\mathbf{k}_{||}} f^{\mathrm{h}}_{\mathbf{k}_{||}} + N^{ex}_{\lambda}\left(\mathbf{q}_{||} = 0\right). \tag{2.44}$$

We see that the carrier contributions are proportional to (i) the product of the electron and hole distributions and (ii) the optically active exciton population with center-of-mass momentum $\mathbf{q}_{||} = 0$, both of which contribute additively to the PL.

When we compare Eqs. (2.43) and (2.36), we note strong similarities. In particular, the frequency dependence of both equations is governed by the same denominator, giving rise to excitonic resonances. Hence, we see that the appearance of these resonances is independent of the detailed structure of the population factor. Since the

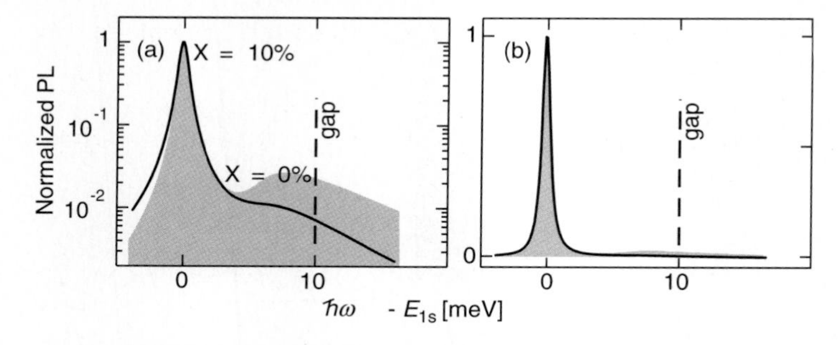

Figure 2.8 Photoluminescence spectrum for a situation without excitonic populations (X = 0%, shaded area) and for a situation where 10% of the electron–hole pairs are bound into excitons (solid line). The results are presented (a) on a logarithmic and (b) on a linear scale. The carrier density is 2×10^9 cm^{-2} and the system temperature is 20 K.

electron–hole plasma and exciton population terms appear additively in N_λ, both contributions can lead to luminescence and therefore also to emission at the excitonic resonances.

Figure 2.8 shows the computed luminescence for a situation without exciton population (gray shaded area) and a configuration where 10% of the electron–hole pairs are bound into excitons (solid line). Panels (a) and (b) present the same data on a logarithmic and linear scale, respectively. We see a prominent peak around the 1s exciton energy in all cases demonstrating the fact that the mere presence of this resonance does not signify excitonic populations in the system. However, the presence of true excitons modifies the ratio between the emission at the 1s resonance relative to the emission from the higher states, that is, 2s all the way to the continuum. A careful evaluation of this ratio makes it possible to determine the actual exciton population in the optically active states [35].

Altogether, we see that excitons and electron–hole pair excitations influence many aspects of the optical semiconductor properties. Excitonic signatures are also relevant in surface systems; see, for example, Ref. [36] and Section 2.4.3.4. In Ref. [37] the SBE have been applied to analyze these features in optical spectra; however, many more experimental and theoretical investigations on surface systems are needed before a comparable level of understanding can be achieved as in the low-dimensional semiconductor systems.

2.2.5
Polarons: Electron–Phonon Coupling in Polar and Ionic Solids

Conducting electrons in solids interact with phonons (quantized modes of vibration occurring in the crystal lattice). As a result of this interaction, a polarization of the lattice acts back on the electron that may be transformed into a quasi-particle called *polaron*, an electron accompanied by a cloud of phonons, that is, an electron coupled to phonons.

In the case of conduction bands of polar semiconductors and ionic solids that have their minimum at the Γ point and have an isotropic effective mass, and assuming that phonons are dispersionless, the electron–phonon interaction can be described by the so-called Frohlich Hamiltonian [38]:

$$H = \frac{p^2}{2m} + \hbar\omega_0 \sum_{\mathbf{q}} a_{\mathbf{q}}^{\dagger} a_{\mathbf{q}} + H_{\text{el-ph}}, \tag{2.45}$$

where

$$H_{\text{el-ph}} = \sum_{\mathbf{q}} \left(M_{\mathbf{q}} a_{\mathbf{k}} e^{i\mathbf{k}\cdot\mathbf{r}} + M_{\mathbf{q}}^{*} a_{\mathbf{k}}^{\dagger} e^{-i\mathbf{k}\cdot\mathbf{r}} \right). \tag{2.46}$$

Here, **r** and **p** represent the conjugate coordinates of the electron (with mass m), $a_{\mathbf{q}}^{\dagger}$ ($a_{\mathbf{q}}$) are creation (annihilation) operators for a phonon with wave vector **q** and energy $\hbar\omega_0$, and $M_{\mathbf{q}}$ represents a coupling constant of the form

$$M_{\mathbf{q}}^2 = \frac{4\pi\alpha\hbar(\hbar\omega_0)^{3/2}}{V(2m)^{1/2}} \frac{1}{q^2}. \tag{2.47}$$

The q^{-2} dependence derives from the 3D Fourier transform of the (bare) Coulomb potential $\propto 1/r$ of the polar semiconductor or ionic crystal. V is a normalization volume and α represents the dimensionless polaron constant. For electrons that are linearly coupled to a system of optical phonons in a polar solid, one finds [39]

$$\alpha = \frac{e^2}{\hbar} \left(\frac{m}{2\hbar\omega_0} \right)^{1/2} \left(\frac{1}{\varepsilon_\infty} - \frac{1}{\varepsilon_0} \right), \tag{2.48}$$

where ε_∞ represents the optical high-frequency dielectric constant (the square of the refractive index) and ε_0 is the static dielectric constant (which includes optical, polar, and electron–electron contributions). Typical values of the Frohlich polaron constant of Eq. (2.48) are in the range 0.01–10 (see, for example, Ref. [40]).

A polaron is characterized mainly by its self-energy (the difference between the actual polaron energy and the energy of the corresponding uncoupled electron) and the effective mass. When the electron–phonon coupling is weak (α small), many-body perturbation theory can be used to find contributions to the polaron self-energy from Feynman diagrams containing one or more phonons. On the mass shell (i.e., by replacing the actual polaron energy by the energy of the corresponding uncoupled electron) and for zero momentum of the electron, one finds [41]

$$\Sigma = -\hbar\omega_0 \left[\alpha + 0.0159\alpha^2 + 0.008765\alpha^3 + O(\alpha^3) \right]. \tag{2.49}$$

For the effective mass, as obtained from an expansion of the polaron self-energy with respect to small momenta of the electron, one finds [41]

$$m^* = m\left[1 - 16\alpha + 0.02263\alpha^2 + O(\alpha^3) \right]. \tag{2.50}$$

When the electron–phonon coupling is strong (α large), one can follow the variational Gaussian wave function approach introduced by Landau and Pekar [42]. Alternatively, assuming that the electron is localized with a Gaussian wave function, the self-energy can be expanded as a power series in $1/\alpha$ to find [43]

$$\Sigma = -\hbar\omega_0\left[0.1085\alpha^2 + 2.836 + O(1/\alpha^2)\right]. \tag{2.51}$$

Feynman introduced a variational method based on path integrals, which has been found to be accurate for all values of the polaron coupling α [44]. A comparison between Eqs. (2.49) and (2.51) and between these equations and the Feynman calculations indicates that for the Frohlich Hamiltonian of Eqs. (2.45) and (2.46) the correct self-energy is given by the weak coupling result of Eq. (2.49) for values up to $\alpha \leq 5$ and by the strong coupling result of Eq. (2.51) for values $\alpha \geq 5$. Furthermore, a variational estimation of the actual size of the Gaussian wave function describing the electron leads to the conclusion that for $\alpha > 5$ or 6 polarons are expected to become localized within the size of the atomic unit shell, which is in contrast with the continuum Frohlich Hamiltonian of Eqs. (2.45) and (2.46) that assumes a continuum theory for the ions and a free particle motion for the unperturbed electron. These strongly localized polarons are the so-called small polarons. Small-polaron theories are based on the use of a Hamiltonian that includes the periodicity of the solid and the assumption that the size of the polaron corresponds to atomic dimensions (see, for example, Ref. [39]).

Electrons that are confined in two dimensions (2D), as occurs in the case of electrons at surface states or semiconductor heterostructures [45], also interact with phonons propagating along the 2D system. This coupling yields polarons in two dimensions, that is, 2D polarons, which can be described by a Frohlich Hamiltonian of the form of Eqs. (2.45) and (2.46) but with the electron–phonon coupling M_{q} of Eq. (2.47) replaced by [46]

$$\left(M_{\mathrm{q}}^{2\mathrm{D}}\right)^2 = \frac{2\pi\alpha^{2D}\hbar(\hbar\omega_0)^{3/2}}{A(2m)^{1/2}}\frac{1}{q}, \tag{2.52}$$

where A represents a normalization area. In particular, for image state electrons that are attracted to the surface of an ionic crystal by its image potential

$$V_{\mathrm{im}}(z) = \frac{e^2}{4}\frac{\varepsilon_{\mathrm{s}}-1}{\varepsilon_{\mathrm{s}}+1}\frac{1}{z} \tag{2.53}$$

and are repeled from the interior of the solid, one finds the following value for the polaron coupling constant [46]:

$$\alpha^{2\mathrm{D}} = \frac{e^2}{\hbar}\left(\frac{m}{2\hbar\omega_0}\right)^{1/2}(E_0 - E_\infty), \tag{2.54}$$

where $E_0 = (\varepsilon_0-1)/(\varepsilon_0+1)$ and $E_\infty = (\varepsilon_\infty-1)/(\varepsilon_\infty+1)$. In this case, image state electrons couple to surface phonons of energy $\hbar\omega_0$, with

$$\omega_0^2 = 12(\omega_{\mathrm{L}}^2 + \omega_{\mathrm{T}}^2), \tag{2.55}$$

where ω_{L} and ω_{T} are the frequencies of the bulk longitudinal and transverse phonons, respectively.

The self-energy and the effective mass of 2D polarons can be obtained from the Frohlich Hamiltonian of Eqs. (2.45) and (2.46) with the 2D electron–phonon coupling of Eq. (2.52). In the weak coupling limit and for zero momentum of the electron, one finds the 2D polaron self-energy to be given by the following expression [47]:

$$\Sigma^{2D} = -\hbar\omega_0^{2D}\left[\frac{\pi}{2}\alpha^2 + 0.06397\alpha^2 + O(\alpha^3)\right]. \tag{2.56}$$

For the effective mass, one finds [48]

$$(m^*)^{2D} = m\left[1 + \frac{\pi}{8}\alpha + 0.1272348\alpha^2 + O(\alpha^3)\right]. \tag{2.57}$$

2.3 Collective Excitations

2.3.1 Plasmons: Electron Density Oscillations

In a pioneering study, Pines and Bohm [49] pointed out that the long-range nature of the Coulomb interaction between valence electrons in metals yields collective plasma oscillations similar to the classical electron density oscillations that had been observed by Tonks and Langmuir in electrical discharges in gases [50]. A simple classical model can be used in order to illustrate the collective nature of the motion of a free electron gas. Let us replace the positive ions by a uniform background of positive charge, and let us suppose that a charge imbalance is established in the plasma by displacing a slab of charge of thickness d by a small distance x ($x \ll d$) (see, for example, Ref. [51]). Hence, the slab behaves like a capacitor in which a constant electric field **E** is set up acting to restore charge neutrality. The magnitude of the field is simply

$$E = -4\pi n_0 e x, \tag{2.58}$$

since the surface charge density on either end of the condenser is simply $n_0 e x$, with n_0 being the unperturbed electron density. The equation of motion of an electron moving under the influence of this field is then

$$m\ddot{x} = -4\pi n_0 e^2 x, \tag{2.59}$$

with the result that electrons will undergo a simple harmonic oscillation at a frequency given by the following expression:

$$\omega_\mathrm{p} = \left(\frac{4\pi n_0 e^2}{m}\right)^{1/2}. \tag{2.60}$$

Typical values of the density of valence electrons in metals are on the order of the Avogadro's number (more than 10 orders of magnitude larger than the density of electrons and positive charges in gaseous discharges). Hence, while in the case of gaseous plasmas the Fermi–Dirac distribution goes over to the Maxwell–Boltzmann distributions, quantum statistics must be employed for valence electrons in metals, that is, the free electron gas at low (and room) temperatures and metallic densities can be regarded as a quantum plasma. The energy $\hbar\omega_p$ is the minimum value of the energy required to excite a simple harmonic oscillator of characteristic frequency ω_p, that is, a *plasmon*. Typical values of the plasmon energy ($\hbar\omega_p$) at metallic densities are in the range $2-20\,eV$. Plasmons are observed directly through energy losses in multiples of $\hbar\omega_p$ when electrons are fired through metallic films [52].

Plasma oscillations in a *uniform* free electron gas can also be described through the knowledge of a frequency-dependent dielectric function. Consider the electric and displacement fields of elementary electrostatics in the absence of external sources. They are known to satisfy the following equations:

$$\nabla \cdot \mathbf{D} = 0 \tag{2.61}$$

and

$$\nabla \cdot \mathbf{E} = 4\pi e \varrho, \tag{2.62}$$

where $\varrho(\mathbf{r}, t)$ represents a polarization charge density. At this point, we take the Fourier transform in space and time of Eqs. (2.61) and (2.62), we assume that the electronic response of the free electron gas is proportional to the applied field $\mathbf{D}$, that is,

$$\mathbf{E}(\mathbf{q}, \omega) = \frac{\mathbf{D}(\mathbf{q}, \omega)}{\varepsilon(\mathbf{q}, \omega)}, \tag{2.63}$$

where $\varepsilon(\mathbf{q}, \omega)$ is a frequency and wave vector-dependent dielectric function, and one finds

$$\varepsilon(\mathbf{q}, \omega)\, i\mathbf{q} \cdot \mathbf{E}(\mathbf{q}, \omega) = 0 \tag{2.64}$$

and

$$i\mathbf{q} \cdot \mathbf{E}(\mathbf{q}, \omega) = 4\pi e \varrho(\mathbf{q}, \omega). \tag{2.65}$$

One trivial solution of Eqs. (2.64) and (2.65) is

$$\mathbf{q} \cdot \mathbf{E}(\mathbf{q}, \omega) = \varrho(\mathbf{q}, \omega) = 0, \tag{2.66}$$

which corresponds to no net charge density or electric field present in the electron gas. However, for frequencies ω such that

$$\varepsilon(\mathbf{q}, \omega) = 0, \tag{2.67}$$

one may have a nonvanishing value of $\mathbf{E}(\mathbf{q}, \omega)$ and $\varrho(\mathbf{q}, \omega)$. This means that for frequencies such that the dielectric function $\varepsilon(\mathbf{q}, \omega)$ vanishes one has a free oscillation of the charge density with no external field. Hence, Eq. (2.67) represents the condition for the existence of plasma oscillations at frequency ω, which correspond to a net longitudinal electric field in the plasma.

In the limit of very long wavelengths ($\mathbf{q} \to 0$), the actual dielectric function of a *uniform* free electron gas coincides with the so-called Drude dielectric function (see, for example, Ref. [1]):

$$\varepsilon(0,\omega) = 1 - \frac{\omega_p^2}{\omega^2}, \tag{2.68}$$

with ω_p being given by Eq. (2.60). Hence, in the long-wavelength limit plasma oscillations exist at the frequency ω_p of Eq. (2.60), which was derived (see Eqs. (2.58)–(2.60)) in an approximation (neglecting altogether the random motion of the electrons) that is justified in the limit of very long-wavelength oscillations.

2.3.1.1 **Surface Plasmons**

In the presence of a boundary, there is a new collective oscillation (that propagates along the interface), which for a planar surface is the so-called surface plasmon predicted by Ritchie [53, 54]. Let us consider a classical model consisting of two semiinfinite nonmagnetic media with local (frequency-dependent) dielectric functions ε_1 and ε_2 separated by a planar interface at $z = 0$. The full set of Maxwell's equations in the absence of external sources can be expressed as follows [55]:

$$\nabla \times \mathbf{H}_i = \varepsilon_i \frac{1}{c} \frac{\partial}{\partial t} \mathbf{E}_i, \tag{2.69}$$

$$\nabla \times \mathbf{E}_i = -\frac{1}{c} \frac{\partial}{\partial t} \mathbf{H}_i, \tag{2.70}$$

$$\nabla (\varepsilon_i \mathbf{E}_i) = 0, \tag{2.71}$$

and

$$\nabla \mathbf{H}_i = 0, \tag{2.72}$$

where the index i describes the media: $i = 1$ at $z < 0$ and $i = 2$ at $z > 0$.

For an ideal surface, if waves are to be formed that propagate along the interface there must necessarily be a component of the electric field normal to the surface. Hence, one seeks conditions under which a traveling wave with the magnetic field $\mathbf{H}$ parallel to the interface may propagate along the surface ($z = 0$), with the fields tailing off into the positive ($z > 0$) and negative ($z < 0$) directions. From Eqs. (2.69)–(2.72), one finds the following system of equations:

$$\frac{\kappa_1}{\varepsilon_1} H_{1_y} + \frac{\kappa_2}{\varepsilon_2} H_{2_y} = 0 \tag{2.73}$$

and

$$H_{1_y} - H_{2_y} = 0, \tag{2.74}$$

which has a solution only if the determinant is zero, that is,

$$\frac{\varepsilon_1}{\kappa_1} + \frac{\varepsilon_2}{\kappa_2} = 0, \tag{2.75}$$

where

$$\kappa_i = \sqrt{q^2 - \varepsilon_i \frac{\omega^2}{c^2}}, \tag{2.76}$$

where q is the magnitude of a 2D wave vector in the plane of the surface. Equation (2.75) represents the condition for the existence of surface–plasmon polaritons. For a metal–dielectric interface with the dielectric characterized by ε_2, the solution $\omega(q)$ of Eq. (2.75) has slope equal to $c/\sqrt{\varepsilon_2}$ at point $q = 0$ and is a monotonic increasing function of q, which is always smaller than $cq/\sqrt{\varepsilon_2}$ and for large q is asymptotic to the value given by the solution of

$$\varepsilon_1 + \varepsilon_2 = 0. \tag{2.77}$$

This is the *nonretarded* surface plasmon condition (Eq. (2.75) with $\kappa_1 = \kappa_2 = q$), which is valid in the nonretarded regime in which the phase velocity ω/q is much smaller than the speed of light. In the case of a Drude semiinfinite metal in vacuum, with ε_1 given by Eq. (2.68) and $\varepsilon_2 = 1$, the nonretarded surface plasmon condition of Eq. (2.77) is fulfilled at the so-called surface plasmon frequency $\omega_s = \omega_p/\sqrt{2}$. For a recent review on surface plasmons and surface plasmon polaritons, see Ref. [12].

2.3.1.2 Acoustic Surface Plasmons

A variety of metal surfaces, such as Be(0001) and the (111) surfaces of the noble metals Cu, Ag, and Au, are known to support a partially occupied band of Shockley surface states within a wide energy gap around the Fermi level (see, for example, Ref. [56]). Since these states are strongly localized near the surface and disperse with momentum parallel to the surface, they form a quasi-2D surface state band with a 2D Fermi energy ε_F^{2D} equal to the surface state binding energy at $\bar{\Gamma}$ point.

In the absence of the 3D substrate, a Shockley surface state would support a 2D collective oscillation, the energy of the corresponding plasmon being given by [45]

$$\omega_{2D} = \left(\frac{2\pi n^{2D} e^2 q}{m}\right)^{1/2}, \tag{2.78}$$

with n^{2D} being the 2D density of occupied surface states, that is,

$$n^{2D} = \frac{\varepsilon_F^{2D} m}{\pi \hbar^2}. \tag{2.79}$$

Equation (2.78) shows that at very long wavelengths plasmons in a 2D electron gas have low energies; however, they do not affect electron–hole and phonon dynamics near the Fermi level, due to their square-root dependence on the wave vector. Much more effective than ordinary 2D plasmons in mediating, for example, superconductivity, would be the so-called acoustic plasmons with sound-like long-wavelength dispersion.

Recently, it has been shown that in the presence of the 3D substrate the dynamical screening at the surface provides a mechanism for the existence of a *new* acoustic collective mode, the so-called *acoustic surface plasmon*, whose energy exhibits a linear dependence on the 2D wave number [57–59]:

$$\omega_{\text{acoustic}} \sim v_F^{2D} q, \tag{2.80}$$

where v_F^{2D} represents the 2D Fermi velocity ($v_F^{2D} = \hbar\sqrt{2\pi n^{2D}}/m$). This *novel* surface plasmon mode has been observed at the (0001) surface of Be, showing a linear energy dispersion that is in very good agreement with Eq. (2.80) and first-principles calculations [60].

2.3.2
Magnons: Elementary Excitations in Ferromagnetic Materials

The notion of *spin waves* is one of the fundamental concepts in the physics of magnetism. In this section, we will focus on ferromagnetic systems. At low temperatures, the ferromagnetic system can be in good approximation considered as the gas of noninteracting bosonic quasi-particles, the so-called *magnons*, which are the quanta of the spin-wave excitations.

In modern applications, where the magnetization dynamics plays a very important role, the spin waves can be employed to transfer information between different parts of nanoscale devices. In most applications dealing with magnetization switching, it is crucial that the magnetic state disturbed by an external stimulus attenuates sufficiently fast to the equilibrium state. Therefore, the damping is an important aspect of spin dynamics. The spectrum of spin-wave excitations governs essentially the thermodynamics of ferromagnets. The interaction of magnons with other quasi-particles like electrons and phonons enriches the physics of condensed matter systems.

This section provides an introduction to the physics of spin waves and demonstrates how the concept of spin waves arises in different physical models. Moreover, we will present the results of the most recent studies of spin-wave dispersion and attenuation to acquaint the reader with current state-of-the-art investigations.

One can point out two limiting cases in the spin dynamics of ferromagnetic systems. One limit is represented by the *Heisenberg model*. Here, one supposes that the atoms in the condensed matter systems, similar to isolated atoms, can be characterized by the magnetic moments, which are well-defined atomic quantities. The energetics are governed by the effective interaction between the moments. The exchange parameters describe the energy associated with the relative change in the moments' directions. The values of the moments are constants of the model. Although the Heisenberg Hamiltonian can be justified under certain assumptions starting with the consideration of the Hamiltonian of interacting electrons, the electronic degrees of freedom are "integrated out" and do not enter the Hamiltonian.

The second limit consists in metallic (itinerant) magnets. One of the adequate approaches to study magnetic excitations in this case is the *linear-response density functional theory* [61]. This approach is based on the calculation of the transverse magnetic susceptibility within the parameter-free calculational scheme. The electron system enters these calculations through detailed information on the energies and wave functions of the electronic states.

The wave functions in metals are delocalized and the states are spread over the whole system. However, the real condensed matter systems are microscopically strongly nonuniform and the electronic states are, in general, very far from simple plane waves. This is particularly true for the 3d electron states of the 3d transition atoms; they feature electronic density that is large within the atomic volumes and much lower in the interstitial regions between the atomic spheres. The large electron density leads to an increase in the exchange–correlation potential at corresponding points and results in strong magnetism of the itinerant electron system. It makes the spin polarization of the electron states energetically advantageous and leads to the nonzero magnetization of the system. The strong intraatomic exchange–correlation potential makes the introduction of the concept of atomic moments useful also in the case of itinerant electron systems. However, this quantity is by no means identical to the rigid atomic moment of the Heisenberg theory.

Next, we consider how spin waves appear in different physical models. We not only show the robustness of the concept of the spin wave with respect to the physical model but also demonstrate important differences between the two limiting descriptions outlined above.

2.3.2.1 Spin Waves in the Heisenberg Model

We begin with the Heisenberg model of classical atomic spins. The localized moment picture associates a spin moment S with each atom. The energetics of the magnet is described by the Heisenberg Hamiltonian

$$H = -\sum_{ij} J_{ij}\mathbf{S}_i \cdot \mathbf{S}_j, \tag{2.81}$$

where J_{ij} are the parameters of the effective interaction between the atomic moments of atoms i and j. A positive sign of the exchange parameters makes the parallel (ferromagnetic) orientation of the atomic moments the energetically preferable ground state.

The precessional dynamics of the atomic moments is described by the Landau–Lifshitz torque equation

$$\frac{d\mathbf{S}_i}{dt} = -\mathbf{S}_i \times \mathbf{h}_i \tag{2.82}$$

The effective field $\mathbf{h}_i$ acting on the moment of the ith atom is given by

$$\mathbf{h}_i = 2\sum_j J_{ij}\mathbf{S}_j. \tag{2.83}$$

By introducing Eq. (2.83) in the Landau–Lifshitz equation (2.81), we obtain the equation for the dynamic variables $\mathbf{S}_i(t)$. We will assume that the deviations of the atomic moments from the ground state's z-direction is small and will keep only terms of first order in the deviation angle. In this approximation, the z component of the spins remains equal to S and the dynamics of the x and y components is described by the equations

$$\frac{dS_{ix}}{dt} = -2S\sum_j J_{ij}(S_{iy} - S_{jy}) \tag{2.84}$$

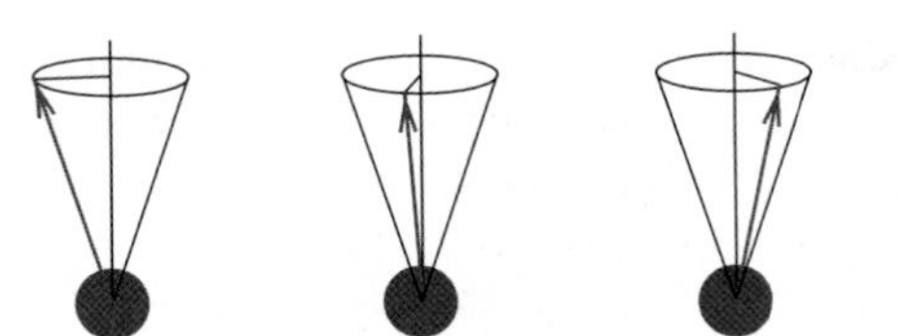

Figure 2.9 Schematic picture of the magnetic configuration corresponding to a spin wave with wave vector **q**. (Please find a color version of this figure on the color plates.)

$$\frac{\mathrm{d}S_{iy}}{\mathrm{d}t} = 2S\sum_j J_{ij}(S_{ix}-S_{jx}) \tag{2.85}$$

We will look for the solution of Eqs. (2.84) and (2.85) in the form

$$\mathbf{S}_i(t) = S(\sin\theta\cos(\mathbf{qR}_i+\omega t),\ \sin\theta\sin(\mathbf{qR}_i+\omega t), \cos\theta), \tag{2.86}$$

where $\mathbf{R}_i$ is the position of atom i, $\mathbf{q}$ and ω are the wave vector and frequency of the spin wave, respectively, and angle θ gives the deviation of the atomic moments from the z-direction. This magnetic configuration is depicted in Figure 2.9. It is easily proven by direct substitution that expression (2.86) is in first order with respect to the angle θ, the solution of Eqs. (2.84) and (2.85) for the frequencies determined by

$$\omega_{\mathbf{q}} = 2S(J(0)-J(\mathbf{q})), \tag{2.87}$$

where

$$J(\mathbf{q}) = \sum_j J_{0j}\exp(i\mathbf{qR}_j). \tag{2.88}$$

Thus, we obtain spin-wave precessional eigenmodes, with their dispersion given by Eq. (2.87), as the solution of the Landau–Lifshitz equation for small deviations of the atomic spins from the ground-state directions.

It is instructive to evaluate the energy of the magnetic configuration (2.86) with respect to the ground-state energy. By substituting (2.86) in (2.81) and by assuming small values of θ, we obtain

$$\Delta E(\mathbf{q},\omega) = \frac{\theta^2}{2}SN\omega_{\mathbf{q}}. \tag{2.89}$$

Since $(\theta^2/2)SN$ gives the reduction of the z-projection of the total spin moment, ΔS_z, Eq. (2.89) can be represented in the form

$$\Delta E(\mathbf{q},\omega) = \Delta S_z\,\omega_{\mathbf{q}}. \tag{2.90}$$

Therefore, the increase in energy is proportional to the change in the total z-projection of S_z. The proportionality coefficient equals the precession frequency found in the solution of the Landau–Lifshitz equation. In other words, the frequency of the spin wave precession equals exactly the energy of the spin wave configuration for a given momentum $\mathbf{q}$, which corresponds to the change in the z-projection of the total spin moment by 1.

If we recall that the spin is a quantum quantity and can only discreetly change its value by a minimal change equal to 1, $\omega_{\mathbf{q}}$ gives the energy of the lowest possible spin excitation for a given $\mathbf{q}$. This consideration leads to the quantization of the spin wave excitations.

Quantum Heisenberg Hamiltonian In the discussion above, we started with the Hamiltonian of the classical vectors and introduced the quantization of the spin waves in the last part of the treatment.

We obtained perfect agreement between the frequency of the eigen precession following from the Landau–Lifshitz equation and the energy consideration on the basis of Eq. (2.81). It is an important physical question how the properties of excited states will change if the atomic spins are from the beginning treated as quantum mechanical operators.

We consider now the quantum Heisenberg model

$$\hat{H} = -\sum_{ij} J_{ij}\hat{\mathbf{S}}_i\hat{\mathbf{S}}_j. \tag{2.91}$$

Here, $\hat{\mathbf{S}}_i$ is the operator of the spin of the ith atom. All atomic spins are characterized by the quantum number S. The spin states of the ith atom can be characterized by the value of the z-projection

$$\hat{S}_{iz}|S_z>_i = S_z|S_z>_i, \tag{2.92}$$

where $S_z = -S, -S+1, \ldots, S-1, S$ and $|S_z>_i$ is the corresponding eigenspinor of the ith atom.

It can be verified that

$$|0> \; = \; \prod_i |S>_i \tag{2.93}$$

is the lowest energy eigenstate of the operator (2.91) giving the ferromagnetic ground state of the quantum Heisenberg model

$$\hat{H}|0> = -\sum_{ij} J_{ij}S^2|0> . \tag{2.94}$$

The ground-state energy is

$$E_0 = -\sum_{ij} J_{ij}S^2 \tag{2.95}$$

On the basis of the experience gained in the previous section, we will assume that the excited spin wave states of the Hamiltonian (2.91) have the form

$$|\mathbf{q}> = \frac{1}{\sqrt{N}}\sum_i \exp\left(i\mathbf{q}\mathbf{R}_i\right)\frac{1}{\sqrt{(2S)}}\hat{S}_{i-}\,|0> \tag{2.96}$$

where operator $\hat{S}_{i-} = \hat{S}_{ix} - i\hat{S}_{iy}$ lowers the value of the z-projection of the spin of the ith atom by one:

$$\hat{S}_{i-}|S_z>_i = \sqrt{(S+S_z)(S+1-S_z)}\,|S_z-1>_i. \tag{2.97}$$

Since in the ground state $S_z = S$, we will need the last equation just for this value of S_z. Therefore, it simplifies to

$$\hat{S}_{i-}|S>_i = \sqrt{2S}\,|S-1>_i. \tag{2.98}$$

By acting with the Hamiltonian (2.91) on the spinor function $|\mathbf{q}>$ given by Eq. (2.96), we obtain

$$\hat{H}|\mathbf{q}>= (E_0 + \omega_{\mathbf{q}})|\mathbf{q}> . \tag{2.99}$$

Therefore, $|\mathbf{q}>$ is indeed the eigenstate of the quantum Heisenberg Hamiltonian with the excitation energy $\omega_{\mathbf{q}}$ that corresponds exactly to the energy we obtained by considering the classical Heisenberg Hamiltonian.

By calculating the expectation values of the atomic spins for the state $|\mathbf{q}>$, we arrive again at the spin configuration shown in Figure 2.9.

Thus, we can conclude that the model of local atomic spins gives a robust picture of the spin wave excitations in a Heisenberg ferromagnet. Considering the precessional motion of the spins along with the energetics of the corresponding magnetic states gives exactly the same magnon energies as the eigenstate solutions of the corresponding quantum mechanical equations.

Despite the robustness of the localized moment picture, one should clearly understand its strong limitations in the description of real materials where the electronic states are delocalized over the whole system and the introduction of the atomic spins as dynamical variables is not straightforward. In the next section, we consider a consistent approach to the spin dynamics of itinerant electron systems referring to the similarities of both descriptions and their important differences.

2.3.2.2 **Itinerant Electrons**

An efficient and reliable approach to deal with magnetic properties of itinerant electron systems is offered by the framework of the density functional theory (DFT). The magnetic ground state is determined as the solution of a self-consistent Kohn–Sham problem where formally noninteracting electrons are placed in the potential that itself depends on the quantum mechanical states of the electrons given by the Kohn–Sham energies $\varepsilon_{n\mathbf{k}\sigma}$ and wave functions $\psi_{n\mathbf{k}\sigma}(\mathbf{r})$. Here, n is the number of electronic bands, $\mathbf{k}$ is the Bloch wave vector of the state, and σ is the spin projection.

We will discuss two different DFT-based approaches to the study of spin waves in itinerant electron magnets. First, we will very briefly introduce the frozen magnon method that basically maps the itinerant electron system on the classical Heisenberg Hamiltonian. Then, we will present in some details a more powerful method to investigate the magnetic excited states in itinerant magnets based on the calculation of the dynamic magnetic susceptibility.

Frozen Magnon Approach The so-called constrained calculations of the DFT [62] combined with the generalized Bloch theorem for spiral magnetic structures [8] allows for performing a direct DFT computation of the energy of the magnetic configurations shown in Figure 2.9. In elemental 3dferromagnets, the energy of these configurations directly provides an estimate of the magnon dispersion. This type of calculation gives reasonably good agreement with experimental spin wave energies for long-wavelength spin waves [63]. For systems with inequivalent magnetic atoms, the calculations become more involved since the deviations of the magnetic moments of inequivalent atoms in a spin wave excitation are different and *a priori* unknown. Therefore, there is no magnetic configuration of the type shown in Figure 2.9 that can be associated with spin waves in complex systems without evaluating the relative deviations between the moments of different atoms. However, also in this case the frozen magnon approach allows us to estimate the energies of the spin wave modes as described, for example, in Ref. [64].

This treatment is close to the consideration performed above, but with the Heisenberg-like exchange parameters determined in an *ab initio* calculation. Similar to the approaches based on the Heisenberg Hamiltonian the frozen magnon calculations are characterized by an infinite lifetime of the spin waves independent of the energy and wave vector of the excitation. In this respect, the calculations fail to capture the strong damping of the spin waves in metallic systems experimentally observed for wave vectors, which are off the Brillouin-zone center.

Calculation of the Dynamic Magnetic Susceptibility A consistent approach to study the spin waves in itinerant electron systems that takes full account of the electronic states is based on the calculation of the magnetic response of the system

$$\mathbf{m}(\mathbf{r},t) = \iint \mathrm{d}\mathbf{r}'\mathrm{d}t'\,\chi\left(\mathbf{r},\mathbf{r}',t-t'\right)\mathbf{h}\left(\mathbf{r}',t'\right) \tag{2.100}$$

to the transversal magnetic field

$$\mathbf{h}(\mathbf{r}) = h(\cos(\mathbf{qr}+\omega t),\sin(\mathbf{qr}+\omega t),0). \tag{2.101}$$

The presence of the spin wave eigenmodes with given $\mathbf{q}$ and ω leads to an enhancement of the susceptibility for just these parameters of the field. This allows us to identify the spin wave excitations.

The calculation of the dynamic magnetic susceptibility in the framework of the DFT consists of two steps. First, a nonself-consistent response of the itinerant electron system is calculated. It gives a noninteracting susceptibility, also called unenhanced or *Kohn–Sham susceptibility*:

$$\begin{aligned}\chi_{\mathrm{KS}}(\mathbf{r},\mathbf{r}') = &\sum_{\mathbf{k}\Omega_{\mathrm{BZ}}}\sum_{nn'}\left(f^{\uparrow}_{n\mathbf{k}-\mathbf{q}}-f^{\downarrow}_{n'\mathbf{k}}\right)\\ &\times\frac{\psi^{\uparrow}_{n\mathbf{k}-\mathbf{q}}(l'\mathbf{r}')\psi^{\uparrow}_{n\mathbf{k}-\mathbf{q}}(l\mathbf{r})^{*}\psi^{\downarrow}_{n'\mathbf{k}}(l\mathbf{r})\psi^{\downarrow}_{n'\mathbf{k}}(l'\mathbf{r}')^{*}}{\omega-\left(\varepsilon^{\downarrow}_{n'\mathbf{k}}-\varepsilon^{\uparrow}_{n\mathbf{k}-\mathbf{q}}\right)+i0^{+}},\end{aligned} \tag{2.102}$$

where $\psi^{\sigma}_{n\mathbf{k}}$ stands for the Bloch state of band n, and $\varepsilon^{\sigma}_{n\mathbf{k}}$ is its Kohn–Sham energy. The response is determined by the available electronic transitions between the occupied ($f = 1$) majority spin states ($\uparrow$), with crystal momentum $\mathbf{k}-\mathbf{q}$, and the empty ($f = 0$) minority spin states ($\downarrow$) with momentum $\mathbf{k}$. These transitions are called the Stoner excitations. The momenta $\mathbf{k}$ belong to the Brillouin zone of the system.

According to the principles of DFT the induced magnetization results in an effective exchange–correlation field that adds to the external field. The problem requires a self-consistent solution: the magnetization induced by the sum of the external and exchange–correlation fields must produce the same exchange–correlation field. This leads to a collective response of all electronic states in the system and the emergence of spin waves.

The self-consistency condition is mathematically formulated in the form of the Dyson-like equation

$$\chi = \chi_{\mathrm{KS}} + \chi_{\mathrm{KS}} K_{\mathrm{xc}} \chi. \tag{2.103}$$

In concrete calculations, all quantities in Eq. (2.103) are presented by matrices in certain functional basis. The exchange–correlation kernel, K_{xc}, determines the effective Kohn–Sham potential. In the calculations discussed below, the form of

$$K_{\mathrm{xc}}(\mathbf{r}) \sim \frac{B_{\mathrm{xc}}(\mathbf{r})}{m(\mathbf{r})} \tag{2.104}$$

corresponds to the adiabatic local density approximation [61, 65]. In Eq. (2.104), $B_{\mathrm{xc}}(\mathbf{r})$ is the value of the exchange-–correlation magnetic field at point $\mathbf{r}$ that is the difference of the spin-up and spin-down potentials at this point, and $m(\mathbf{r})$ is the magnetization.

The solution of Eq. (2.103) is the true *enhanced magnetic susceptibility*. It can be written in the following form:

$$\chi = (1 - \chi_{\mathrm{KS}} K_{\mathrm{xc}})^{-1} \chi_{\mathrm{KS}}. \tag{2.105}$$

The spin waves appear as the peaks in the imaginary part of the susceptibility and signify strong absorption of energy by the system. The frequency corresponding to the maximum of the peak, $\omega_0(\mathbf{q})$, is identified as the magnon energy, whereas the full-width at half-maximum (FWHM) of the magnon peak is interpreted as the inverse lifetime Γ of the excitation. The finite lifetime of the spin waves is explained by their hybridization with the continuum of the single-electron Stoner excitations that enter the theory as poles of the Kohn–Sham susceptibility χ_{KS} (Eq. (2.102)). The Stoner transtion correspoding to a given momentum $\mathbf{q}$ and energy ω is schematically presented in Figure 2.10. This attenuation mechanism is commonly referred to as *Landau damping*. The calculations for the elemental ferromagnetic metals, in particular bcc Fe and fcc Ni, show that the decay becomes so strong that the spin wave excitation cannot be considered as well defined.

In the Heisenberg model, the number of spin wave modes is equal to the number of magnetic atoms in the unit cell. The spectral function presenting the energy distribution of the excited states takes in this case the form of a number of delta functions. In the itinerant electron systems, the situation is fundamentally different.

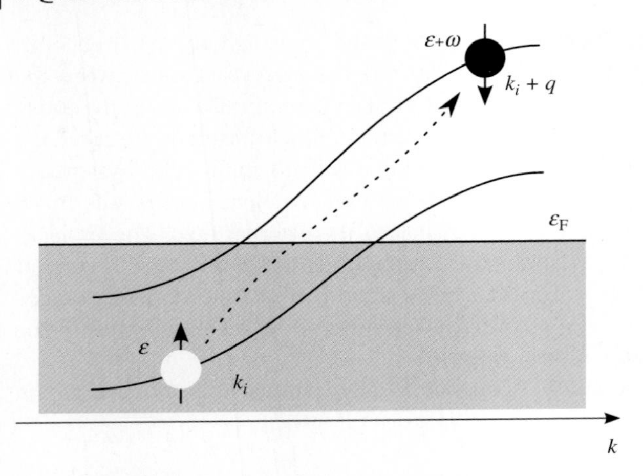

Figure 2.10 Schematic representation of the Stoner excitation. An electron is excited from an occupied majority spin state to an empty minority spin state. The states differ by energy ω and monetum **q**.

The energy position of a spin wave excitation is associated with the root of the real part of the denominator of Eq. (2.105)

$$\det\left(1-\mathbf{Re}\,\chi_{\mathrm{KS}}K_{\mathrm{xc}}\right)=0 \tag{2.106}$$

Since only the reactive (real) part of the unenhanced susceptibility enters the equation, it can be satisfied for the energy for which no Stoner transitions are available. In this case, the spin wave excitation is represented by a delta function peak in the spectral density of spin-flip excitations and has an infinite lifetime. In other words, the spin wave excitation appears as the eigenstate of the many-electron quantum mechanical problem.

However, if there also exist Stoner transitions at the energy associated with the spin wave state, the imaginary part of the Kohn–Sham susceptibility is nonzero. In this case, the delta function-like feature is replaced by a peak of finite width. The width is determined by the spectral density of the Stoner excitations in the corresponding energy region. This is the manifestation of the Landau damping. Depending on the wave vector and energy of the spin wave excitation, the influence of the Stoner transitions can vary from relatively weak to very strong. This scenario is confirmed by calculations performed for the 3d ferromagnets and by inelastic neutron scattering experiments. As an example, in Figure 2.11 we present the energies and lifetimes of magnons in bulk bcc iron based on the evaluation of the dynamic susceptibility and compare them with experimental neutron scattering data. Apart from the accurate prediction of magnon energies, the susceptibility reveals the experimentally observed sudden decrease in the magnon lifetime as the spin wave band enters the continuum of Stoner excitations.

An important physical question arises: are the spin wave excitations obtained by analyzing the dynamic susceptibility of itinerant electrons related to the precessional motion of the atomic moments in the Heisenberg model? The atomic moments do

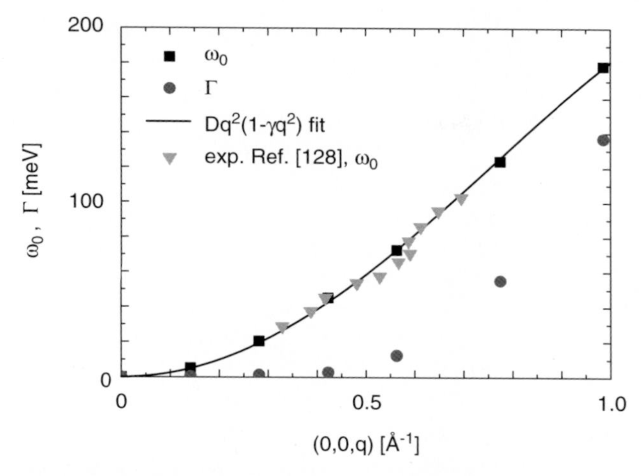

Figure 2.11 Energies (ω_0) and inverse lifetimes Γ of spin waves in bulk bcc Fe for wave vectors parallel to the (001)-direction of the reciprocal space.

not enter the calculation of the magnetic susceptibility directly. Instead, the system is physically described by the itinerant electron energies and wave functions. Nevertheless, the answer about the relevance of the atomic moments in the spin wave formation is positive also in this case. The reason for this is again the strong intraatomic exchange interaction mediated by the exchange–correlation kernel of the Dyson equation. The strong intraatomic exchange interaction has its origin in the strong spatial localization of the 3d states in the atomic spheres. The large electron density at a certain space point **r** results in a large exchange–correlation potential at this point and makes an alignment of the electron spins energetically preferable. Therefore, in the self-consistent precessional motion corresponding to spin waves, the electron spins in the intraatomic regions tend to order parallel to each other at each time instant. On the other hand, the Stoner transitions with the wave vector and energy equal to the wave vector and energy of the spin wave compete against the trend of preserving the atomic moment.

This competition between the formation of the spin wave as a precessional motion of well-defined atomic moments and the disturbance of atomic moments by single-electron Stoner excitations is the important mechanism for spin wave attenuation.

The question of the relevance of the atomic moments for the spin wave excitations in the itinerant electron systems is directly related to the question of the number of magnon modes in the excitation spectrum of the system. In the Heisenberg model, the number of modes is equal to the number of magnetic atoms in the unit cell. In the calculation of the dynamic susceptibility, the information about the system enters through the electronic band structure and, therefore, the number of the degrees of freedom is not restricted by the number of atoms. However, neither the experiment nor the calculations of the magnetic susceptibility detect an increase in the number of spin wave modes with respect to the Heisenberg model with the one exception of fcc-Ni where a number of investigations report the formation of a mode different from the low-energy acoustic mode. Since the Ni spin moment per atom is the smallest from the 3d elements, it is expected to be less robust than the spin moments of Fe and Co, and the formation of a mode different from the Heisenberg one is most probable for nickel. The question of the presence of the "optical" mode in the spin wave spectrum of Ni is not solved (see, for example, Ref. [66]) and it will most probably remain the topic of detailed research work in the near future.

In systems with several different magnetic atoms in the unit cell, it is of strong interest to address the question whether the number of calculated magnon modes is equal to the number of atoms. Because of the complexity of the calculations of the dynamic susceptibility, there are only a small number of such studies for systems of this type. Here, we briefly consider a recent calculation by Buczek *et al.* [65] for the full Heusler alloy Co_2MnSi.

Because of the attenuation effects described above, peaks characterizing the individual spin wave modes are broadened. In the total susceptibility, these peaks can overlap and a broad common feature may arise that is difficult to analyze. Therefore, it is important to develop methods for the theoretical analysis that give access to individual spin wave modes. This approach is based on the diagonalization of the loss matrix and the analysis of the frequency dependence of its eigenvalues.

An example of the calculated loss tensor spectrum is presented in Figure 2.12, where three clear spin wave peaks can be discerned. The energies and inverse lifetimes of the spin waves along main directions in the Brillouin zone are presented in Figure 2.13a and b. The system has three magnetic atoms per unit cell and features three modes of the spin wave excitations. This shows that the concept of well-defined atomic moments remains useful also in the case of 3d compounds. The width of the lowest mode is zero, which is a consequence of the half-metallic character of the

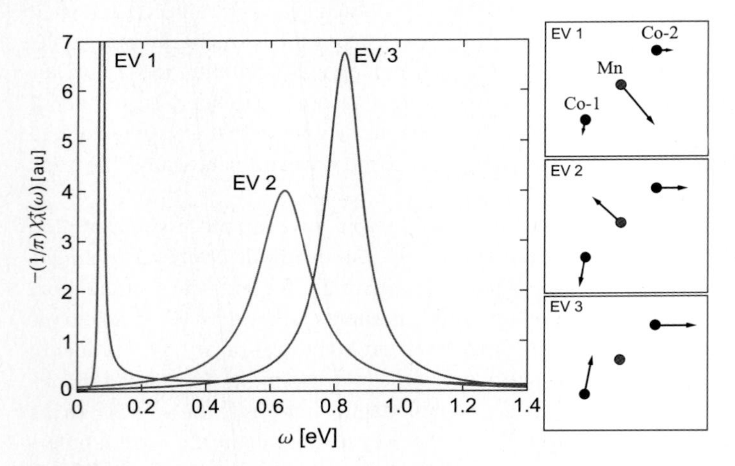

Figure 2.12 Enhanced susceptibility of Co_2MnSi, an example of the spectrum of the loss matrix for $\mathbf{q} = 0.28(1, 1, 0)2\pi/a$. a stands for the lattice constant. The three largest eigenvalues are shown, and other eigenvalues are of vanishing magnitude. Three clear peaks (EV 1, 2, and 3 corresponding to the labels in Figure 2.13) can be discerned. The panels show corresponding eigenvectors; arrows indicate the deviations of magnetic moments. Reprinted with permission from Ref. [65]. Copyright (2009) by the American Physical Society. (Please find a color version of this figure on the color plates.)

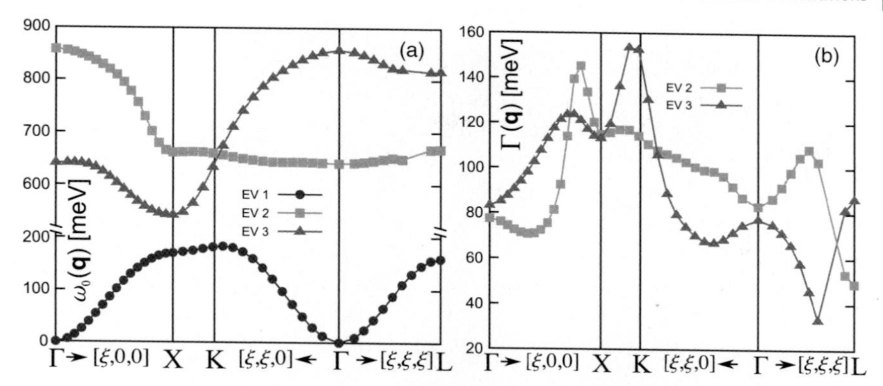

Figure 2.13 Energies (a) and inverse lifetimes Γ (b) of three SW modes in Co_2MnSi. Reprinted with permission from Ref. [65]. Copyright (2009) by the American Physical Society. (Please find a color version of this figure on the color plates.)

electron structure (cf. Figure 2.14). On the contrary, the "optical" modes (EV 2 and 3) appear where the continuum density is not small and have significant width, which depends strongly and nonmonotonously on $\mathbf{q}$. The shapes of the excitations corresponding to the three spin waves are characterized by an almost rigid rotation of the magnetization around the atomic sites and the number of spin wave modes corresponds to the number of magnetic atoms in the primitive unit cell. Therefore, Co_2MnSi – up to the damping – behaves as a typical Heisenberg–Hamiltonian system. In the EV 1 mode, the atomic moments oscillate in phase, roughly given by $e^{i\mathbf{q}\cdot\mathbf{r}_i}$, where $\mathbf{r}_i$ stands for the position of atom i. This justifies the use of the customary term "acoustic." For the optical modes, some of the moments align themselves roughly in the antiphase relation to the others, leading to quite intricate shapes of the excitations. As an example, let us consider the direction ΓK. The almost dispersionless mode EV 2 formed by Co and Mn moments (cf. Figure 2.12) can be regarded as an optical counterpart to EV 1, where Mn moments acquire an additional phase of π. Mode EV 3 involves only Co atoms.

Our last example is an iron film of three monolayer (ML) thickness. In Figure 2.15, we show the calculated spin-flip spectrum of the 3 ML Fe film on Cu(001). There are

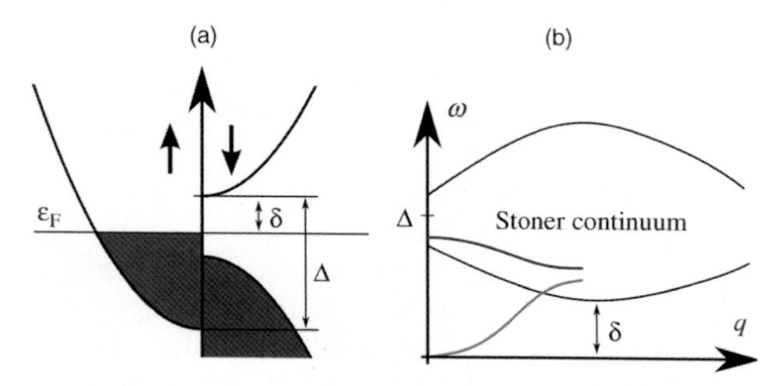

Figure 2.14 Schematic density of electronic states (a) and Stoner excitations (b) in a half metal. Because of the gap in the minority spin channel, there is a finite energy δ necessary to excite a Stoner pair. The energy in Co_2MnSi is on the order of 150 meV. Magnons with energies below this value cannot decay via the Landau mechanism.

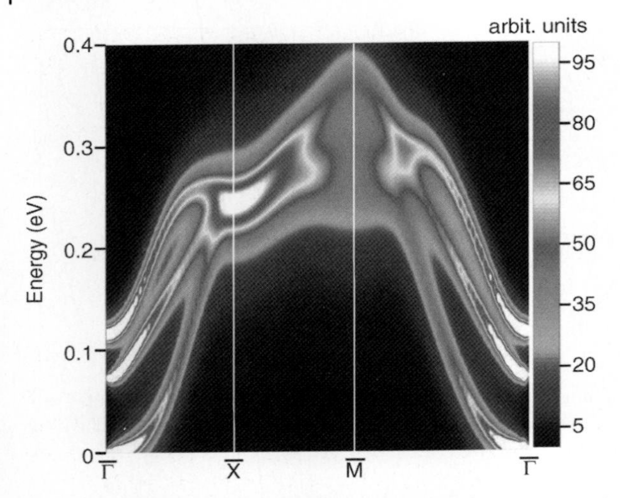

Figure 2.15 Spectral power $\mathrm{Im}\chi(q_{\|}, \hbar\omega)$ of spin-flip excitations in three-monolayer iron film on Cu (100) obtained from LRDFT calculations. Reprinted with permission from Ref. [17]. Copyright (2010) by the American Physical Society. (Please find a color version of this figure on the color plates.)

three spin-wave branches: one acoustic branch starting with zero energy and two optical branches with activation energies around 0.1 eV. The characteristic form of the spectrum with the parabolic dispersion of acoustical magnons is a consequence of the exchange interactions in the film that are stronger within than between the atomic planes. Because of the nearly half-metallic character of the film, the number of Stoner states in the spin wave energy region is small and the magnon peaks are well defined. Time-resolved photoemission experiments, discussed in detail in Section 2.4.3.5, show that spin wave emission in this iron film constitutes a significant source of inelastic electron decay.

2.3.2.3 **Conclusions**

In this chapter, we have shown that in the ferromagnetic Heisenberg model of interacting atomic moments the spin waves can be derived as the eigenmodes of the precessional motion of the atomic moments or as eigenstates of the Heisenberg Hamiltonian. Both classical and quantum treatments give the same dispersion relation between the energy and the wave vector of the spin waves. Thus, the description on the basis of the atomic moment picture appears robust and transparent.

However, in the itinerant electron magnets the primary physical quantities are the energies and wave functions of the itinerant electron states. The consequent approach to the study of the magnetic excitations in itinerant electron systems is the evaluation of the dynamic magnetic susceptibility. We have shown that also in this case the concept of the well-defined atomic moments remains important for the interpretation of the results of the calculations. Although the atomic moments do not enter the theory as dynamic variables, the physical trend to their formation is present and originates from the strong intraatomic exchange–correlation potential.

A very important aspect of the study of the dynamic magnetic susceptibility is the account for the attenuation of the spin waves because of the hybridization with single-particle Stoner excitations. These excitations compete with the trend to the formation of robust atomic moments and lead to the finite lifetime of magnons.

2.4 Experimental Access to Quasi-Particle and Collective Excitations

2.4.1 Coherent Phonons

Electrons and holes exchange energy and momentum with the crystalline lattice by emitting or absorbing phonons (e–ph scattering), and so do phonons among themselves (ph–ph scattering). Phonons involved in these scattering events are *incoherent*. By contrast, illumination of a crystal with an ultrashort optical pulse can impulsively create *coherent* phonons, if the pulse duration is sufficiently short compared to the inverse phonon frequency. Like their incoherent counterpart, coherent phonons can be classified into the optical and acoustic branches.

2.4.1.1 Coherent Optical Phonons

Coherent optical phonons are in principle Raman active modes with wave vector $\mathbf{q} \simeq 0$, which are in phase over a macroscopic area illuminated by the laser pulse. They are observed as periodic modulations in the optical constants and other material properties at Raman phonon frequencies, typically in the THz (10^{12}–10^{13} Hz) range, as shown in Figure 2.16a for diamond.

Theoretical Description of Coherent Optical Phonons In most situations involving coherent optical phonons, a classical description is adequate. The classical equation of motion for a small nuclear displacement Q is that of a driven harmonic oscillator [68]

$$\mu\left[\frac{\partial^2 Q(t)}{\partial t^2} + 2\gamma\frac{\partial Q(t)}{\partial t} + \omega_0^2 Q(t)\right] = F(t), \tag{2.107}$$

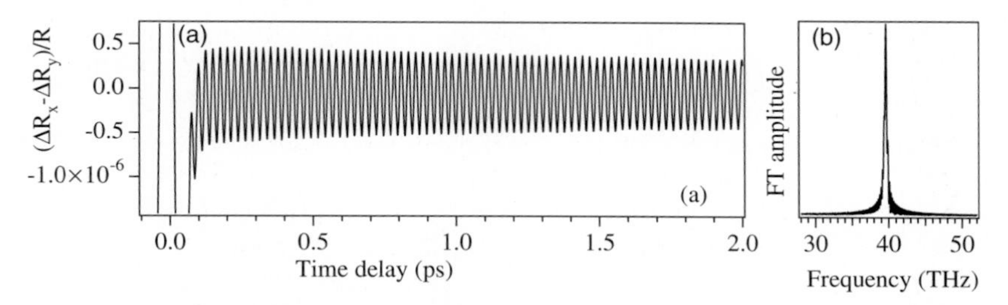

Figure 2.16 (a) Coherent optical phonon observed as a periodic modulation in the reflectivity of a diamond single crystal. (b) Fourier-transformed spectrum of the reflectivity trace in (a). The peak frequency is 39.8 THz (1330 cm^{-1}) See Ref. [67] for further details.

with the reduced lattice mass μ, the phonon damping γ, and the frequency ω_0. Solving the equation of motion gives the trajectory of a damped harmonic oscillator:

$$Q(t) = Q_0 \exp(-\gamma t) \sin(\omega_0 t - \varphi), \tag{2.108}$$

where φ is the initial phase. The trajectory deviates from the harmonic oscillator (2.108), for example, when the time-dependent interaction with photoexcited electron–hole plasma cannot be neglected [69–71].

While the classical model captures the essential physics, on a microscopic level the dynamics of the electrons and phonons should be described in terms of quantum mechanical operators. One way to microscopically describe coherent optical phonons is to do so with a synchronous motion of *many* different modes with locked phases. A finite optical penetration depth in an opaque solid would allow for a finite range of phonon wave vectors $\mathbf{q}$ to be coupled with the incident light and create a wave packet of optical phonons. This picture is convenient to describe coherent lattice oscillations in real space. Scholz *et al.* developed a microscopic theory for Ge [72], in which they wrote the Hamiltonian for electron–phonon coupling as a phonon-assisted hopping between lattice sites. The driving force of coherent phonons in Ge was given by the coherent part of the anisotropic hole hopping between the valence bands.

An alternative description of coherent optical phonons is given as a *single* zero-wave vector mode. This was achieved by Kuznetsov and Stanton [73] by considering deformational electron–phonon coupling, in which only the $\mathbf{q} = 0$ optical phonon mode is directly coupled to the total number of photoexcited electron–hole pairs. A simplified picture for a two-band semiconductor is given by a Hamiltonian

$$\hat{H}_{\mathrm{el}} = \sum_{\mathbf{k},\alpha} \varepsilon_{\alpha\mathbf{k}} c^{\dagger}_{\alpha\mathbf{k}} c_{\alpha\mathbf{k}} + \sum_{\mathbf{q}} \hbar\omega_{\mathbf{q}} b^{\dagger}_{\mathbf{q}} b_{\mathbf{q}} + \sum_{\alpha,\mathbf{k},\mathbf{q}} M_{\mathbf{kq}} \left(b_{\mathbf{q}} + b^{\dagger}_{-\mathbf{q}} \right) c^{\dagger}_{\alpha\mathbf{k}} c_{\alpha\mathbf{k}+\mathbf{q}}, \tag{2.109}$$

where $c, c^{\dagger}$ are the electron second quantization operators in k space, $\varepsilon_{\alpha\mathbf{k}}$ is the energy dispersion in the band $\alpha = \{c, v\}$ (conduction or valence), $\omega_{\mathbf{q}}$, $b^{\dagger}_{\mathbf{q}}$ and $b_{\mathbf{q}}$ are the phonon dispersion and the phonon creation and annihilation operators. The deformational coupling is described by the third term in the Hamiltonian. The coherent amplitude of the $\mathbf{q}$-th phonon mode is defined by the statistical averages of the *single* phonon operators

$$D_{\mathbf{q}} \equiv \langle b_{\mathbf{q}} \rangle + \langle b^{\dagger}_{-\mathbf{q}} \rangle \equiv B_{\mathbf{q}} + B^{*}_{-\mathbf{q}}, \tag{2.110}$$

so that it is proportional to the Fourier component of the displacement given by

$$\langle \hat{u}_{\mathbf{q}} \rangle \equiv \left\langle \frac{1}{V} \int d^3\mathbf{r}\, \hat{u}(\mathbf{r}) e^{-i\mathbf{q}\cdot\mathbf{r}} \right\rangle = \sqrt{\frac{\hbar}{2\varrho V \omega_{\mathbf{q}}}} D_{\mathbf{q}}. \tag{2.111}$$

Here, ϱ is the reduced mass density and V is the system's volume. The averages in Eq. (2.110) will vanish when the phonon oscillator is in one of its energy eigenstates. In this case, there is a certain number, $\mathcal{N}_{\mathbf{q}} = \langle b^{\dagger}_{\mathbf{q}} b_{\mathbf{q}} \rangle$, of *incoherent* phonons in the mode. However, if the wave function of the oscillator is a *coherent* superposition of different eigenstates, the displacement will not average to zero. An extreme case of such a superposition are *coherent states* that are widely used in quantum optics to describe quantum states of the electromagnetic field

$$\Psi^{\text{coh}} = |z\rangle = \sum_n \frac{z^n}{\sqrt{n!}} e^{-z^2/2} |n\rangle. \tag{2.112}$$

For these states, the "fully coherent" phonon amplitude and number can be defined by

$$B_{\mathbf{q}}^{\text{coh}} \equiv \langle z|b_{\mathbf{q}}|z\rangle = z, \tag{2.113}$$

$$\mathcal{N}_{\mathbf{q}}^{\text{coh}} = \langle z|b_{\mathbf{q}}^\dagger b_{\mathbf{q}}|z\rangle = |z|^2 = |B_{\mathbf{q}}|^2. \tag{2.114}$$

To have a nonzero amplitude in Eq. (2.110), however, the mode does not have to be in a fully coherent state. In the general case, a mode can have a certain number of both coherent and incoherent phonons

$$\mathcal{N}_{\mathbf{q}} = |B_{\mathbf{q}}|^2 + (\langle b_{\mathbf{q}}^\dagger b_{\mathbf{q}}\rangle - \langle b_{\mathbf{q}}^\dagger\rangle\langle b_{\mathbf{q}}\rangle) = \mathcal{N}_{\mathbf{q}}^{\text{coh}} + \mathcal{N}_{\mathbf{q}}^{\text{incoh}}. \tag{2.115}$$

The equation of motion for the coherent amplitude is obtained as

$$\frac{\partial^2}{\partial t^2} D_{\mathbf{q}} + \omega_{\mathbf{q}}^2 D_{\mathbf{q}} = -2\omega_{\mathbf{q}} \sum_{\alpha,\mathbf{k}} M_{\mathbf{kq}}^{\alpha} n_{\mathbf{k},\mathbf{k}+\mathbf{q}}^{\alpha}, \tag{2.116}$$

where $n_{\mathbf{k},\mathbf{k}+\mathbf{q}}^{\alpha} \equiv \langle c_{\alpha\mathbf{k}}^\dagger c_{\alpha\mathbf{k}+\mathbf{q}}\rangle$ is the electronic density matrix, which is directly related to the photoexcited carrier density. Within this picture, the equations of motion for different modes are completely uncoupled. The coherence in the phonon system is the result of simultaneously occurring multiphonon processes within the same mode. For $\mathbf{q} = 0$ and assuming that the electron–phonon coupling constant M does not depend strongly on $\mathbf{k}$, Eq. (2.116) can be simplified to

$$\frac{\partial^2}{\partial t^2} D_0 + \omega_0^2 D_0 = 2\omega_0 \sqrt{\frac{\hbar}{2\omega_0 \varrho V}} [C^v - C^c] N(t) \equiv W N(t), \tag{2.117}$$

with C the deformation potential and N the total number of photoexcited electron–hole pairs. If the decay of N is neglected and N is approximated by a step function of time, Eq. (2.117) can be solved by Fourier transform in the limit $t \to +\infty$:

$$D_0(t) = \frac{W}{\omega_0^2} N_0 [1 - I(\omega_0) \cos \omega_0 t] \tag{2.118}$$

with the normalized pump intensity I. Assuming $\varrho = 5\,\text{g/cm}^3$, $V = 1\,\text{cm}^3$, $\omega_0 = 10$ THz, and $C^v - C^c = 10^9$ eV/cm, this would lead to the coherent lattice displacement of $\sim 3 \times 10^{-4}$ nm, which is 10^{-3} of a typical lattice constant and the number of coherent phonons $\mathcal{N}_0^{\text{coh}} = 10^{19}\,\text{cm}^{-3}$, for excitation density of $10^{19}\,\text{cm}^{-3}$. Incoherent emission of $\mathbf{q} = 0$ phonons by photoexcited hot electrons is forbidden by energy and momentum conservation, and thus the quantum mechanical fluctuation of the lattice displacement,

$$\frac{\langle(\hat{u}-\langle u\rangle)^2\rangle}{\langle u\rangle^2} = \frac{2\mathcal{N}^{\text{incoh}}+1}{2\mathcal{N}} \simeq \frac{1}{2\mathcal{N}}, \tag{2.119}$$

is negligible. Therefore, as a result of the optical excitation, the $\mathbf{q}=0$ phonon mode acquires a macroscopically large number of coherent phonons. The macroscopic occupation of a single $\mathbf{q}=0$ quantum state would lead to an equivalent of Bose–Einstein condensation [70, 73].

The coherent states in Eq. (2.112) are a set of minimum-uncertainty states as noiseless as the vacuum state. Hu and Nori [74] proposed phonon squeezed states, in which the quantum noise is reduced below the zero-point fluctuation level, based on this single mode picture.

Generation of Coherent Optical Phonons The generation mechanism of the coherent optical phonons can be classified into two limiting cases according to the temporal profile of the driving force $F(t)$ in the equation of motion (2.107).

In transparent materials, impulsive stimulated Raman scattering (ISRS) is the only possible generation mechanism with a single pump pulse. A femtosecond, broadband optical pulse can offer multiple combinations of two photon difference frequencies required for the stimulated Raman process, $\hbar\omega_1 - \hbar\omega_2 = \hbar\omega_0$ (Figure 2.17a). This gives a δ function-like driving force defined by the Raman

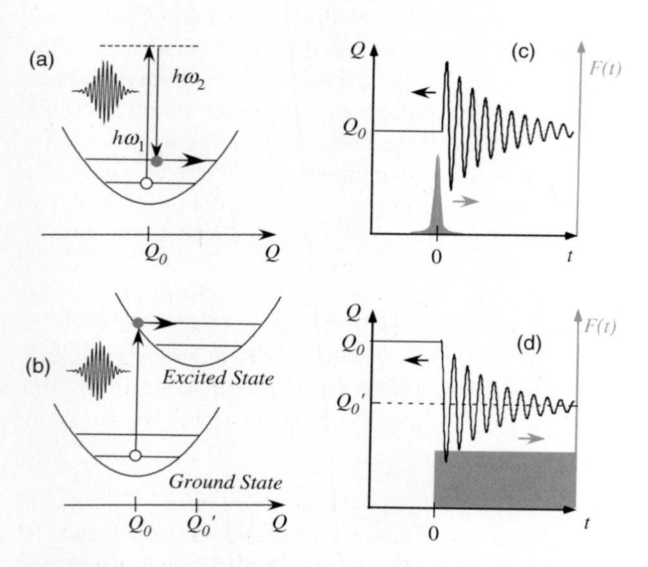

Figure 2.17 Schematic illustration of the two limiting cases for coherent optical phonon generation. (a) nonresonant ISRS. (b) DECP. Graphs (c) and (d) display the time evolution of the driving force (gray areas) and that of the displacement (solid curves) for nonresonant ISRS and DECP, respectively.

tensor R_{jkl} and the time profile of the excitation pulse [75, 76]:

$$F_j^{\mathrm{ISRS}}(t) = R_{jkl} E_k E_l = \left(\frac{\partial \chi}{\partial Q}\right)_{kl} E_k E_l. \tag{2.120}$$

The resulting nuclear oscillation is a sine function of time (i.e., zero amplitude at $t=0$), as illustrated in Figure 2.17c. In case of a resonant photoexcitation, ISRS can undergo an enhancement in the oscillation amplitude [77] and, for polar phonon modes, the breakdown of the Raman selection rule [78]. In the limit of a vanishing decay rate of the electronic excited state ($\Gamma \to 0$), the driving force can be given by a step function and should lead to an oscillation with a cosine function of time (i.e., maximum amplitude at $t = 0$) [77, 79], as illustrated in Figure 2.17d. For a finite value of Γ, the initial phase of the coherent optical phonons can vary between a sine and a cosine function of time [80].

When the electronic band structure is sensitive to the atomic motion of a particular phonon mode, photoexcition of carriers with ultrashort pulses can shift the vibrational potential surface suddenly and thus offer a step function-like driving force (Figure 2.17c). The resulting nuclear oscillation is a cosine function of time, as illustrated in Figure 2.17d. The displacive excitation of coherent phonons (DECP) was first proposed for the A_{1g} phonon of bismuth [81] and confirmed by a later time-resolved X-ray diffraction measurement [82] as well as a theoretical simulation [83].

At a semiconductor surface, the Fermi-level pinning by intrinsic or extrinsic surface states can lead to the formation of a depletion/accumulation layer. When such a surface is excited with ultrashort optical pulses, photogenerated electrons and holes are swept to opposite directions within 100 fs, as illustrated in Figure 2.18. The polarization buildup by such ultrafast drift-diffusion currents J can offer a step function-like driving force for *polar* phonons [68, 76]

$$F_j^{\mathrm{TDFS}}(t) = -\frac{e^*}{\varepsilon_\infty \varepsilon_0} \int_{-\infty}^{t} \mathrm{d}t' J_j(t'), \tag{2.121}$$

where ε_∞ and ε_0 are the high frequency and the vacuum dielectric constants, respectively. The transient field screening in the depletion layer (transient depletion field screening, TDFS) dominates the generation of coherent LO phonons in III–V and II–V semiconductors, as well as in their heterostructures.

The generation mechanism cannot be determined solely from the initial phase of the coherent oscillation, because resonant ISRS, DECP and TDFS would all result in an oscillation with a cosine function of time. To distinguish the different generation mechanisms experimentally, the dependence of the coherent amplitude on the pump polarization angle, on the applied surface voltage, on the doping level, etc. should be examined carefully [84].

Detection of Coherent Optical Phonons With the recent development of ultrashort radiation and particle sources, a variety of detection techniques are available to investigate ultrafast dynamics of coherent optical phonons of different nature.

Linear optical detection, such as transient reflectivity and transmission measurements, have been the most conventional, standard experimental technique. In a first-order approximation, the coherent nuclear displacement Q induces a

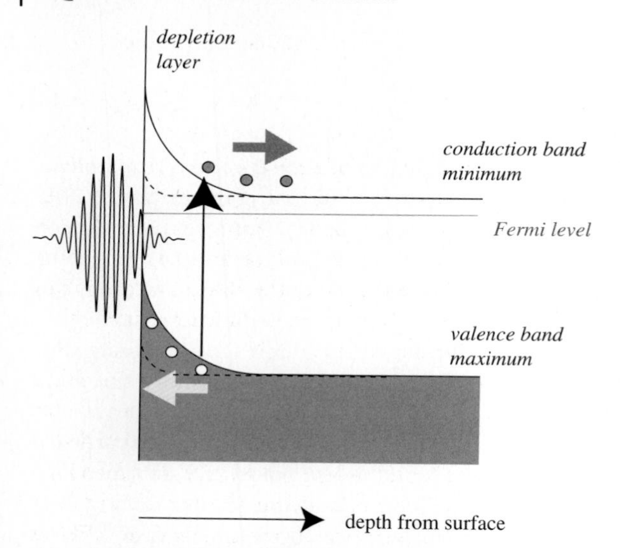

Figure 2.18 Schematic illustration of the TDFS in an n-type semiconductor. The band bending magnitude before and after photoexcitation is plotted by solid and broken curves, respectively. Because of the built-in depletion field, photoexcited holes are swept toward the surface and reach there the maximum density, whereas electrons are swept into the bulk. The ultrafast drift diffusion current screens the depletion field, and thus kick-starts coherent polar phonons.

change in the optical properties (e.g., reflectivity R) through the refractive index n and susceptibility χ [68]

$$\Delta R = \frac{\partial R}{\partial n}\Delta n \approx \frac{\partial R}{\partial \chi}\frac{\partial \chi}{\partial Q}Q(t), \tag{2.122}$$

which makes ΔR to be a quantitative measure for Q. Since $\partial\chi/\partial Q$ is a first-order Raman tensor, only Raman active modes can be observed by linear optical detection, provided that detection geometry and optical polarization satisfy the Raman selection rule. When the probe wavelength is at resonance with the optical absorption of the crystal, however, the breakdown of the Raman selection rule occurs. Dipole-forbidden Raman scatterings can then participate, in addition to allowed Raman scatterings, in the detection process of polar phonon modes [84]. The forbidden and allowed scatterings can be experimentally distinguished by the probe polarization dependence of the coherent amplitudes.

By using a nonlinear optical process such as second harmonic generation (SHG), one can probe coherent surface phonons and adsorbate-related vibrations exclusively [85, 86]. The second harmonic electric field is driven by the nonlinear polarization $P(2\omega)$ at the surface, which can be expanded into the nuclear displacement $Q_{n,q}$ for the n-th phonon mode with wave vector q [85]:

$$P_i(2\omega) = \left[\chi^{(2)}_{ijk}(2\omega) + \sum_{n,q}\left(\partial\chi^{(2)}_{ijk}(2\omega)/\partial Q_{n,q}\right)_0 Q_{n,q}\right] \times E_j(\omega)E_k(\omega). \tag{2.123}$$

The SH intensity is proportional to $|P|^2$ and can be approximated to be a linear function of Q by ignoring the second-order term.

Time-resolved X-ray diffraction (TRXRD) using a visible pump and X-ray probe scheme can directly monitor the structural dynamics of crystals far from the equilibrium. TRXRD can be applied for the detection of coherent optical phonons [87, 88]. However, in contrast to the TRXRD detection of acoustic phonons described in Section 2.4.1.2, the atomic motions associated with optical phonons do not modify the Bragg peak position because they do not change the barycentric positions of the crystal lattice. Instead, coherent optical phonons modify the Bragg peak intensity periodically if the structure factor can be approximated by a linear function of displacement in the vicinity of the equilibrium position [87].

In strongly correlated materials, the nuclear motion of particular phonon modes can couple strongly with the electronic system to affect the electronic conductivity in the THz frequency region. Time-resolved THz spectroscopy [89] can thus simultaneously monitor the coherent lattice and the coupled electron–hole dynamics. Time-resolved photoemission (TRPE) spectroscopy [90, 91] constitutes an alternative technique to detect the coupled phonon–electron dynamics as a periodic modulation in the binding energy of the electronic states.

Decay of Coherent Optical Phonons Coherent optical phonons decay via similar processes as incoherent phonons do, typically with a decay rate γ of 10^{10}–$10^{11}\,\mathrm{s}^{-1}$. In good-quality crystals and at finite temperature, anharmonic coupling between normal modes provides the main decay path. The lowest order anhamonic term in the crystal potential is the third-order anharmonicity [92]. The corresponding phonon broadening $\gamma_A^{(3)}$ includes both the downconversion term, where the initial ω_0 phonon with wave vector $q \sim 0$ decays into two or more lower energy phonons, and the upconversion term, where the initial phonon is scattered by a thermal phonon into a higher energy phonon. If there is no phonon mode of higher frequency than ω_0, no upconversion mechanism is allowed, and the decay occurs only by the downconversion. Assuming that the high-energy phonon decays into two low-energy phonons such that $\omega_i(\mathbf{q}) + \omega_j(-\mathbf{q}) = \omega_0$, the linewidth can be written as

$$\gamma_A^{(3)}(\omega_0) = \sum_{ij} \gamma_{ij}(\omega_0)[1 + n(\omega_i) + n(\omega_j)]d_{\omega_i+\omega_j}(\omega_0), \tag{2.124}$$

where $n(\omega)$ is the phonon occupation number and γ_{ij} is the average coupling coefficient. The scattering efficiency is proportional to the density of states of the final two-phonon band $d_{\omega_i+\omega_j}(\omega_0)$, and therefore the decay channels involving zone edge phonons give the dominant contributions. Discrimination between the different allowed channels can be realized through their different behavior with temperature. For a high-frequency optical phonon decaying into two isoenergetic acoustic phonons ($\omega_i = \omega_j = \omega_0/2$), for example, the temperature dependence is given by [92, 93]

$$\gamma_A(T) = \gamma_0[1 + 2n(\omega_0/2)] = \gamma_0\left[1 + \frac{2}{\exp(\hbar\omega_0/2k_B T) - 1}\right]. \tag{2.125}$$

Crystalline defects, such as atoms with different isotopic masses and impurity atoms, also contribute to the decay of optical phonons at low temeprature and/or in defective crystals. The rate for the mass disorder-induced scattering γ_I is given by [94]

$$\gamma_I = g_2 \frac{\pi\omega_0^2}{12} N_d(\omega_0), \tag{2.126}$$

where $N_d(\Omega_0)$ is the phonon density of states, and

$$g_2 \equiv \sum_i x_i[\Delta M_i/\bar{M}]^2, \tag{2.127}$$

is a measure of the isotopic disorder, with x_i the concentration of the i-th isotope whose mass differs from the mean atomic mass $\bar{M}$ by the amount ΔM_i. Equation (2.126) can also be applied to describe scattering by single vacancies quantitatively by putting $\Delta M_i = \bar{M}$ and x_i to be the vacancy concentration [95].

In low-dimensional conducting systems, electron–phonon coupling can be the major decay channel for high-frequency optical phonons. For graphite, the decay rate of the E_{2g} mode (Raman G mode) can be computed by [96]

$$\gamma_g = \frac{\pi^2\omega_0\alpha'}{2c}\left[f\left(-\frac{\hbar\omega_0}{2}-\varepsilon_F\right)-f\left(\frac{\hbar\omega_0}{2}-\varepsilon_F\right)\right], \tag{2.128}$$

where c is the speed of light, f is the Fermi–Dirac distribution, and $\alpha' = 4.39 \times 10^{-3}$. Equation (2.128) implies that the shift of the Fermi level ε_F can lead to a *decrease* in the phonon decay rate through breakdown of the adiabatic approximation [96]. Such an anomalous decrease in γ_g can also be realized by the photoexcited nonequilibrium electronic distribution, that is, without affecting the charge neutrality, as was experimentally demonstrated in transient reflectivity measurements of coherent E_{2g} phonons of graphite [71]. Electron–phonon coupling can also dominate the phonon decay in three-dimensional crystals in the presence of an extremely nonequilibrium electronic distribution, leading to a Fano-type lineshape that is indicative of interference effects [97, 98].

Coherent Phonon-Quasiparticle Coupled Modes Like their incoherent counterparts, coherent optical phonons can couple with other quasiparticles and collective excitations in solids to form coherent coupled modes.

In polar semiconductors, LO phonons couple with the collective charge density oscillations to form LO phonon–plasmon coupled (LOPC) modes. The frequency of the LOPC modes can be obtained by solving the equation for the complex dielectric constant [99]

$$\varepsilon(\omega) = \varepsilon_\infty\left[1+\frac{\omega_{LO}^2-\omega_{TO}^2}{\omega_{TO}^2-i\gamma\omega-\omega^2}-\frac{\omega_p^2}{\omega^2+i\Gamma\omega}\right] = 0, \tag{2.129}$$

where ω_{LO} and ω_{TO} are the LO and TO phonon frequencies, γ and Γ, the phonon and plasmon damping rates, respectively. The plasma frequency $\omega_p = \sqrt{ne^2/m^*\varepsilon_0\varepsilon_\infty}$ depends on the carrier density n and the carrier effective mass m^*. For the LOPC

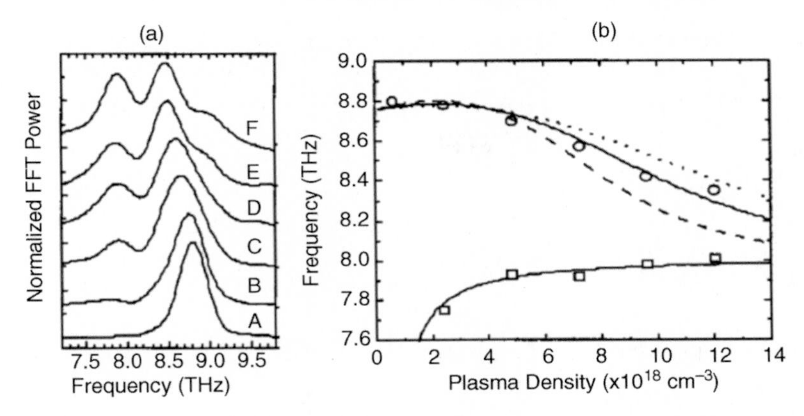

Figure 2.19 (a) Normalized Fourier power spectra for a TRSHG measurement on n-type GaAs. The pump-injected plasma densities are (A) 6×10^{17}, (B) 2.4×10^{18}, (C) 4.8×10^{18}, (D) 7.2×10^{18}, (E) 9.6×10^{18}, and (F) $1.2 \times 10^{19}\text{cm}^{-3}$. (b) Plot of the measured frequencies of the LO-hole (○) and LO-electron (□) coupling modes as a function of injected plasma density. Curves denote the calculated dielectric response function with different damping constants. Reprinted with permission from Ref. [102]. Copyright (2002), American Institute of Physics.

mode coupled with the *electron* plasma, the frequencies can be reproduced by the undamped ($\gamma = \Gamma_e = 0$) solutions, one of which appears above ω_{LO} (L^+ branch) and the other below ω_{TO} (L^- branch). Both branches of the LOPC mode were observed as coherent oscillations in transient reflectivity measurements on n-doped GaAs [100, 101]. Their frequencies were determined by the sum of doped and photo-injected carrier densities, indicating the coupling of the LO phonon with both the doped and the photo-injected electron plasma.

By contrast, a TRSHG study on similar n-doped GaAs [102] observed a coherent oscillation at a frequency between ω_{LO} and ω_{TO}, in addition to that below ω_{TO}, as shown in Figure 2.19. Because of the surface sensitivity, the SHG detection monitored the depletion layer of n-doped GaAs exclusively. The higher frequency oscillation was attributed to the LO phonon coupled with the photo-injected *hole* plasma, which was swept toward the surface due to the strong built-in field. The observed frequency was reproduced by Eq. (2.129) by assuming a heavy damping for the hole plasma ($\Gamma_{\text{h}} = 25$ THz). The lower frequency oscillation was likewise assigned to the LO phonon coupled with the photo-injected electron plasma, which has drifted toward the bulk.

Phonon–polaritons arise from the coupling of far infrared light to infrared-active phonons in crystals. They offer a good probe for the properties of low-frequency optical phonons, which are crucially related to the ferroelectric phase transition. Coherent phonon–polaritons [103] are generated in ferroelectrics such as $LiTaO_3$ and $LiNbO_3$ by four-wave mixing technique, which employs two pump pulses with wave vectors $\mathbf{k}_1$ and $\mathbf{k}_2$ crossing at the sample and thus creates a transient grating of phonons with wave vector $\mathbf{k}_{\text{p}} = \mathbf{k}_1 - \mathbf{k}_2 \neq 0$. By detecting the diffraction of the delayed probe light at the transient grating, one can monitor the coherent oscillation at the selected vector in the time domain. The light-phonon coupling leads to the formation of two polariton dispersion branches (polariton frequency ω_p as a function of

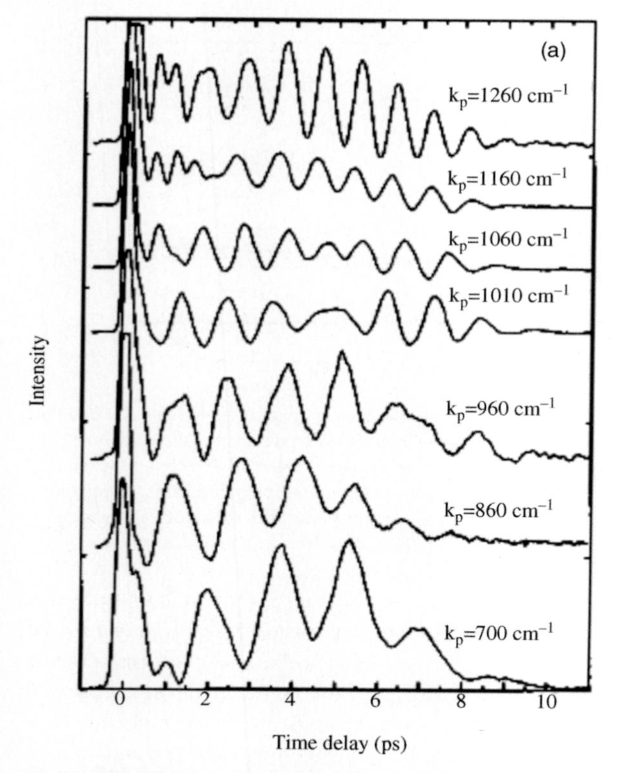

Figure 2.20 Phonon polaritons in $LiTaO_3$ at 300 K for different wave vectors. Reprinted with permission from Ref. [103]. Copyright (1998) by the American Physical Society.

polariton wave vector k_p) that avoid crossing at the resonance frequency. Time-resolved measurements on phonon–polaritons often show strong beating (polariton beats) in particular wave vector regimes, as shown for $LiTaO_3$ in Figure 2.20, due to the avoided crossings at the low-frequency resonances in the polariton dispersion.

In ferromagnetic metals, a coherent phonon–magnon coupled mode can be induced at the surface via spin-flip scattering. Transient reflectivity and TRSHG studies of the Gd(0001) surface observed such an coherently coupled oscillation in the presence of a magnetic field, whereas time-resolved photoemission measurements revealed that the same coherent vibration leads to a periodical shift of the surface-state energy via the exchange coupling between surface state and bulk bands [104].

2.4.1.2 Coherent Acoustic Phonons

Coherent acoustic phonons can be generated by the absorption of a laser pulse at a solid surface in the form of an ultrasonic wave packet with a spatial extent on the order of the penetration depth of the light. Uniaxial stress exceeding tens of MPa develops on a timescale much shorter than the acoustic propagation time across the laser penetration depth. As a result, the strain comprises a broad spectrum with frequencies extended from GHz up to a THz. The propagation of high-frequency (0.1–1 THz range) acoustic phonons through crystals at low temperature is characterized by both ballistic and diffusive transport. Phonons can live long enough for multiple round-trips in a μm thick film, with typical periods of ns to μs depending on the crystal temperature.

Generation of Coherent Acoustic Phonons At semiconductor and metal surfaces, absorption of a laser pulse leads to an abrupt heating within the optical penetration depth through a variety of mechanisms including the volume deformation potential interaction, optical–phonon emission and subsequent decay, and Auger heating. Both the electron–hole plasma (through the deformation potential) and the heated lattice (through thermal expansion) contribute to a buildup of stress.

We consider a laser pulse of duration <1 ps and energy Q incident on the free surface of a film of thickness d [105]. The total energy deposited per unit volume at a distance z into the film is

$$W(z) = (1-R)\frac{Q}{A\varsigma}e^{-z/\varsigma}, \tag{2.130}$$

where R is the reflectivity, ς is the optical absorption depth ($\varsigma \ll d$), and A is the area illuminated by the light pulse ($\sqrt{A} \gg d, \varsigma$). This heating gives a temperature rise $\Delta T(z) = W(z)/C$, where C is the specific heat per unit volume. This sets up an isotropic thermal stress given by $-3B\beta\Delta T(z)$, where B is the bulk modulus and β is the linear expansion coefficient. The stress σ_{zz} is given by

$$\sigma_{zz}(z,t) = -3B\beta\Delta T(z,t) - B\frac{\partial E_g}{\partial P}\Delta n(z,t) + \varrho v^2 \eta_{zz}(z,t). \tag{2.131}$$

The second term represents the photoexcited electron–hole contribution to the stress, with Δn denoting the photoexcited carrier density, E_g is the band gap, and P is the pressure. The last term is the elastic response of the crystal, with ϱ denoting the mass density, v the longitudinal speed of sound, and η_{zz} the resultant strain. The stress and strain are related to the atomic displacement u_z by

$$\frac{\partial \sigma_{zz}}{\partial z} = \varrho\frac{\partial^2 u_z}{\partial t^2}, \tag{2.132}$$

$$\eta_{zz} = \frac{\partial u_z}{\partial z}. \tag{2.133}$$

The strain $\eta_{zz}(z,t)$ obtained by solving these equations is shown in Figure 2.21a. The solution contains two separate components: a static thermal layer proportional to the instantaneous stress and a bipolar coherent acoustic pulse that travels away from the surface at the speed of sound. The width of this pulse is on the order of twice the absorption length ς. After propagating across the film, the pulse will be reflected at the boundary with the substrate. The acoustic pulse is detected after each round trip in the film as a series of equally spaced pulses, as shown in the right panel of 21.

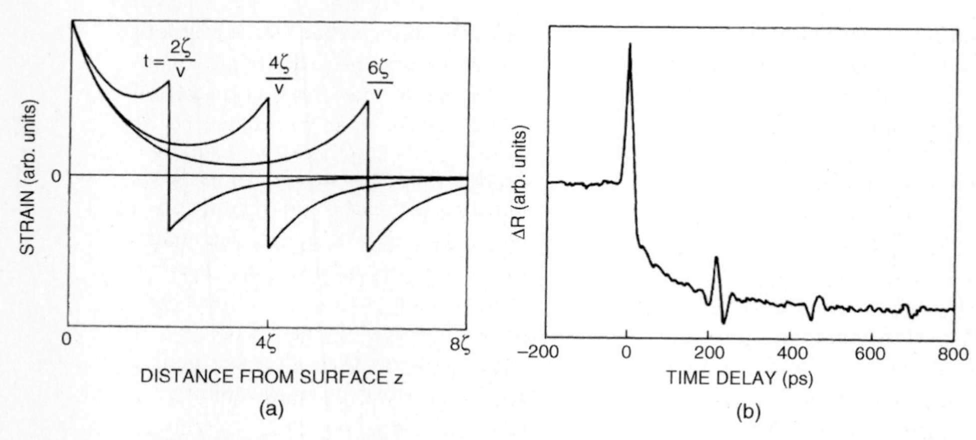

Figure 2.21 (a) Calculated spatial dependence of the elastic strain at different times after the pump pulse has been absorbed. (b) Photoinduced changes in reflectivity of a 2200 Å film of As_2Te_3 on a sapphire substrate. Reprinted with permission from Ref. [105]. Copyright (1986) by the American Physical Society.

The successive echoes in Figure 2.21(b) are inverted with respect to each other. This is consistent with the boundary conditions at the two surfaces of the film: at the free surface the phase changes by π and at the substrate interface there is no phase change because of the higher acoustic impedance of sapphire.

In semiconductor superlattices, the bulk phonon dispersion is backfolded into the mini-Brillouin zone of the artificially periodic heterostructure. The zone folding results in a series of Raman active acoustic phonon branches with $\omega \neq 0$ at wave vectors $q \simeq 0$ [106]. For this reason, the generation of coherent zone-folded acoustic phonons in superlattices is dubbed as ISRS [68, 76], which was described in Section 2.4.1.1.

In transparent materials, the generation and detection of coherent acoustic phonons is based on impulsive stimulated Brillouin scattering (ISBS) in a transient grating scheme [107]. ISBS utilizes two crossing laser pulses to generate counterpropagating acoustic phonons of selected wave vectors through photoelastic effects. The third light pulse, a probe, is diffracted by the interference fringe and detected by a photodetector. The intensity of the diffracted light oscillates according to the propagation of the acoustic phonons that exist as a standing wave.

Detection of Coherent Acoustic Phonons The time-resolved detection of coherent acoustic phonons in transparent solids has been based on the ISBS technique described above [107].

For opaque crystals at low temperature ($<$10 K), detection of the heat pulse with a bolometer has been a standard technique for relatively slow ($>$ ns) timescales. By spatially scanning the exciting laser spot with respect to the detector, one can visualize the highly anisotropic transport of the acoustic wave (phonon imaging) [108]. With an intense fs laser as a light source, pump–probe optical detections such as transient reflectivity measurements can achieve much faster time resolution, which is required for a very thin film or nanostructured sample, without cooling. Scanning the probe pulse with respect to the pump pulse on a thin film enables to image the snapshots of surface acoustic wave fronts propagating laterally like ripples [109].

TRXRD in a visible pump and X-ray probe scheme can directly monitor the coherent acoustic phonon dynamics [88, 110]. Acoustic phonons excited by an intense femtosecond laser pulse modulate the translational periodicity of a crystal, which is probed with an X-ray pulse by recording the rocking curve (X-ray reflectivity as a function of incident angle) at different time delays [88, 110]. Time-resolved electron diffraction (TRED) offers a highly surface sensitive technique to monitor coherent acoustic phonons as a periodic shift in the Bragg peak position [111].

Decay of Coherent Acoustic Phonons Mass defect scattering by impurities or isotopes [112] and anharmonic decay [113, 114], together with phonon focusing effects [108], dominate the dynamics of high-frequency phonon transport in non-metallic solids. In metals, interaction with photoexcited nonequilibrium electrons has also to be taken into account [115].

In high-quality crystals at low temperature (<10 K), scattering occurs mainly at mass defects, which typically are impurities and naturally occurring isotopes. Tamura [112] demonstrated that the scattering rate of acoustic phonons by substitutional isotopic atoms in a Bravais lattice can be written as

$$\gamma_I(\omega) = \frac{\pi}{6} V_0 g_2 \omega^2 D(\omega), \tag{2.134}$$

where V_0 is the volume per atom, D is the phonon density of states per unit volume, and g_2 is a measure of the isotopic disorder given by Eq. (2.127). Equation (2.134) implies that the scattering rate depends only on the frequency of the initial phonon, not its wave vector or polarization. In the long-wavelength limit ($\omega \ll \omega_D$), Eq. (2.134) reduces to [116]

$$\gamma_I = \frac{V_0 g_2 \omega^4}{4\pi v_D^3}, \tag{2.135}$$

with v_D the Debye velocity, which exhibits a ω^4 dependence. For scattering of phonons in the high-frequency dispersive region, a stronger frequency dependence than ω^4 is expected [112].

At finite temperature, anharmonic coupling, in which the initial high-frequency phonon decays spontaneously into two or more lower-frequency phonons, is also important. As the phonons downconvert by spontaneous decay, their mean free path increases dramatically. The anharmonic decay rate of acoustic phonons cannot be expressed in a simple way but varies with propagation direction and phonon polarization [116]. In the regime $\omega \gg k_B T/\hbar$, the lifetime of LA phonons against anharmonic three-phonon processes is very short, whereas TA phonons have anomalously long lifetime. This is because the TA phonons cannot decay spontaneously except through the collinear processes [85]. By contrast, the LA phonons can split spontaneously into two phonons, whose decay rate is be proportional to ω^5 from in weakly dispersive, isotropic crystals [67, 81]. For strongly dispersive, anisotropic crystals, however, the spontaneous decay rate exhibits a stronger frequency dependence than ω^5 [83]. Calculations indicate that the dominant decay channel for cubic crystals is LA $\rightarrow$ TA + TA, and total dacay rates at 1 THz is from 10^5 to 10^6 s^{-1} [113].

2.4.2
High-Resolution Angle-Resolved Photoemission

2.4.2.1 Photoemission Spectral Function of Quasi-Particles

The formation of quasi-particles in a many-body solid can conveniently be accessed by angle-resolved photoelectron spectroscopy. In the previous section, it has been addressed how quasi-particles can form by means of interaction between electrons on the one hand and excitations, such as, for example, phonons on the other hand. In now turning to photoemission, which probes the occupied states of the solid, one may naively think that it provides an account of the electronic band structure of the solid. However, there is a fundamental caveat: conventional band theory traditionally only accounts for the interaction of electrons with the static ion lattice. Yet, coupling to further microscopic degrees of freedom can alter the electron dynamics and lead to new many-body ground states not foreseen in that picture. A prominent example is the interaction with lattice vibrations (phonons) [19] that enhances the effective electron mass on the corresponding energy scale. A possible consequence is that it may give rise to (conventional) superconductivity. Moreover, the impact on the electrons from interaction with other excitations such as spin excitations in magnetic materials should also exist, yet it is not nearly as well established. This problem has received new attention due to the suggestion that high-temperature superconductivity in cuprate materials may result from electronic coupling to spin fluctuations [117]. Yet another mechanism, which is a notorious problem even for “simple” band structure calculations in an effective one-electron picture, is the existence of electron–electron correlations, which are – per definition – not treated exactly in standard approaches such as density functional theory. Therefore, by exploiting ARPES as an experimental probe, one must expect to get an account of all these interactions simultaneously. Let us consider what will happen (and which corrections will apply) to the picture of noninteracting electrons. Most importantly, interaction of conduction electrons with elementary excitations leads to a *renormalization* of electronic energies, that is, to deviations of their band dispersion from that expected for the noninteracting case. At low binding energies, the electrons become dressed by excitations, thereby forming dressed *quasi-particles* of increased mass (note that here the term “quasi-particle” will be used in conjunction with dressing by bosonic excitations, while strictly the concept of quasi-particles can be applied to all electron states in a many-body electron system [1]). As illustrated in Figure 2.22a, this is reflected in a reduced slope of the electron band which is inversely proportional to the electron mass. Beyond the characteristic energy scale ω_0 of the coupled excitations, the electrons resume their noncoupling band dispersion. Quasi-particle formation due to interaction with phonons as depicted in Figure 2.22b has been reported experimentally by ARPES, and we shall examine examples below [118, 119].

The analogy to electron–phonon coupling would be the coupling to spin waves (magnons) illustrated in Figure 2.22c. The highest spin wave mode to which the

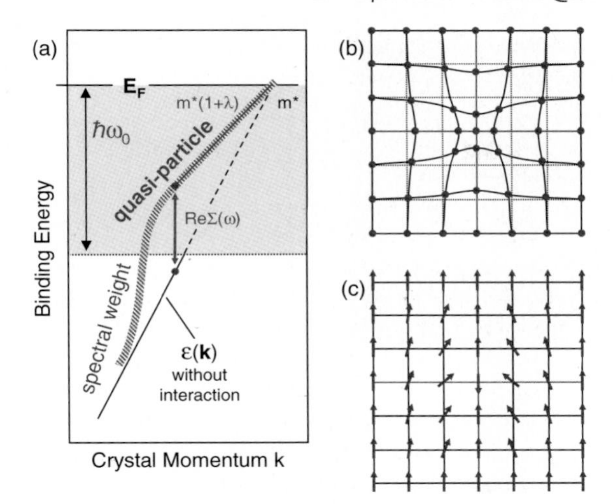

Figure 2.22 Schematic display of quasi-particle formation. (a) The electron is dressed with the excitation (phonon or spin excitation) up to an energy ω_0 below the Fermi energy, leading to a mass enhancement. (b) Electron–phonon coupling implies a distortion of the crystal lattice surrounding the electron. (c) Electron–magnon coupling implies spin scattering mediated by magnetic interactions. (Please find a color version of this figure on the color plates.)

electrons couple accordingly defines the energy ω_0. Unfortunately, similar energy scales of phonon and spin excitations are a serious hindrance for determining relative interaction strengths in the cuprates [120, 121]. In turn, it seems rather insightful to study magnetic metals. For example, spin wave energies in the elemental ferromagnets such as Fe, Co, and Ni are exceptionally high, well exceeding those of phonons and making it thus a good candidate for such studies. A very important property of the ARPES experiment is that it probes the solid (including the many-body interactions) in terms of an *excited* state. In particular, the incoming photon removes an electron from the many-body system, and an excited hole state remains. The spectrum recorded in ARPES, based on a Green's function algorithm (which reflects the response of the many-body system), is given by the so-called *spectral function*. The spectrum obtained in the presence of interactions in the electron system can be written as [122]

$$A(k,\omega) \propto \frac{\mathrm{Im}\,\Sigma(\omega)}{[\hbar\omega - \varepsilon_k - \mathrm{Re}\,\Sigma(\omega)]^2 + [\mathrm{Im}\,\Sigma(\omega)]^2} \qquad (2.136)$$

Here, ε_k is the (hypothetical) noninteracting dispersion. The real and the imaginary part of the so-called self-energy $\Sigma(\omega)$ reflect the contribution of the interaction to the spectrum. It is an important consequence that the photoemission spectrum described by $A(k,\omega)$ is no longer a delta function (as it was in the noninteracting case, marking the corresponding band dispersion), yet it is an intensity landscape when plotted upon energy ω and momentum k. $\mathrm{Im}\,\Sigma(\omega)$ incorporates the scattering processes and is given by an integral over the density of states of those bosonic

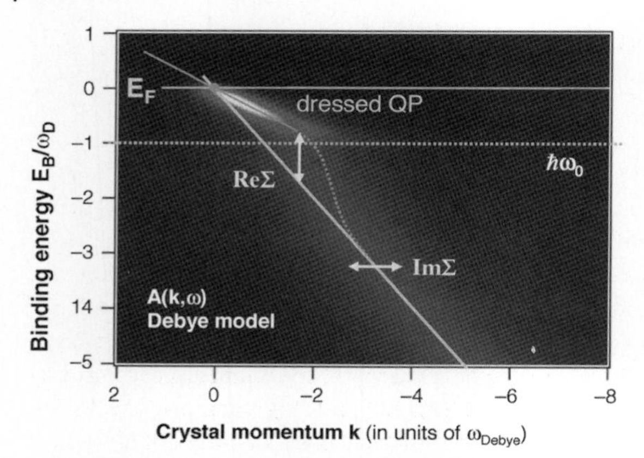

Figure 2.23 Spectral function generated numerically ($T = 20$ K, $T_{\text{debye}} = 340$ K, impurity scattering 10 meV). The shift of spectral weight near the Fermi level and a reduced slope become obvious. The self-energy Σ implies a broadening (given by ImΣ) and an energy shift (given by ReΣ) of the spectrum compared to the noncoupling case. (Please find a color version of this figure on the color plates.)

modes the electron is coupled to. This is, for example, the phonon density of states in the case of electron–phonon coupling:

$$\text{Im}\,\Sigma(\omega) \propto \lambda \int_0^{\omega} \varrho_{\text{Phonon}}(\Omega)\text{d}\Omega \tag{2.137}$$

where λ is the coupling constant. Note that $\text{Im}\,\Sigma(\omega)$ relates to the lifetime of the excited state, as explained in Section 2.2.3 in connection with Eq. (2.12). The phonon density of states can, for example, be described by the Debye model as in Eq. (2.17) for bulk phonons. $\text{Im}\,\Sigma(\omega)$ leads to a *broadening* of the spectral function $A(k,\omega)$. The real part $\text{Re}\,\Sigma(\omega)$ is then obtained from a Kramers–Kronig transformation, similar to relations known from optics. It leads to an *energy shift* of the spectrum. As an example, in Figure 2.23 the spectral function is plotted from a computer simulation for coupling of the electrons to a Debye-type phonon dispersion with a maximum energy of $T_{\text{Debye}} = 340$ K. The coupling constant λ can be extracted from the slope of the dressed quasi-particle band, which is renormalized by $1/(1+\lambda)$.

2.4.2.2 Experimental Considerations for Photoelectron Spectroscopy

Experimentally, studies of electron–phonon and electron–magnon coupling are very commonly performed at synchrotron light sources. Their advantage is twofold. For one, they provide high brightness that allows studies on samples (such as metal surfaces) with limited lifetime before degradation, and second, the tunability of the photon energy allows one to choose rather high energies not available with helium discharge lamps. In particular, in going to photon energies near ≈ 100 eV, the assumption of free-electron final states becomes justified, so that final state effects do not modulate the spectral function $A(k,\omega)$ of the initial state in unwanted manner. The ARPES experiment also usually requires very high energy and angular resolution, to achieve the best possible account of the spectral function $A(k, \omega)$. With the advent of modern electron analyzers with imaging detectors, energy and angular resolution values on the order of $\approx 5-10$ meV and $0.1°$ have become feasible. In addition, the beamline optics may contribute notably to the finite resolution. A total experimental resolution of $\approx 10-20$ meV seems practical today.

Best observation of self-energy effects in ARPES is achieved on surface states and resonances [119, 122], as will be illustrated in the experimental examples below. The reason is that these states have little or no dispersion in direction perpendicular to the surface. Since photoemission has a finite electron escape depth on the order of 10–20 Å for common photon energies in the range of 20–130 eV, this implies a small broadening of the k-perpendicular component, which in turn might affect or slightly distort the spectral function. Thus, surface states without this influence usually are the best candidates to look even for small renormalization effects. In photoemission from *bulk* states, one has thus to precisely consider the energy broadening of the spectra due to escape depth of the photoelectron. It leads to a broadening of the final state, showing a Lorentzian distribution in $k_\perp$ with a full-width at half-maximum of $k_\perp = w^{-1}$, where w is the mean free path of the electron. This results in a total energy broadening of the spectra with contributions from the hole state and the photoelectron [123]:

$$\Gamma_{\text{tot}} = \Gamma_{\text{h}} + \frac{|\nu_{\text{h}}|}{|\nu_{\text{e}}|}|\Gamma_{\text{tot}} \tag{2.138}$$

where $\nu_{\text{e/h}} = \partial\varepsilon_{\text{e/h}}/\partial k_\perp$ are the velocities of electrons and holes, respectively. Hence, the final state might influence the photoemission spectrum significantly, unless the dispersion ν_{h} of the initial state perpendicular to the surface is vanishing. This is fulfilled, for example, for surface states. Yet even for bulk states, the crystal symmetry can be favorably exploited. Selecting the photon energy such that k lies in a *high-symmetry plane* implies an extremum in the band topology. Thus, ν_{h} will vanish, leaving Γ_{h} as the desired quasi-particle contribution of interest. This is illustrated in detail in Ref. [124].

2.4.2.3 **Quasi-Particles from Electron–Phonon Interaction**

An interesting realization of coupling of electron states to phonon modes is achieved at the surface of metals. Not only does this provide good observation conditions for the renormalization but also special vibrational modes solely observed for surface adsorbates can be studied. Studies on free metal surfaces have, for example, been performed on Mo(110) and Be(0001) surfaces [118, 119]. Here, a kink on the phonon energy scale was observed near the Fermi level, as expected according to Figure 2.23. The physics of electron–phonon interaction at surfaces with adsorbate phonons can nicely be studied by using a W(110) substrate, where one can, for example, prepare a monolayer coverage with hydrogen, as illustrated in Figure 2.24a. Its vibrational properties have been studied by scattering methods in the past [125]. Also, the surface states of the clean substrate are well known [126]. On the Fermi surface of H:W(110) on the $\bar{\Gamma}$–$\bar{S}$ line, metallic surface state are well separated from the bulk state. Here, quasi-particle renormalization is observed very clearly, as shown in

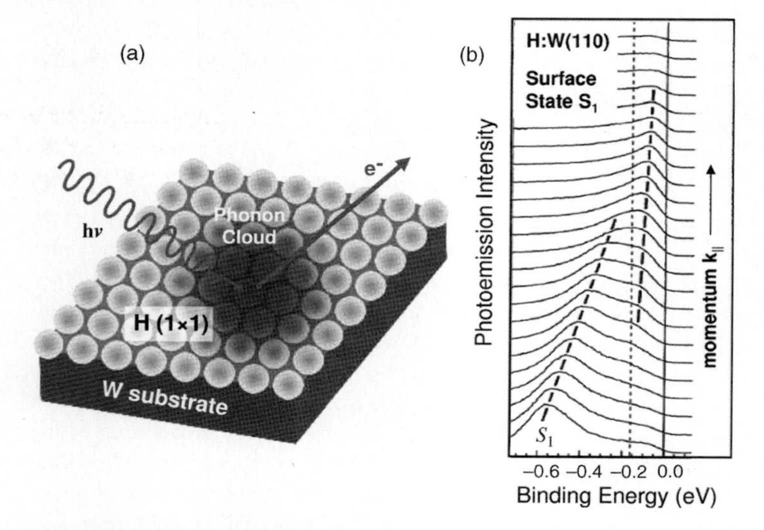

Figure 2.24 The H:W(110) system allows the study of adsorbate vibrations in the presence of interactions with a metallic electron bath. (b) Spectra of the surface state band (T = 150 K, $h\nu$ = 100 eV). They reflect the electron–phonon coupling in an energy window of ≈ 160 meV that corresponds to the H stretching vibration. (Please find a color version of this figure on the color plates.)

Figure 2.23b. The energy window of ≈ 160 meV in which the renormalization occurs is in excellent agreement with the symmetric stretch vibration of the H atoms. The coupling constant of $\lambda \approx 1.4$ is rather large, and the question arises whether superconductivity might be achieved at low temperatures. Confirmation of the interpretation as due to adsorbate vibrations was obtained by performing an *isotope replacement* experiment of the adsorbate. For this purpose, the adsorption was done with deuterium instead of hydrogen. This reduces the oscillator frequency by a factor of $\sqrt{2}$ as expected, reflecting the mass increase of the adsorbate atoms [127]. These results help to elucidate the aspect of energy dissipation of adsorbates by electronic excitations in metallic substrates, which also has important implications for surface chemistry. Such nonadiabatic processes contribute to the dampening of adsorbate vibrations and thus are essential to controlling surface reaction mechanisms and kinetics.

2.4.2.4 **Quasi-Particles from Electron–Magnon Interaction**

Surface States of Magnetic Materials A prime candidate to explore the interaction of electrons with spin waves is ferromagnetic iron. Here, the energy scale of the spin waves [128, 129] is approximately one order of magnitude higher than that of the phonons. Furthermore, the recent observation of superconductivity in the nonmagnetic high-pressure phase of Fe has been related to electronic coupling with spin fluctuations [130]. Using high-resolution ARPES of the Fe(110) surface, two metallic surface state bands can be identified that display spectroscopic signatures

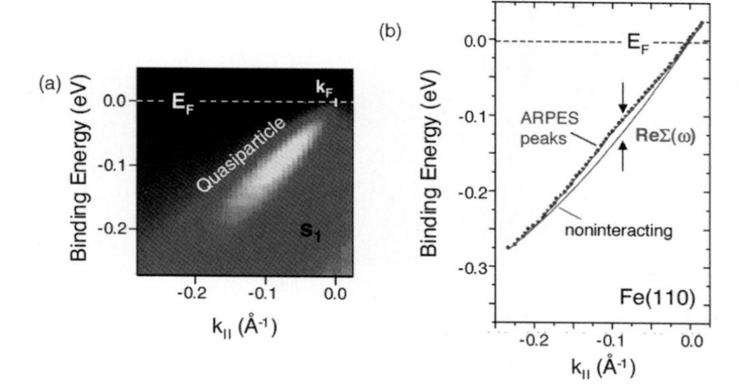

Figure 2.25 (a) ARPES data of the metallic Fe (110) surface state near $\bar{S}$ at $T = 85$ K. It is a surface resonance overlapping with the projection of bulk bands of opposite spin. (b) Peak position extracted from the band map using a fitting procedure (thick dots). The smooth line is a parabolic interpolation for the noninteracting band. A deviation from this band extends over a large energy scale below the Fermi level. Reprinted with permission from Ref. [131]. Copyright (2004) by the American Physical Society. (Please find a color version of this figure on the color plates.)

of quasi-particle renormalization. For ferromagnetic iron, the (110) face exhibits metallic surface states s_1 and s_2 around the $\bar{N}$ and the $\bar{S}$ points of the Brillouin zone, respectively. Both are of minority spin character, as explained in Ref. [131]. Spin interaction becomes possible because they overlap in energy with the surface-projected bulk bands of opposite spin.

The experimental dispersion of these surface states near the Fermi level extends along $\bar{\Gamma}$–$\bar{S}$ for s_1 and along $\bar{\Gamma}$–$\bar{N}$ for s_2. In the raw data of surface state s_1 in Figure 2.25a, the dressed quasi-particle region shows up with high intensity extending beyond 0.1 eV binding energy. The accurate peak position and width are obtained from a fit [119] of the momentum spectra at constant energy shown in Figure 2.25b. One typically uses a convoluted Lorentzian–Gaussian lineshape where the Gaussian part takes care of the experimental broadening. The Lorentzian width is used for the self-energy analysis. The dispersion displays a weak "kink" in the 100–200 meV region below E_F, which bears much resemblance to quasi-particle renormalization effects observed for electron–phonon coupling [118, 119]. The dispersion anomaly becomes obvious when comparing it to the nominally "undressed" dispersion. The undressed band is obtained here by parabolic interpolation between the lowest data points and the experimental Fermi vector, as included in Figure 2.25b. The experimentally determined dispersion is significantly offset from the noncoupling band interpolation. This suggests the interpretation that the observed kinks are indeed caused by many-body effects. The real part of the self-energy $\mathrm{Re}\Sigma(\omega)$ is the *difference* between interpolated and observed band dispersion, as marked in Figrue 2.25b. For both surface bands, it increases toward a maximum around 125 ± 10 meV and then gradually approaches zero again. The imaginary part of the self-energy, $\mathrm{Im}\,\Sigma(\omega)$, as displayed in Figure 2.26, reflects the scattering processes that increase with increasing binding energy. It is given by the Lorentzian

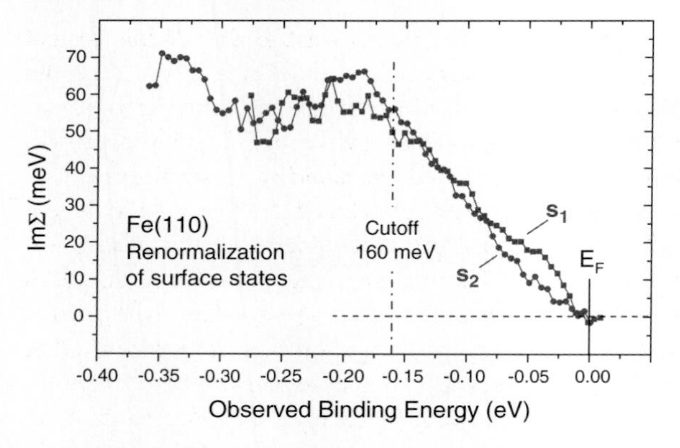

Figure 2.26 Imaginary part of the self-energy of the Fe(110) surface state, determined from the width of the ARPES momentum spectra. The increasing broadening reflects the interaction that the electrons are experiencing. The characteristic energy of $\approx$ 160 meV (which marks the saturation) coincides with known magnon energies.

half-width of the momentum spectra, applying the energy scale of the undressed band. Impurity scattering [119] adds a small offset of 30 meV that one has to subtract. The pronounced increase in $\mathrm{Im}\Sigma(\omega)$ with binding energy saturates at 160 ± 20 meV. The large energy scale clearly rules out electron–phonon coupling as origin. The phonon spectrum of iron has a maximum energy of $\approx$ 30 meV in the bulk [132] and also at the (110) surface [133]. This energy scale may be faintly visible in the data for s_1, yet cannot be considered relevant on a larger binding energy scale. Electron–electron scattering through Coulomb interaction [119] cannot account for characteristic structure either since in iron the total electronic density of states is smooth in the relevant energy region. Thus, we are left to consider coupling of the electrons with magnetic excitations.

Magnons in ferromagnetic Fe are well known from inelastic neutron scattering [128, 129]. Between approximately 100 meV and 200 meV both experiment [129] and theory [134, 135] find an "acoustic" and an "optical" magnon branch, separated by a gap where sharply defined spin waves do not exist. An acoustic spin wave branch has also been inferred from magnetization studies, locating its cutoff at 166 meV [136]. This energy scale is remarkably close to the characteristic energy observed in ARPES. It suggests the interpretation in terms of electron–magnon quasi-particle renormalization. Independent evidence comes from spin-polarized electron energy loss spectroscopy on the Fe(110) surface [137]. A loss structure at 170–200 meV is interpreted as exchange scattering by spin waves. It has been argued that spin waves at larger energies will be dampened by the onset of Stoner excitations [138]. We also note that a study on Co films observed strong electronic interaction only with the acoustic magnon branch [139], in very good agreement with our ARPES data. The compatibility of the ARPES self-energies with an electron–magnon coupling scenario can be tested using a simple model in analogy to electron–phonon coupling. The model uses the magnon density of states for the interaction, as described in Ref. [131]. It must be mentioned that a precise theoretical treatment of modified quasi-particle spectra due to spin excitations does not exist to date. While several approaches are available to describe magnons per se, as laid out in Section 2.3.2.2, the interaction of hole states with magnons (both of electronic nature) will require an advanced many-body approach. Thus, the bosonic magnon coupling model used here must be considered a simplification, albeit justified *a posteriori* by its effectiveness. The functional form of $\mathrm{Im}\Sigma(\omega)$ and $\mathrm{Re}\Sigma(\omega)$ is well reproduced. Notably, one finds that the peak in $\mathrm{Re}\Sigma(\omega)$ occurs before the maximum mode energy, that is, at 0.79 ω_0, beyond which it decreases asymptotically. The analysis yields a coupling constant of $\lambda = 0.20 \pm 0.04$, indicating a relatively moderate coupling. It may relate to the fact that scattering occurs from spin-down *surface* states into spin-up *bulk* states with a rather small spatial overlap. Alternatively, the electrons may also couple to *surface* magnons. These have been found recently on the very same energy scale ($\approx$ 160 meV) by spin-polarized electron energy loss spectroscopy of the Fe(110) surface [140].

Interaction with Magnons in Bulk Magnetic Metals The question of electron energy renormalization in bulk magnetic materials has important bearing on correlated materials in general. They often exhibit an (anti-)ferromagnetically ordered phase. For access to these many-body interactions, a three-dimensional Fermi liquid in the ferromagnetic state seems a suitable model system. The energy scales for the lattice and spin wave excitations in typical ferromagnets such as Ni differ by approximately an order of magnitude, and hence will affect the quasi-particles at different binding energies [132, 141]. Furthermore, it is established that the valence band states are strongly correlated [142, 143], which is proven by a concomitant photoemission satellite. This allows to directly address the interplay of correlation physics and quasi-particle formation in the presence of distinct spin excitations. Here, we discuss an ARPES study of quasi-particle states in Ni(110) [124]. Despite the ubiquitous problem of perpendicular momentum broadening, suitable observation conditions can be found for *bulk* bands. An electronic self-energy analysis is carried out by comparison with a Gutzwiller calculation that describes quasi-particle renormalization more reliably than state-of-the-art density functional theories [144, 145].

A clean Ni(110) crystal was prepared by repeated Ar sputtering and annealing of a crystal *in situ*. During ARPES, the sample was kept at $T = 10$ K, and to ensure validity of the free electron final-state approximation, a high photon energy of 100 eV was chosen. For the purpose of studying quasi-particle effects in a wide energy range, the minority spin band at the K-point (labeled Σ_2) was selected. The ARPES data in Figure 2.27 show the band of interest, which is unaffected by neighboring states for binding energies up to 800 meV. A reliable method for determining the self-energy Σ (in terms of $\mathrm{Re}\Sigma$ and $\mathrm{Im}\,\Sigma$) is from the peak positions ω and widths Δk of the momentum distribution curves. Then, $\mathrm{Re}\Sigma(\omega) = \varepsilon(k) - \omega$, where $\varepsilon(k)$ is the bare band dispersion, and $\mathrm{Im}\,\Sigma(\omega) = |v(k)|\Delta k$, where $|v(k)|$ is the slope of the bare band and Δk the half-width at half-maximum of the peak [124]. For extraction of the self-energy contribution, a Gutzwiller calculation for Ni was used as reference [146] because in the Gutzwiller theory, electron–electron correlation effects are accounted for in a much better way than in density functional theory calculations. Hence, deviations of the experimental dispersion should result essentially from coupling of

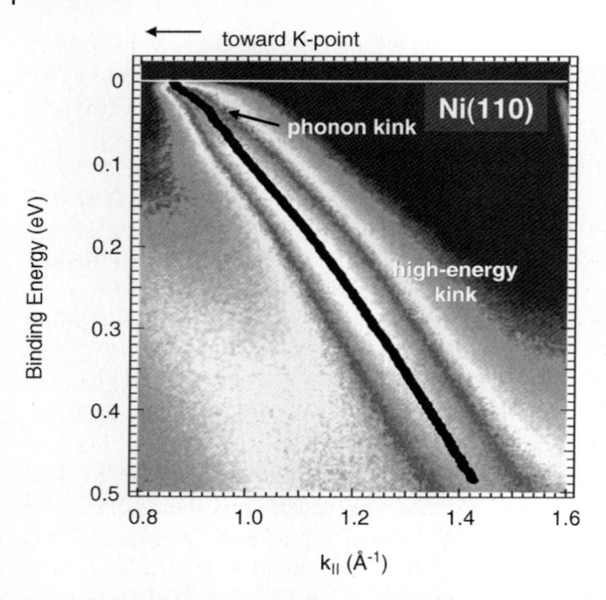

Figure 2.27 ARPES data of Ni(110) recorded at $h\nu = 100$ eV (corresponding to a high-symmetry plane). The band map shows the bulk band near the *K*-point. The momentum distribution curves are fitted to yield the peak positions. At low and high binding energy (as indicated), indications for kinks in the dispersion are discernible, corresponding to the phonon and magnon energy scale, respectively. Reprinted with permission from Ref. [124]. Copyright (2009) by the American Physical Society. (Please find a color version of this figure on the color plates.)

the electrons to bosonic excitations. The results for ReΣ and ImΣ as a function of binding energy are plotted in Figure 2.28a and b, respectively. Comparing experimental data and calculation, a first-mass renormalization can be found around $E_B \approx 30$ meV, which corresponds to the kink expected for electron–phonon coupling [147]. The visibility of this feature proves the high quality of the data. It demonstrates that many-body interactions are indeed observable in bulk bands at high-symmetry planes. A second deviation from the calculation is found in the binding energy region from 250–300 meV. This behavior is reflected by the real part of the self-energy in Figure 2.28a, which is showing a second maximum, corresponding well to the structure of the imaginary part ImΣ in Figure 2.28b. Hence, there is strong indication of a second many-body interaction effect at higher binding energies in both parts of the self-energy. A plausible explanation in a ferromagnet with correlated electrons like Ni is the coupling of the electrons to spin wave excitations.

Magnons in Ni are well known from inelastic neutron scattering, which yields energies up to 250 meV [141]. However, this method is problematic for scattering at large wave vectors, which makes it difficult to obtain data near the zone edges. Therefore, one expects an even larger energy for the maximum of the spin wave spectrum. Cutoff energies obtained by various calculations are 270–370 meV [134, 148, 149]. In an attempt to model the observed characteristics of the self-energy, the simple relation known from electron–phonon coupling as used above is analogously applied here: $\mathrm{Im}\,\Sigma(\omega) \propto \lambda \int_0^{\omega} \varrho_{\mathrm{bos}}(\Omega)\mathrm{d}\Omega$, where ϱ_{bos} is the boson density of states and λ is the coupling constant. From the linear dispersion for bulk phonons and the quadratic one for bulk magnons, up to their respective cutoffs ω_0, it follows $\varrho_{\mathrm{pho}} \propto \omega^2$ and $\varrho_{\mathrm{mag}} \propto \omega^{1/2}$, respectively. These results can be used to calculate $\mathrm{Im}\Sigma$ contributions from electron–boson interaction with two parameters, the coupling strength λ and the cutoff energy ω_0. The corresponding real parts are calculated directly employing the Kramers–Kronig relation. The results for $\mathrm{Re}\Sigma$ and $\mathrm{Im}\Sigma$ are laid out in Figure 2.28a and b, respectively.

For a complete description of the imaginary part, one has to include the electron–electron scattering contribution, for which a quadratic energy dependence as obtained from Fermi liquid theory is assumed (effects such as impurity scattering and experimental resolution can be accounted for by a constant offset). These contributions add up linearly to the full imaginary part. Note that an electron–electron scattering contribution does not need to be considered explicitly for $\mathrm{Re}\Sigma$ because it is already included in the Gutzwiller reference band. With this model, the main features of both self-energy parts are very well reproduced; see Figures 2.24b and 2.25b. The parameters in the model for the cutoff energies are $\omega_{0,\mathrm{ph}} = 35$ meV and $\omega_{0,\mathrm{mag}} = 340$ meV. The latter is compatible with the available bulk spin wave data. The analysis also yields the coupling constants $\lambda_{\mathrm{ph}} = 0.3$ and $\lambda_{\mathrm{mag}} = 0.19$. The electron–magnon coupling is somewhat weaker than the electron–phonon coupling

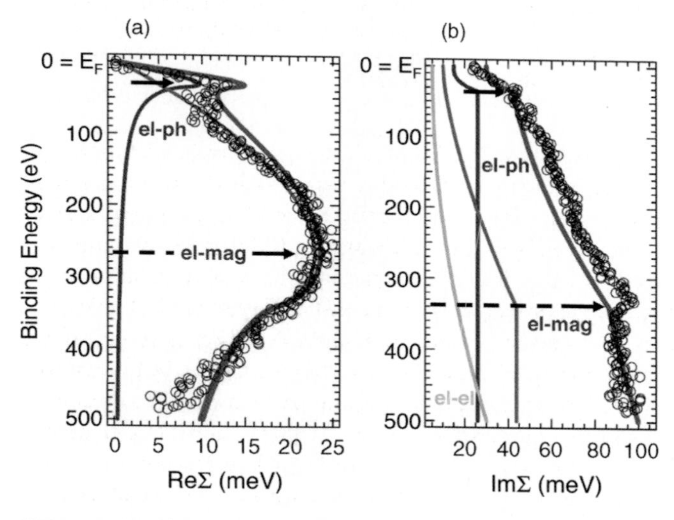

Figure 2.28 Self-energy analysis of the *K*-point band in Ni(110) derived from ARPES. (a) Real part $\mathrm{Re}\,\Sigma$ and (b) imaginary part $\mathrm{Im}\,\Sigma$. In both panels, peaks in the self-energy are seen at energy scales that correspond to the phonons and magnons: in terms of $\mathrm{Im}\,\Sigma$, this occurs at ≈ 30 meV and ≈ 340 meV, respectively. Note that in (a) a Gutzwiller reference band is used, which includes the electron correlation effects, so that they are effectively removed from $\mathrm{Re}\,\Sigma$. In (b) for $\mathrm{Im}\,\Sigma$, electron correlations have to be explicitly considered. The model curves for $\mathrm{Re}\,\Sigma$ and $\mathrm{Im}\,\Sigma$ are Kramers–Kronig transformable, consistent with the quasi-particle picture. Reprinted with permission from Ref. [124]. Copyright (2009) by the American Physical Society. (Please find a color version of this figure on the color plates.)

in this particular case. It should be added that the spectral features and magnitudes for Re Σ and Im Σ obey a Kramers–Kronig relation and *simultaneously* describe the experimental data very closely. This is a stringent criterion for the validity of the electron–magnon coupling picture, and proves that this model can well explain the observed phenomenon.

2.4.2.5 **Conclusions and Implications**

While electron–phonon coupling is well established and has been reported extensively in the past, including ARPES, electron–magnon coupling studies are rather novel. While first demonstrated for a surface state [131], experiments on bulk magnetic materials show that it is feasible to deduce quasi-particle interactions even in three-dimensional solids, using a technique based on high-symmetry planes. The self-energy extracted with high precision including a Gutzwiller reference satisfies the Kramers–Kronig criterion. It reveals characteristic structure in the quasi-particle dispersion that is well compatible with known spin wave energies. In the context of correlated materials, kinks have also been observed in high-temperature superconductors [121]. It is being discussed whether such kinks are derived from coupling to phonons or to spin fluctuations, while their similar energy scales in those materials make them difficult to separate. Particularly close resemblance to magnetic metals is found in the newly discovered iron-based pnictides [150]. Their parent compounds are metals with an antiferromagnetic spin density wave. A pairing mechanism based on spin fluctuations has been suggested [151]. The ARPES observations on elementary metals strongly suggest that renormalization effects reported to date for the cuprates [120] might relate to coupling to magnetic fluctuations. This implies important consequences for models of high-temperature superconductivity based on spin-mediated pairing. Such interpretation would be in line with a number of theoretical treatments, for example, on cuprates, in which the emergence of kinks has been attributed to the coupling of quasi-particles and bosonic (antiferromagnetic) spin excitations [152–154]. Regarding the detailed microscopic nature of the quasi-particle renormalization relating to spin excitations, a different explanation was suggested recently for strongly correlated electron systems, which does not require electron–boson coupling. In extended model calculations for pure electron–electron interaction, it was found that two well-separated regimes of quasi-particle renormalization can result [155]. While near the Fermi level well-defined quasi-particles exist according to Fermi liquid theory, at higher energies the self-energy changes abruptly, resulting in reduced quasi-particle lifetimes. In the transition between these situations, a dispersion anomaly is expected to emerge [155]. In this picture, kinks appear due to the frequency dependence of the local self-energy within the dynamical mean field theory. The results in Ref. [155] allow an estimate of the binding energy at which the kink would be expected. For the parameters of the Ni bandstructure, a coarse approximation yields $\approx$ 330 meV [10], which is in the same range as the observed kink energy in ReΣ of 270 meV. However, that kink scenario [155] was developed for a paramagnetic single-band model in infinite dimensions with strong quasi-particle renormalization. In contrast, Nickel is a ferromagnetic multiband system with only moderate renormalization of quasi-particle masses.

Therefore, it remains open whether or not the approach in Ref. [155] is of relevance for the kinks observed in Nickel. Yet, intriguingly, probably the boundary between coupling to a well-defined spin excitation on the one hand and coupling between electrons on the other hand might not sharply be drawn. This does not question the existence of well-behaved magnons seen in scattering experiments [141], but rather pertains to the nature of the quasi-particle. The interaction model of Byczuk *et al.* [155] may be seen as a different methodical ansatz for the same quasi-particle phenomenon. In extending this approach, a recent theoretical study [156] argues that this kink formation can, in fact, be viewed as resulting from emerging *internal* collective excitations – which are addressed as spin excitations or magnons by the experimentalists.

2.4.3
Time-Resolved Photoelectron Spectroscopy

In this section, we briefly discuss what can be learned from time-resolved photoelectron spectroscopy about the excitation and the decay of quasi-particles. For these purposes, we will focus on a few model cases studying the lifetime of photoexcited electrons, electron–phonon coupling, surface exciton formation, and magnon emission. There are a number of elaborate reviews considering time-resolved photoemission spectroscopy and quasi-particle dynamics (see, for example, Refs [4, 9]). For further reading, we also refer the reader to Volume 1 of this series.

2.4.3.1 Experiment

As illustrated in Figure 2.29a and b, there are two major schemes applied in time-resolved photoemission.

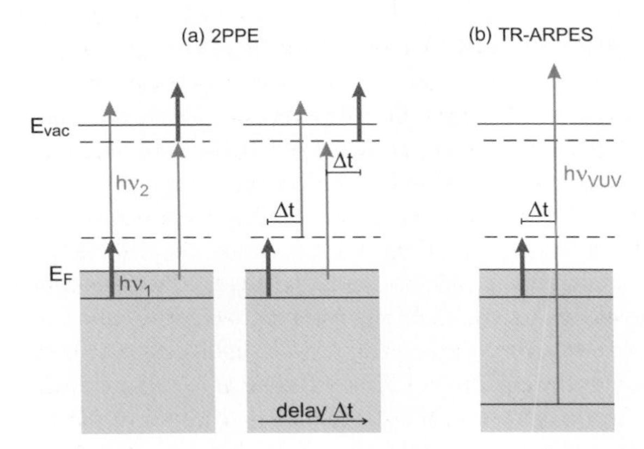

Figure 2.29 Excitation scheme of (a) bichromatic time-resolved two-photon photoemission (2PPE) spectroscopy and (b) time- and angle-resolved photoemission spectroscopy (TR-ARPES). Varying the delay Δt between pump and probe pulses allows us to follow the dynamics of excited states and ensembles, and thereby the creation of quasi-particles and collective excitations.

Two-Photon Photoemission In what is usually called two-photon photoemission, two femtosecond laser pulses with photon energies below the work function (metal) or ionization potential (semiconductor) are applied to populate and probe unoccupied states between the Fermi level E_F (metal) or valence band maximum (VBM) (semiconductor) and the vacuum level E_{vac}. Laser pulses with photon energies in the visible to ultraviolet (UV) range ($\approx$ 2–6 eV) are sufficient to span this energy range for various materials. In bichromatic two-photon photoemission, different photon energies $h\nu_1$ and $h\nu_2$ are used. Thereby, direct photoemission is largely suppressed and laser pulse intensities I can be made sufficiently high to optimize the count rate $\propto I_1 \cdot I_2$ without running into space charge problems.

Common laser sources are tunable Ti:sapphire oscillators (730–870 nm, that is, $h\nu = 1.69-1.42$ eV) or regenerative amplifiers with 80 MHz or 300 kHz repetition rates, respectively. While the fundamental of an oscillator is either frequency doubled ($2h\nu$) or tripled ($3h\nu$), the higher output power of amplifier setups allows us to widely tune the photon energy via optical parametric amplification at the cost of the lower repetition rate [157]. Typical pulse energies range between 0.1 nJ and 1 μJ at pulse durations of 20–100 fs focused on spots of 50–200 μm diameter. To detect the photoelectrons, either electrostatic hemispherical analyzers or time-of-flight spectrometers (typical repetition rates ≤ 1 MHz) are used. For both types of analyzers, multidimensional detectors have been developed, which record arrays of energies and emission angles. As the kinetic energy of the photoemitted electron is only a few electron volts, special diligence has to be applied to obtain reliable energy versus angle maps and to derive band dispersions $E(k_{||})$. The momentum parallel to the surface $k_{||}$ is conserved in the 2PPE process. In a time-resolved measurement, such maps are recorded as a function of the pump–probe delay. In the simplest case, we observe a buildup of the population and its exponential decay. From the latter, we obtain the lifetime of a specific surface state, for example, at $E(k_{||} = 0)$.

Time- and Angle-Resolved Photoemission Spectroscopy When photon energies above the ionization energy are employed, the technique is usually termed time-resolved and angle-resolved photoemission spectroscopy (TR-ARPES) [158, 159]. Sources are either the fourth harmonic of a Ti:saphire laser system ($4h\nu$) [104], picosecond pulses of synchrotron radiation or femtosecond pulses of a free electron laser, both synchronized to a laser pump pulse [160–162], or higher harmonics generated by focusing mJ-pulses on a rare-gas target [158, 163–165]. In contrast to 2PPE experiments, in TR-ARPES the pump pulse is usually used to create a hot electron population and the response of the system to this rather massive perturbation is followed by photoemission spectroscopy. Vacuum ultraviolet light pulses (VUV) allow us to map the valence bands throughout the Brillouin zone or to probe at least shallow core levels. Space charge problems may either occur due to the intense pump pulse or due to the direct photoemission (for a detailed discussion, see Ref. [166, 167]). With TR-ARPES, laser-induced phase transitions have been studied, which reveal collective excitations, such as coherent phonons [91, 168, 169], and the coupling among quasi-particles, for example, electrons, phonons, and magnons belonging to the interacting heat baths of a ferromagnet [104, 170, 171].

2.4.3.2 **Electron Lifetimes**

In metals it is primarily a consequence of the electron–electron interaction that photoexcited carriers, that is, electrons and holes, have a finite lifetime. Here, we must distinguish between the lifetime of an individual state and the relaxation time of a photoexcited electron population, also referred to as *hot* electrons. *Ab initio* theories calculate the self-energy of a single state. On Cu(001) the image potential states constitute a well-defined Rydberg-like series of unoccupied states with binding energies converging toward the vacuum level $E - E_{\mathrm{vac}} = 0.85\ \mathrm{eV}\,(n+a)^{-2}, n = 1, 2, \ldots$ As these states are localized in front of the surface, they exhibit a lifetime much longer than bulk states at comparable energies above E_F. Time-resolved 2PPE measurements reveal that after being populated with an ultrashort laser pulse the image potential states on Cu(001) show an exponential decay of the intensity over four orders of magnitude [172, 173]. There is now excellent agreement between experimentally determined lifetimes and the *ab initio* calculations described in Ref. [9]. For Cu(001) decay is dominated by electron–electron scattering. Phonon emission contributes negligibly to the decay of the image potential states since the bulk penetration of these states and thus the coupling to the lattice is small [9, 174, 175].

The decay of an excited electron or photohole is accompanied by excitation of a secondary electron–hole pair to conserve energy and momentum in inelastic electron–electron scattering. In the time-resolved 2PPE experiment, the count rate in a certain energy and momentum window is recorded as a function of the delay between the pump and the probe pulses. If a photoexcited state gets repopulated from higher excited states or by secondary electrons, the lifetime of the individual state is no longer measured but the population dynamics of the hot electron population. This can be nicely illustrated by 2PPE measurements of the population dynamics of the first image potential state in iron shown in Figure 2.30. When the photon energy of the UV pump pulse is just sufficient to excite the $n = 1$ state from occupied states at E_F ($3h\nu = 4.43$ eV, Figure 2.30a), the measured lifetime of the minority and majority components is $\tau_\downarrow = 11 \pm 2$ fs and $\tau_\uparrow = 16 \pm 2$ fs. These spin-dependent lifetimes are attributed to the spin-dependent density of states in the ferromagnetic iron film [176–179]. The spin resolution in the measurement will become important for discussing magnon emission in Section 2.4.3.5. At higher photon energies ($3h\nu = 4.85$ eV, Figure 2.31b), the pump pulse excites electrons along the parabolically dispersing image potential band up to E_{vac}, and intraband decay leads to a repopulation of the band bottom. As a consequence, the apparent spin-dependent lifetimes of the $n = 1$ image potential state at $k_{\|} = 0$ increase to $\tau_\uparrow = 19 \pm 2$ fs and $\tau_\downarrow = 17 \pm 2$ fs. Intraband scattering involves the excitation of a secondary electron–hole pair close to the Fermi level (cf. Figure 2.31) and opens a significant decay channel for electrons above the bottom of the image potential-state band [180].

A related situation is found for the decay of hot electrons in bulk copper [182]. Figure 2.31a summarizes measured relaxation times and calculated lifetimes for Cu (001) and Cu(111) as a function of the excitation energy $E - E_F$ above the Fermi level. For these monochromatic 2PPE measurements, the surface workfunction was lowered by Cs adsorption. The experimental data sets were recorded for photon

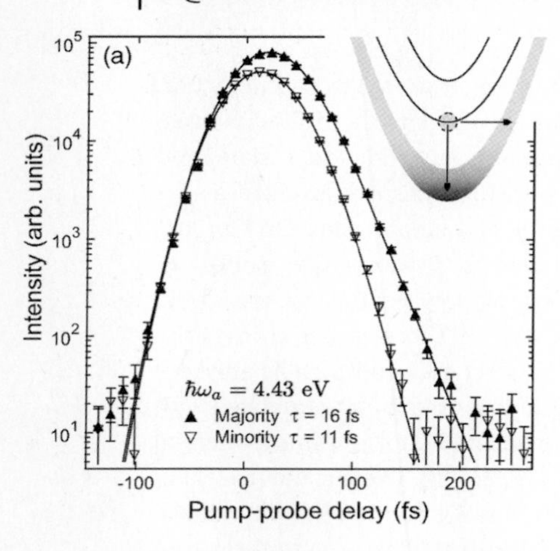

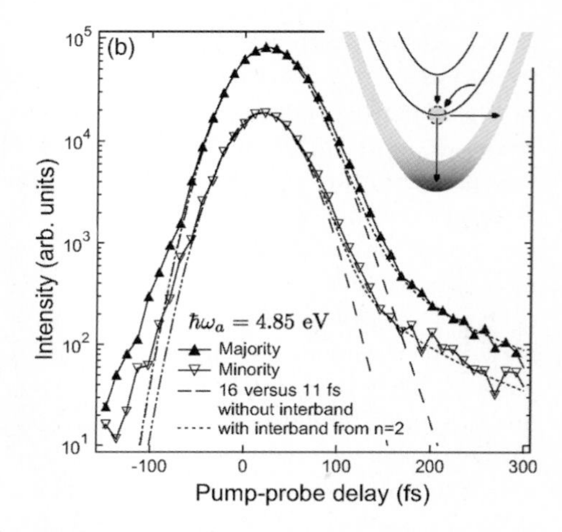

Figure 2.30 Time- and spin-resolved measurements of the $n = 1$ image potential-state lifetime for a 3 ML iron film deposited at room temperature on Cu(001). (a) A UV pump pulse with a photon energy of 4.43 eV populates only the band bottom. (b) For the higher photon energy of 4.85 eV, the image state band is excited up to E_{vac}. In the latter case, inter- and intraband decay leads to a repopulation of the band bottom and apparently longer lifetimes at $k_{||} = 0$. Reprinted with permission from Ref. [181].

energies of about 3.0–3.4 eV [183–185] and 1.63 eV [185] and show differing data in the overlapping energy range. By tuning the photon energy, later experiments revealed, for an intermediate-state energy of $E - E_F = 1.65$ eV, relaxation times of 13, 67, and 42 fs for $h\nu = 3.81$, 3.4 and 3.6 eV, respectively [186]. This difference can be explained by considering the schematic band structure of copper in Figure 2.31b. The processes A and B illustrate sp-intraband transitions (Drude absorption), including phonon or defect scattering, and the much stronger direct transitions from d- to sp-bands. In the latter process, d-band holes are photogenerated and their decay leads to secondary electrons. Due to the high probability of direct transitions B, d-band holes and thus secondary electrons are generated in large numbers. Taking into account the longer relaxation time and localization of the d-band holes [186, 187] substantiates the dependence of the relaxation time on the photon energy; the Auger-like decay of d-band holes (process C in Figure 2.31b) with maximum energy $E_F - h\nu$ leads to a delayed repopulation of states above the Fermi level with a maximum energy of $E_F + h\nu$.

Typical Fermi velocities in metals are on the order of 10^6 m/s [189]. In the case of copper, electrons at the Fermi level have a velocity on the order of 15 Å/fs [190]. Therefore, ballistic transport of the delocalized sp-electrons will contribute to their relaxation time. Since the group velocity and the orientation of the surface-projected band gaps depend on the crystal direction, relaxation times are expected to vary with surface orientation. Small changes in the relaxation time have been demonstrated, for example, for the copper crystal faces (100), (110), and (111); they do, however, not correlate with the Fermi velocity [183].

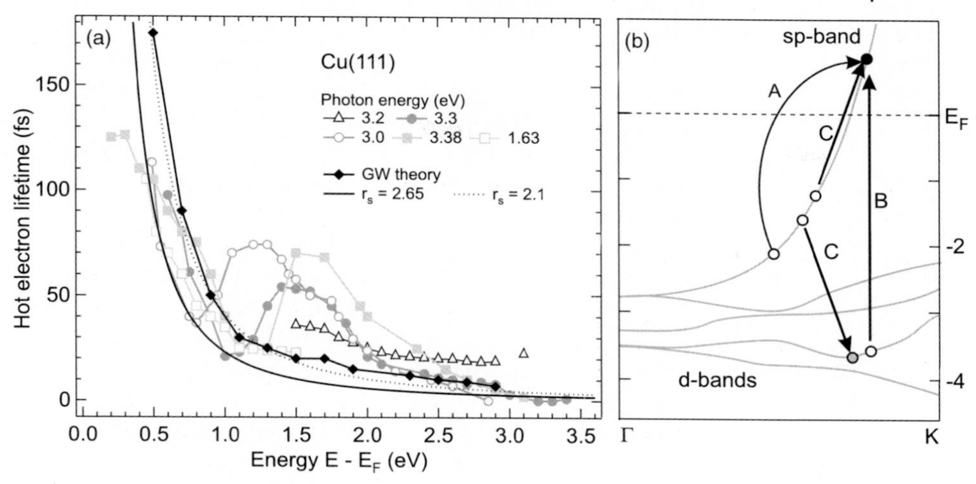

Figure 2.31 (a) Lifetimes of hot electrons in copper as a function of the excitation energy $E-E_F$. All data were taken with pump and probe pulses of the same photon energy $h\nu_1 = h\nu_2 = h\nu$. To lower the surface workfunction alkali metals, such as K and Cs, were adsorbed. Open triangles (Ref. [183]), $h\nu = 3.2$ eV; filled circles (Ref. [184]), Cu(111) $h\nu = 3.3$ eV; open circles (Ref. [184]), Cu(111) $h\nu = 3.0$ eV; filled squares (Ref. [185]), Cu(100) $h\nu = 3.38$ eV; open squares (Ref. [185]), Cu(100) $h\nu = 1.63$ eV; solid diamond (Refs [182, 188]), Cu(111) LMTO GW calculation; and solid and dashed line (Eq. (2.140)) for $r_s = 2.65$ and 2.1. (b) Schematic bulk band structure of copper and various excitation pathways. (A) Drude intraband absorption, (B) interband absorption, and (C) sp-band refilling via Auger decay of long-lived d-holes.

In conclusion, measured electron relaxation times do not represent the pure electronic lifetimes, although the energy-dependent hot electron lifetime for inelastic e–e scattering is the dominant quantity in the relaxation dynamics. This is confirmed by the reasonable agreement of the experimental results obtained for $h\nu = 1.63$ eV with *ab initio* GW calculations for a plane wave [191] or a linear muffin-tin orbital (LMTO) basis set [188] (diamonds in Figure 2.31a). The GW calculations are far from the lifetimes $\tau(E)$ obtained using the free electron gas theory.

$$\tau(E) = 263 \cdot r_s^{-5/2}(E-E_F)^{-2} \tag{2.139}$$

$$n_0 = \frac{3}{4\pi}(r_s a_0)^{-3}. \tag{2.140}$$

Here, a_0 is the Bohr radius and n_0 is the electron density. Good agreement can be obtained only for the radius $r_s = 2.1$ (dashed line in Figure 2.31a) that differs significantly from the value $r_s = 2.65$ obtained from the number of copper s electrons and the lattice parameter (solid line) [182]. The shorter relaxation times observed at low excitation energies are attributed to transport processes, which are not included in the calculations [186, 188]. Besides transport processes, *ab initio* calculations generally neglect secondary electrons, which are always excited in inelastic electron–electron scattering and lead to a repopulation of states above E_F [179].

2.4.3.3 Electron–Phonon Coupling

So far, we have compared the measured relaxation rate of photoexcited electrons and their calculated lifetime. We learned that even at low excitation densities, that is, in the regime of single excitation events, electron decay is influenced by repopulation and transport. 2PPE measures the lifetime only for individual states, such as surface states, and only with a proper choice of the excitation energy. In addition, transport into the bulk is disabled for a state localized at the surface.

The study of the decay of excited electron ensembles, that is, the regime of high excitation density, was pioneered by Bokor and coworkers [192, 193]. For metals it can be investigated by monitoring the electron distribution function above and below the Fermi edge as a function of the pump–probe delay [193, 194]. In a semiconductor, the transient population, for example, near the conduction band minimum of silicon, is recorded [195–197].

Let us assume a constant density of states in the vicinity of the Fermi level. A schematic of the electron density $n(E)$ after laser excitation is sketched in Figure 2.32. Optical absorption repopulates states in the energy interval $[E_F - h\nu, E_F + h\nu]$. During the excitation with the femtosecond laser pulse, the electron distribution is strongly nonthermal. Electron–electron scattering leads to thermalization of the electron gas after which its density can be modeled by a Fermi distribution $f(E, T_e) = (\exp{(E - E_F)/(k_B T_e)} + 1)^{-1}$ at an elevated electronic temperature T_e. In a metal, this equilibration within the electronic system takes about 100 fs. In semiconductors, carrier–carrier scattering depends on the excitation density and becomes important at a density of about $10^{17}\,\text{cm}^{-3}$, where similar timescales are observed [197, 198].

The excess energy in the electron system decays by interaction with the lattice through phonon emission. In a magnetic sample, the spin system adds an additional heat bath and dissipation channel. Both electrons and phonons can couple to magnons and thereby create a wealth of coherent and incoherent quasi-particle and collective excitations. Depending on the excited and probed sample volume, transport processes may also add to the decay of T_e measured in photoemission.

To illustrate the above discussion for a metal and a semiconductor, we next highlight recent studies of gadolinium from Bovensiepen (see Ref. [104] and references therein) and of silicon from Ichibayashi and Tanimura [197].

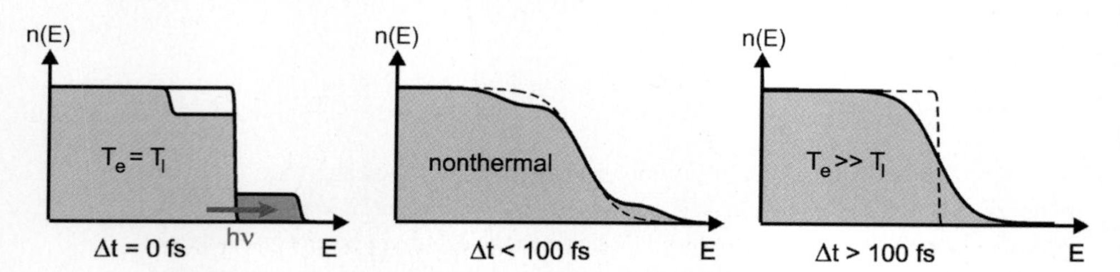

Figure 2.32 Schematic of the electron distribution $n(E)$ as a function of energy E after laser excitation. Ultrafast thermalization occurs via electron–electron scattering. Some 100 fs after laser excitation, the electron distribution is described by a Fermi function with an electron temperature T_e of a few thousand Kelvin, while the lattice temperature T_l is close to the initial temperature.

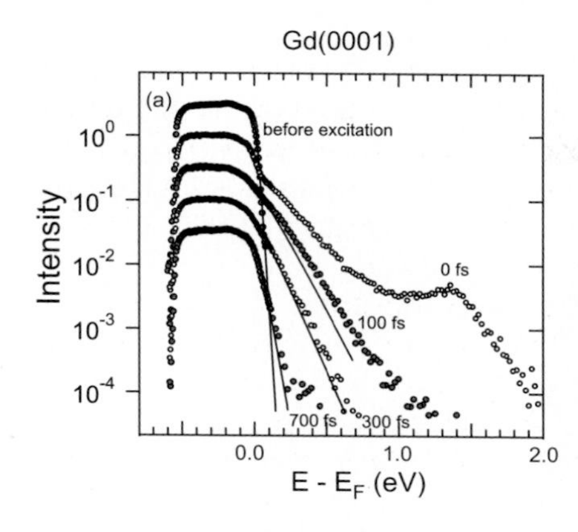

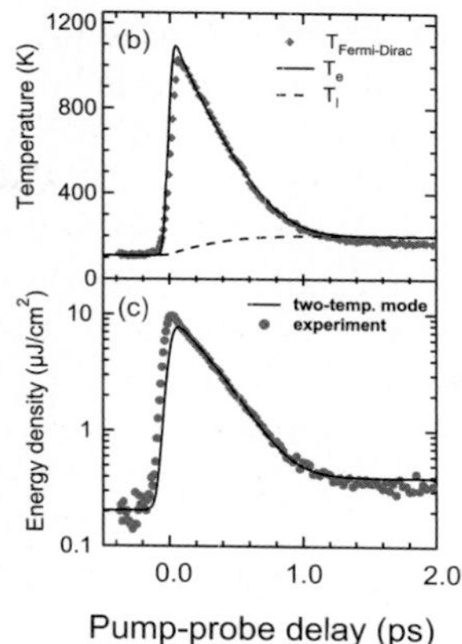

Figure 2.33 (a) Spectra of hot electrons at the Gd(0001) surface recorded for different pump–probe delays. The black lines are fits assuming a constant DOS and a Fermi–Dirac distribution with electron temperature T_e. (b) Transient temperature and (c) energy density of the electronic system. The adsorbed fluence is 0.25 mJ cm^{-2}. Solid lines show simulations based on the two-temperature model. Reprinted with permission from Ref. [104]. Copyright (2007) by the Institue of Physics.

Figure 2.33a depicts the transient electron population close to the Fermi level for a 10 nm Gd(0001) film grown epitaxially on W(110) (work function 3.4–3.7 eV). The sample was excited by an IR pump pulse ($h\nu = 1.5$ eV, 55 fs FWHM duration) with a fluence of 0.25 mJ/cm^2 and probed by the frequency-doubled signal of an OPA ($h\nu = 4.2$ eV, sub-50 fs pulse duration).

In the vicinity of the Fermi level, the gadolinium bulk DOS is nearly constant and the photoelectron signal is described by a Fermi–Dirac distribution. At zero delay $\Delta t = 0$ fs, the electron distribution is nonthermal and electron–electron scattering leads to excitation energies exceeding the pump pulse photon energy. After 100 fs the distribution is thermalized. The transient temperature of the electron gas (circles in Figure 2.33b) is extracted by fitting a Fermi–Dirac distribution to the photoemission data in Figure 2.33a. It is well described by the two-temperature model introduced by Anisimov *et al.* [199] for the case of laser excitation.

$$\begin{aligned} C_e(T_e)\frac{\partial T_e}{\partial t} &= S(z,t) - H(T_e, T_l) + \frac{\partial}{\partial z}\left(\kappa\frac{\partial T_e}{\partial z}\right) \\ C_l(T_l)\frac{\partial T_l}{\partial t} &= H(T_e, T_l) \end{aligned} \tag{2.141}$$

These equations model the interaction of electrons and lattice with their specific heat capacities $C_e = \gamma T_e$ and C_l. The Debye model was used to include the temperature dependence of the lattice-specific heat. Since γ is small, T_e rises to 1000 K (at the moderate fluence of 0.25 mJ/cm^2) before the lattice grows warm $C_l \gg C_e$. The electronic part of the thermal conductivity κ describes the diffusive transport of energy assuming a linear temperature dependence $\kappa = \kappa_0(T_e/T_l)$ [200].

S is the absorbed energy density per time. Its temporal and spatial profiles are determined by the pump pulse and the optical penetration depth.

The quasi-particle interaction between electrons (momentum k' and k) and phonons (momentum $\pm q = (k' - k) \pm G$) is hidden in the term $H(T_e, T_l)$ that describes the energy transfer rate between the electron and the lattice subsystems (phonon emission and absorption, see Eq. (2.13) in Section 2.2.3).

$$H(T_e, T_l) = -\frac{2}{(2\pi)^3} \sum_q \hbar\omega_q \int d^3k' W_{kk'} \delta(E_k - E_{k'} - \hbar\omega_q) \times [(n_q + 1) f_{k'}(1 - f_k) + n_q f_k (1 - f_{k'})]. \tag{2.142}$$

Here, n_q and f_k are the Bose–Einstein and Fermi–Dirac distribution functions of phonons and electrons, respectively. The energy transfer rate H has been calculated assuming one phonon emission and absorption of acoustic phonons in the Debye model ($\omega \propto q$) [201, 202].

$$H(T_e, T_l) = g(T_e) - g(T_l)$$
$$g(T) = 4 g_\infty \Theta_D \left(\frac{T}{\Theta_D}\right)^5 \int_0^{\Theta_D/T} dx \frac{x^4}{e^x - 1} \tag{2.143}$$

The electron–phonon coupling constant g_∞ and the Debye temperature Θ_D depend on the material. Solving Eq. (2.14) yields the solid lines in Figure 2.33b. The femtosecond pump pulse heats a small fraction of the electrons to a temperature of about 1100 K. Within 1 ps the electron and lattice temperatures equilibrate and the (lattice) temperature increases from 110 to 170 K. The 1–2 ps is typical for the equilibration time between a laser excited electron distribution and the lattice. The hot electrons (and holes) can induce desorption [203], chemical reactions [204], or configuration switching [205] in molecules adsorbed at the surface. They also drive ultrafast laser-induced demagnetization [206, 207] and magnetic switching [208] in thin-film ferromagnets.

There is an obvious difference between hot electron relaxation in semiconductors and metals. When the photon energy of the pump pulse is smaller than twice the gap energy, electrons excited to the conduction band (CB) cannot decay into empty CB states and excite a secondary electron across the gap. Electron–hole recombination (Auger-like decay) scales with the third power of the excited density n [209]: $dn/dt = -(C_e + C_h)n^3$. For bulk silicon the measured Auger-decay constants are $C_e(C_h) = 2.8(0.99) \times 10^{-31}\ \mathrm{cm^6/s}$ [210]. For densities of $n = 10^{17}\ \mathrm{cm^{-3}}$ electron–electron scattering among the excited carriers becomes important [198]. However, at this density the effective decay constant amounts to $(C_e + C_h)^{-1} n^{-2} = 260\ \mu s$. Therefore, energy relaxation and population decay are dominated by electron–phonon scattering and defect – as well as surface –recombination [211]. Figure 2.34a shows 2PPE spectra of Si(111) 7×7 as a function of pump–probe delay. The intensity distribution near the silicon CBM is highlighted in the inset. Since the photo-generated carrier density of $10^{18}\ \mathrm{cm^{-3}}$ is much less than the effective CB-DOS, the band shape was simulated by a Boltzmann distribution function with an

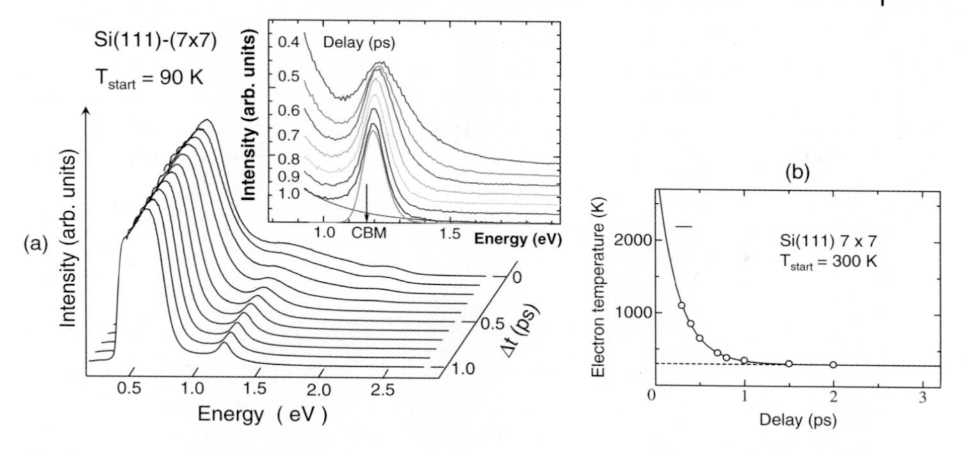

Figure 2.34 (a) Series of 2PPE spectra measured for Si(111) 7×7 for s-polarized 2.21 eV pump pulses (fluence ≈ 0.5 mJ cm^{-2}) and p-polarized 4.95 eV probe pulses. The inset shows the temporal evolution of the CB peak in an expanded scale from 0.4 to 1 ps delay. (b) The electron temperature at the CBM as a function of pump–probe delay. The solid curve is the fit of a single exponential decay component and a constant value of 296 K. The time constant is 240 fs. Reprinted with permission from Ref. [197]. Copyright (2009) by the American Physical Society. (Please find a color version of this figure on the color plates.)

effective electron temperature T_e, convolved with a 70 meV instrument resolution [197]. The decrease of T_e is shown in Figure 2.34b. The electron temperature can be described by an exponential curve with a time constant of $\tau_{e-p} = 240 \pm 20$ fs. This value fits nicely to the relaxation dynamics observed in previous all-optical studies [212]. On changing the sample temperature from 296 to 90 K, the time constant becomes 310 ± 10 fs. As the electron–phonon interaction scales with $2n_q + 1$ (Eq. (2.142)), the temperature dependence of τ_{e-p} reveals scattering with optical phonons of silicon with $\hbar\omega_q \approx 50$ meV. The 2PPE data demonstrate that the energy relaxation of hot electrons in silicon occurs via deformation potential scattering with optical phonons. This process leads to an equilibration of electron and lattice temperature on the timescale of 1 ps, comparable to gadolinium.

In Section 2.4.1, coherent phonons have been extensively discussed. Evidently coherent vibrations couple to the electronic system. A periodic variation in the bond distance leads to an oscillation in the binding energy, which can be followed in a pump–probe photoemission experiment with appropriate energy resolution. This allowed for the identification of phonon modes that are particularly important in phonon–magnon coupling [213] and for separating the relevant modes in charge density wave systems [91, 169].

2.4.3.4 **Surface Exciton Formation**

In the following paragraphs, surface exciton formation and its signature in 2PPE are described by monitoring the carrier dynamics at the silicon (100)-surface [211, 214]. The signatures of coherent and incoherent excitons in terahertz spectroscopy and photoluminescence studies of III–IV quantum well structures have been discussed

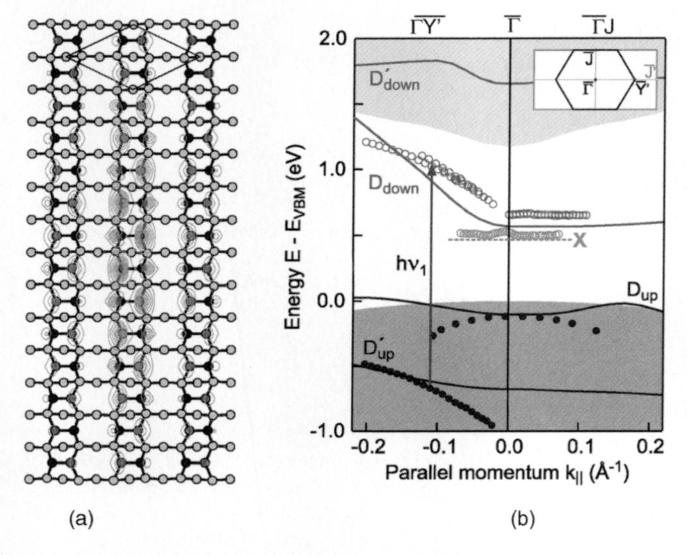

Figure 2.35 (a) The $c(4 \times 2)$ reconstruction of the Si(100) surface. Dimers are arranged in rows along the $[0\bar{1}1]$-direction and are alternately tilted along and perpendicular to the rows. The unit cell (solid rhombus) of the $c(4 \times 2)$ stacking comprises two dimers. The surface exciton is visualized by the contour plot of the probability density of the electron bound to the hole, located in the center of the depicted area. Reprinted with permission from Ref. [214]. Copyright (2005) by Springer. (b) Measured (symbols) and calculated (solid lines, shaded areas) surface-projected band structure of the Si(100) $c(4 \times 2)$ surface. $\overline{\Gamma Y'}$ and $\overline{\Gamma J}$ correspond to the directions along and perpendicular to the dimer rows. The presence of two dimers per unit cell leads to two occupied (D_{up} and D'_{up}) and two unoccupied (D_{down} and D'_{down}) dangling bond bands. X is the signature of a surface exciton in the single-particle band structure. The silicon valence and conduction bands are indicated by the gray-filled areas. Reprinted with permission from Ref. [211]. Copyright (2004) by the American Physical Society. (Please find a color version of this figure on the color plates.)

in Section 2.2.4. For a review on excitons in molecular layers, the reader is referred to Chapter 15 of Volume 1 of this series.

Figure 2.35a depicts the $c(4 \times 2)$ ground-state reconstruction of the Si(100) surface [215]. The lower coordinated atoms at the surface rebond to buckled dimers. They are arranged in rows along the $[0\bar{1}1]$-direction and alternately tilted (black D_{up} and dark gray D_{down} atoms) along and perpendicular to the rows. This structure is observed below 200 K in low-energy electron diffraction (LEED) [216] and predominates in scanning tunneling topographs [217]. Thermal activation of the dimer-rocking mode destroys long-range order [217] and a (2×1) LEED pattern is observed above 200 K.

The half-filled dangling bond states of the dimer split and form the occupied and unoccupied surface bands D_{up} and D_{down}. In Figure 2.35b, a GW-calculation of the surface quasi-particle band structure (solid lines, shaded areas) is compared with the band dispersion (symbols) obtained from 2PPE [211, 214]. The band structure calculation implements many-body exchange and correlation effects among the

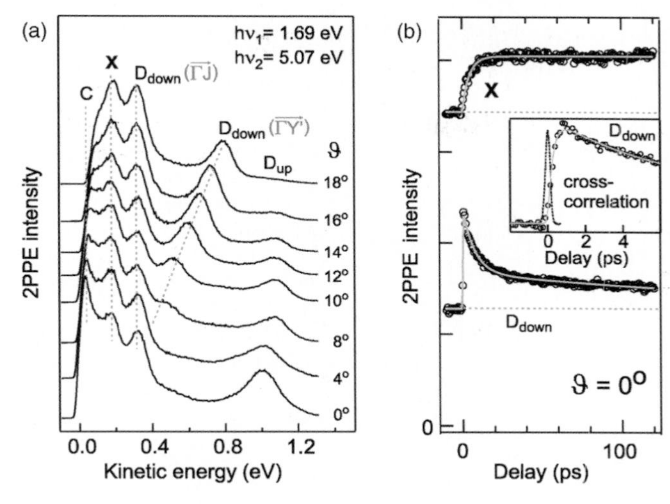

Figure 2.36 (a) Angle-resolved 2PPE spectra of the Si(100) $c(4 \times 2)$ surface recorded at 90 K for in time overlapping pump and probe pulses. (b) Time-resolved 2PPE measurements. Photon energies of the pump and probe pulses are $h\nu_1 = 1.69$ eV and $h\nu_2 = 5.07$ eV. Measurements at $\bar{\Gamma}$ ($\vartheta = 0°$) recorded with the electron analyzer tuned to the peak maxima of X and the D_{down} state. The inset highlights the dynamics of the D_{down} population on an enlarged timescale. Reprinted with permission from Ref. [214]. Copyright (2005) by Springer.

electrons in terms of the dynamical self-energy (see Section 2.2.1) [218–220]. The final and intermediate states of 2PPE (hole in D_{up} and excited electron in D_{down}) are accurately described and the silicon bulk and surface band gaps are reproduced within 100 meV. A set of angle-resolved 2PPE spectra for in-time overlapping pump ($h\nu = 1.69\,\text{eV}$) and probe pulses ($h\nu = 5.07\,\text{eV}$) is shown in Figure 2.36a. The assignment of the D_{down} state is based on the pronounced dispersion along the dimer chain ($\overline{\Gamma Y'}$) and the flat band perpendicular to it ($\overline{\Gamma J}$) (see Figure 2.35b). While the measured and calculated binding energies of the occupied D_{up} dangling bond band agree very well, the dispersion does not. This disagreement was confirmed by subsequent experiments and calculations [221, 222] and remains an open question.

Up to now we have assigned the transitions expected for single-particle excitations. Besides the dangling bond bands the spectra in Figure 2.36a reveal additional intensity at the low-energy cutoff C and a nearly nondispersing peak X about 130 meV below the D_{down} band. Both X and C correspond to long-lived surface excitations in the band gap, which are usually attributed to (nearby) surface defects [223]. Adsorption of activated hydrogen or sputtering–annealing cycles support this assignment for tail C, but not for peak X. The latter parallels the behavior of the D_{down} state, that is, gradually loses intensity with H-exposure and degrading surface quality.

Including the interaction between electrons and holes in the many-body theory [224] suggests that X corresponds to emission of an electron from a bound surface exciton state formed between a hole in the D_{up} band and an electron in the D_{down} band. To understand the signature of an excitonic state in 2PPE spectra, we consider energy conservation. Prior to the photoemission event, the total energy of the system amounts to

$$E_{tot} = E_0 + \Omega^X + h\nu \\ \simeq E_0 + E_{gap}^{surf} - E_{bind}^X + h\nu \tag{2.144}$$

where E_0 is the ground-state energy and Ω^X is the energy required to form the exciton. After the photoemission event, the total energy reads

$$E_{tot} = E_0 + E_{kin} - \langle E_{hole} \rangle, \tag{2.145}$$

where E_{kin} and $\langle E_{hole} \rangle$ are the kinetic energy of the photoemitted electron and the average dispersion of the final state hole, respectively. Energy conservation finally yields

$$E_{kin} = \langle E_{hole} \rangle + \Omega^X + h\nu \\ \simeq \langle E_{hole} \rangle + E_{gap}^{surf} - E_{bind}^X + h\nu \\ = E_{D_{down}} - E_{bind}^X + h\nu. \tag{2.146}$$

Since Ω^X and $h\nu$ are fixed, 2PPE measures the initial state, that is, the photohole $E_{kin} \propto \langle E_{hole} \rangle$. Thus, the flat dispersion of X reflects mainly the dispersion of the D_{up} dangling bond band, but not the degree of localization of the exciton. Consequently, the dispersion of peak X is flat in all directions, in clear contrast to the strong dispersion of the D_{down} band along the dimer chain. We note that peak X is a single quasi-particle feature and only the shift between X and D_{down} at $\bar{\Gamma}$ corresponds to the binding energy of the surface exciton E_{bind}^X. The calculated binding energy of 100 meV is in good agreement with the measured difference of 130 meV between X and D_{down}. Besides the Si(100) $c(4 \times 2)$ surface reconstruction, Figure 2.35a represents the calculated probability density (contour plot) of an electron in the presence of a hole fixed at the center of the depicted area. The electron's wave function is delocalized along the center dimer row. This quasi one-dimensional localization at the surface is in accord with the large binding energy of the surface exciton.

Before closing this section, we address the dynamics of exciton formation at the Si(100) surface. Referring to the discussion in Section 2.2.4, we note that 2PPE directly probes the (incoherent) population dynamics of the intermediate state. Figure 2.36b compares time-resolved measurements recorded at the bottom of the D_{down} band and at peak X. The maximum population of D_{down} occurs after 1.5 ps. Electrons excited to the D_{down} band decay via phonon emission to the band bottom. This intraband decay leads to a delayed rise of the population at the band bottom. The subsequent dynamics reveals a double-exponential decay of the D_{down} signal with time constants of 7.5 ± 3.5 and 220 ± 30 ps, respectively. Compatible with the shorter time constant, exciton formation (buildup of X) occurs in 5 ± 2.5 ps. The increase in signal X in this time span is almost equal to the decrease in the D_{down} signal (Figure 2.36b). Therefore, nearly all excited electrons, which reach the bottom of the D_{down} band are finally trapped in state X; exciton formation is the dominant surface decay channel. The excitation energy of X amounts to about 0.5 eV, which is much smaller than the photon energy of 1.69 eV of the initial excitation step. Thus, before an exciton can form, excited electrons and excited holes need to relax via phonon emission toward the band edges. Energy relaxation within surface bands occurs within 1.5 ps via phonon emission. Carrier relaxation involves an energy difference of 130 meV and thus requires simultaneous emission of at least two phonons. This can explain why *X* is observed only after a relaxation time of about 5 ps. The 220 ps decay constant of the D_{down} signal cannot originate from depopulation of the D_{down} band, but must reflect the decay constant of a reservoir, from where the D_{down} band is repopulated. Electrons excited to unoccupied bulk bands by the infrared laser pulse rapidly relax to the CBM. From there bulk electrons scatter to the D_{down} surface band [211]. The 220 ps decay constant can thus be attributed to the lifetime of the bulk electrons in the CBM. Since peak *X* does not decay on the picosecond timescale accessed here, the excitonic state must have a lifetime of at least nanoseconds. Peak *X* and its dynamics have been confirmed by two independent studies [221, 222].

2.4.3.5 **Magnon Emission**

Thermally excited spin waves are the fundamental excitation of ferromagnets, lowering the magnetization by $2\mu_B$. These collective excitations have been described in detail in Section 2.3.2. Here, we concentrate on magnon emission upon inelastic electron scattering in the itinerant ferromagnet iron. Understanding the generation of such a collective excitation in nonthermal equilibrium is important for developing a microscopic picture of relaxation processes in ferromagnets. The signature of magnons has been found in spin-polarized electron energy loss spectroscopy [139, 225], high-resolution photoemission spectra (Section 2.4.2.4 [131]) and inelastic tunneling spectroscopy [226]. However, these measurements do not address the fundamental problem of the time required to generate a magnon. Based on the small energy of magnons, that is, the dispersion relation $\omega \propto q^2$, spin wave emission is commonly viewed as a slow process, occurring within picoseconds [227].

In the following, the spin-dependent lifetime of the first image potential state at the surface of a ferromagnetic thin iron film on Cu(100) is discussed. It is demonstrated that intraband decay (scattering along the $n = 1$ parabola) is twice as strong for minority spin electrons as compared to their majority counterpart. This can be explained only by magnon emission, which must significantly contribute to inelastic decay on a femtosecond timescale.

The $n = 1$ image potential band was probed by energy-, angle-, and (most important) spin-resolved 2PPE [17]. Details of the experimental apparatus and the spin detector can be found in Refs [228, 229]. Measuring the energy as a function of the momentum parallel to the surface $E(k_{||})$ yields the first and second image potential-state bands ($n = 1, 2$), which show a two-dimensional parabolic dispersion as depicted in Figure 2.37a. Their exchange splitting is a signature of the ferromagnetic order in the iron film at the measurement temperature of 90 K ($T/T_C = 0.24$), where T_C is the Curie temperature [178, 230]. The lifetime τ of electrons photoexcited to the first image potential-state bands is obtained by shifting the time delay between the UV pump and the IR probe pulses. The three possible electronic decay channels

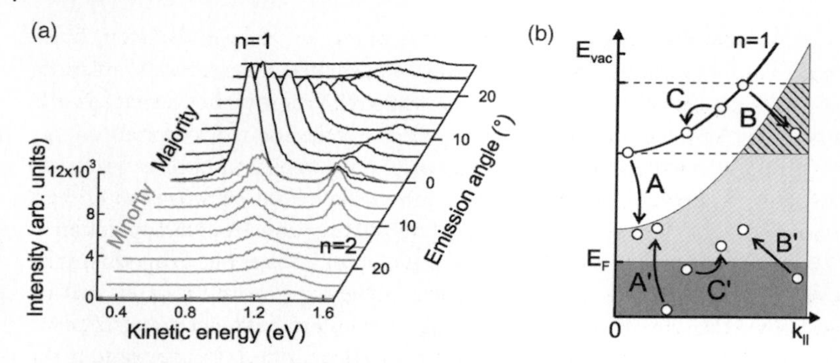

Figure 2.37 (a) Dispersing $n = 1, 2$ image potential-state bands as a function of emission angle parallel to the surface. Majority and minority spin channels are exchange split (data taken from Ref. [17]). (b) The three inelastic decay processes A, B, and C of the $n = 1$ image potential state in the gap of the surface-projected bulk bands (shaded area). In the inelastic decay process, energy and momentum are conserved via electron–hole pair creation A′, B′, and C′.

are sketched in Figure 2.37b. Inelastic decay into empty bulk states is indicated by process A. In this inelastic scattering process, momentum and energy are conserved by an electron–hole pair excited in the iron film A′. For the iron film, the d-bands in the majority spin channel are almost fully occupied, while the d-states in the minority spin channel form an only partially occupied band, exchange split from the majority bands by about 2.5 eV [17]. This spin-dependent density of d-states determines the difference in majority and minority decay rates [176, 177] at the bottom of the image potential-state bands (processes A and A′ in Figure 2.37b).

As illustrated in Figure 2.38a, with increasing energy E above the band minimum E_0, both decay rates increase linearly. However, the minority spin slope is twice as steep as the slope for majority spin electrons ($d\Gamma^{\downarrow}/dE = 0.25 \pm 0.04$ (eV fs)$^{-1}$ versus $d\Gamma^{\uparrow}/dE = 0.12 \pm 0.02$ (eV fs)$^{-1}$).

The increase in the decay rate is due to interband scattering (process B in Figure 2.37b) and intraband scattering (process C) [175, 180]. Energy and momentum are conserved by electron–hole pair creation (processes B′ and C′). The electrons gain additional phase space for decay, which comprises the sp-bands at energies $E > E_0$ (hatched area in Figure 2.37b) and the two-dimensional image potential band (constant density of states). Since the additional phase space for decay is only weakly spin dependent, it *cannot* explain the large difference observed in the slope $d\Gamma/dE$ between minority and majority electrons. This strong spin dependence must therefore stem from spin-flip processes.

A direct spin-flip transition requires strong spin orbit coupling and is therefore inefficient. However, in the exchange spin-flip processes indicated in the Feynman diagrams of Figure 2.37b and c, the primary electron does not change its spin angular momentum but excites substrate electron–hole pairs in the opposite spin channel. Since Stoner excitations have characteristic energies on the order of the exchange splitting of 2.5 eV, they will not play a role in the decay processes B and C with energy transfers below 0.5 eV. In this energy range, magnons play a dominant role in the lifetime of minority electrons [231]. The magnon spectrum of the 3 ML iron film on Cu(100) is shown in Figure 2.15 and discussed in detail in Section 2.3.2.2. It starts at zero energy and contains two additional branches. With increasing energy of the electron above the band bottom $E > E_0$, these *optical* magnon branches can be excited. At the low temperature of our experiment, magnon absorption by electrons is negligible. We conclude that only for minority spin electrons intraband scattering occurs via magnon emission since the angular momentum of an emitted magnon (reduction by $2\,\mu_B$) is compensated by a flip of a minority to a majority spin electron (increase by $2\,\mu_B$). This process corresponds to magnon emission and leads to an increase in the available phase space for decay. It explains the larger slope $d\Gamma/dE$ for minority compared to majority electrons (Figure 2.38a). As the lifetime of the $n = 1$ minority image potential state on iron is only 11 fs, magnon emission must take place on this timescale and is therefore an ultrafast process contrary to the accepted opinion [227].

2.4.3.6 Magnon–Phonon Interaction

In a ferromagnet, the spin system can be considered as the third heat bath interacting with hot electron and lattice subsystems after laser excitation [206]. The emission of magnons, as described in the last section, does not lead to a change in the total

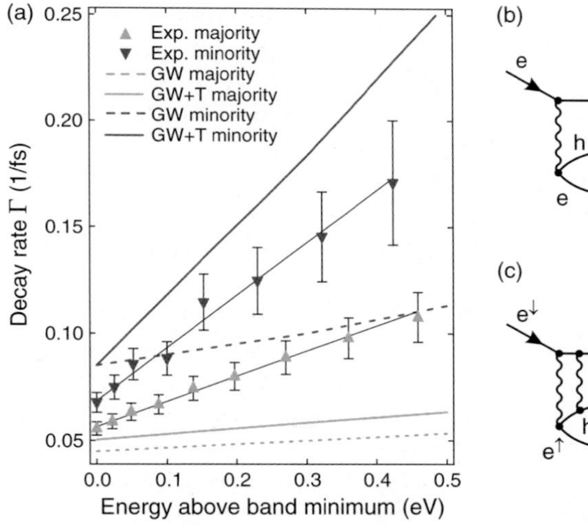

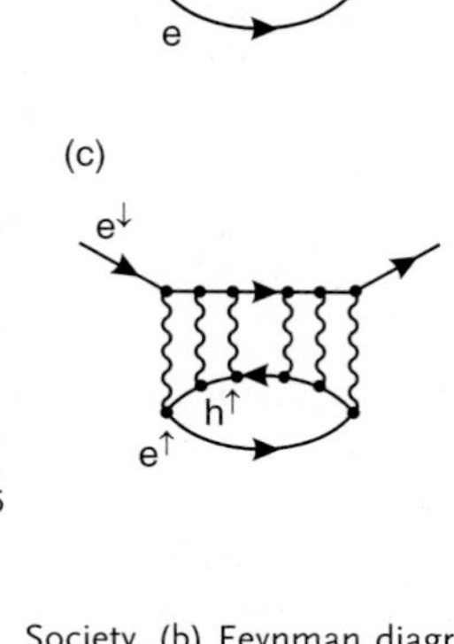

Figure 2.38 (a) Decay rate $1/\tau$ of the $n = 1$ image potential state on the iron film in the majority and minority spin band for increasing energy above the band minimum. The measured decay rates (symbols) are compared with the decay rates of bulk electrons calculated in the GW (dashed lines) and GW + T matrix (solid lines) formalisms. Reprinted with permission from Ref. [17]. Copyright (2010) by the American Physical Society. (b) Feynman diagram for a Stoner excitation, that is, the contribution of electron–hole pair creation in the spin majority channel to the self-energy of a minority spin electron. (c) Magnon emission as a collective excitation can be described as multiple scattering between the primary electron and the hole of an electron–hole pair in the opposite spin channel. Reprinted with permission from Ref. [181].

magnetic momentum *per se*. Admittedly, a change in the magnetization requires angular momentum transfer, for example, to the lattice (Einstein–de Haas effect). This is likewise true for ultrafast laser-driven demagnetization [206, 207] and magnetic switching [208]. To study the spin–lattice interaction, gadolinium is a very suitable model system. In this prototypical Heisenberg ferromagnet (see Section 2.3.2.1), the magnetic moment per atom of $\mu = 7.55\,\mu_B$ is mainly localized at the half-filled 4f core shell ($7.0\,\mu_B$). Below $T_C \simeq 293$ K [232], ferromagnetic order occurs by indirect exchange interaction (Ruderman–Kittel–Kasuya–Yosida) mediated by the spin-polarized $(5d6s)^3$ valence electrons. Since the 4f shell is half-filled, its angular momentum is zero (neglecting core valence hybridization) and the spin-orbit-mediated interaction between the localized 4f magnetic moment and the lattice is weak, as was already established by the pioneering work of Vaterlaus *et al.* [233].

This magnon–phonon coupling in gadolinium was further analyzed in a TR-ARPES experiment at the synchrotron facility BESSY II (Helmholtz-Zentrum Berlin) [170]. A 100 Å thick Gd(0001) film grown epitaxially on W(110) was magnetized with the easy axis in-plane. Figure 2.39a shows angle-resolved Gd 4f core-level spectra recorded at a photon energy of 60 eV with a sample temperature of 120 K. The magnetic linear dichroism (MLD), that is, the contrast between the spectra recorded for reversed magnetization $M\uparrow$ and $M\downarrow$ (see Figure 2.39b), is a measure of the film magnetization. This is proven by the temperature dependence of the MLD,

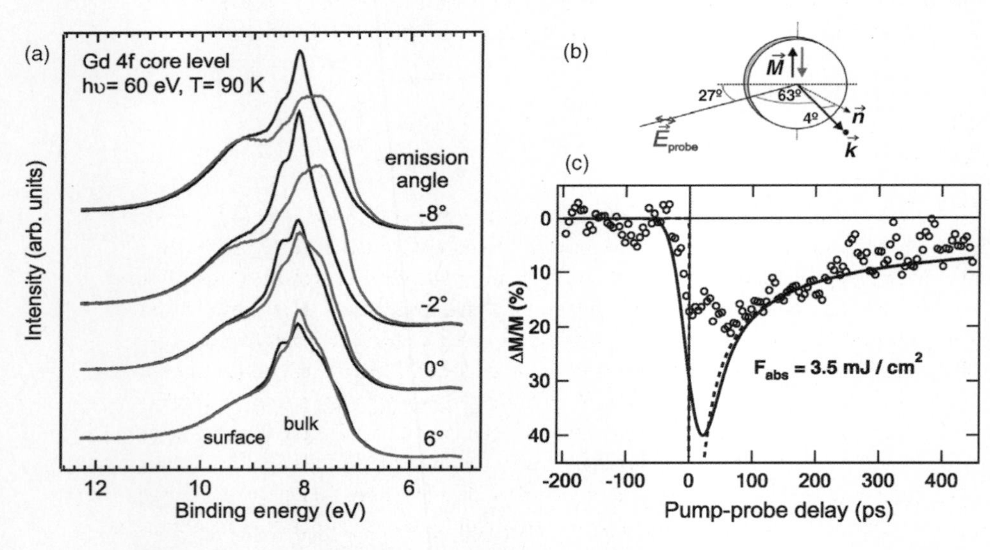

Figure 2.39 (a) Linear magnetic dichroism of the 4f photoemission line (surface and bulk components) of Gd(0001) (adapted from Ref. [234]). (b) Sketch of the measurement geometry. Spectra are recorded close to normal emission (0°) for opposite directions of the in-plane magnetization ($M\uparrow$ and $M\downarrow$). (c) Transient, relative change in the magnetization: measured 4f dichroism (open circles), calculated from lattice temperature and magnetization (dashed line), and convolved with a Gaussian of 50 ps FWHM duration (solid line). Reprinted with permission from Ref. [170]. Copyright (2008) by the American Physical Society. (Please find a color version of this figure on the color plates.)

which was shown to follow the spontaneous magnetization $M(T_l)/M(0)$ [170]. In the TR-ARPES experiment, the Gd electronic system is pumped by an IR laser pulse (synchronized to the BESSY master clock [160]) and the 4f magnetization is probed by the time-delayed synchrotron pulse (50 ps pulse duration). The resulting transient magnetization $M(\Delta t)$ is shown in Figure 2.39c (open circles). A 20 % drop is followed by a recovery of M within 1 ns. A calculation of $M(\Delta t)$ from the transient lattice temperature T_l (see Section 2.4.3.3, Figure 2.33b) and the temperature-dependent magnetization $M(T_l)$ [170, 232] gives the transient magnetization under the assumption that thermal equilibrium is established at all delays. This thermal scenario of $M(\Delta t)$ is plotted in Figure 2.39c by the dashed line. As the lattice reaches a temperature close to T_C at 1.5 ps after the pump pulse, the magnetization drops by 80%. To account for the time resolution governed by the X-ray pulse duration, this temperature dependence is convolved with a Gaussian of 50 ps FWHM and depicted by the solid line in Figure 2.4. Starting at about 80 ps, the measured 4f MLD and the calculated magnetization agree nicely, which substantiates our modeling of $T_l(\Delta t)$. However, the measured drop in the magnetization at earlier times is smaller than expected from the thermal modeling (solid line) by a factor of 2. This confirms that equilibrium between the Gd spin system and the lattice is not established in this time regime. The 4f spin lattice relaxation thus takes much longer than the electron lattice equilibration (1–2 ps). This result was confirmed in a recent X-ray magnetic circular dichroism experiment at the BESSY femtoslicing facility of the Helmholtz-Zentrum Berlin that establishes a spin lattice relaxation time of 40 ps [171].

2.5 Summary

The physical properties of a solid-state system are determined not only by its geometrical and electronic structure. Various elementary excitations also have a crucial role to play. Every response of a solid to external stimuli involves the creation and annihilation of quasi-particles and/or collective excitations. These excitations and their mutual interactions are responsible for most of the fundamental properties of the matter, such as heat and electronic transport, and the optical and magnetic responses. Moreover, the delicate coupling among quasi-particles rules the physics and functionality of correlated systems and can stabilize distinct phases. It is, therefore, of utmost importance to understand the dynamics of elementary excitations, that is, the timescale on which quasi-particle excitations build up and decay.

In this chapter, we have presented an overview of elementary excitations in solids and at solid surfaces and interfaces. Thereby, we have introduced the following quasi-particle and collective excitations: electrons, phonons, excitons, polarons, plasmons, and magnons. We outlined the theoretical concepts to describe their creation and annihilation and discussed various couplings among these excitations. Recent calculations and experiments are highlighted to give an impression of the state of the art of the field.

Neither from a theoretical nor from an experimental viewpoint are the problems presented in this chapter solved. One of the most prominent examples is the pairing mechanism in high-temperature superconductors, which is still under active debate. For an *ab initio* calculation of the elementary excitations in a solid, we need detailed information on the electronic structure and states, which may by itself become a substantial theoretical problem. This knowledge is then fed into various schemes to calculate the dynamic response. Obviously, we need to know the electronic structure of a material quite accurately to arrive at a reliable description of its dynamics. This problem becomes even more complex in layered structures, and at surfaces and interfaces, where the response function changes quite abruptly and novel one- and two-dimensional electronic states can appear. Since quasi-particle excitations and couplings are often enhanced in low-dimensional systems, the study of surfaces and nanostructures remains a formidable challenge for the future.

References

1 Ashcroft, N. and Mermin, N. (1976) *Solid State Physics*, Saunders.
2 Fürst, C., Leitenstorfer, A., Laubereau, A., and Zimmermann, R. (1997) *Phys. Rev. Lett.*, **78**, 3733.
3 Betz, M., Göger, G., Laubereau, A., Gartner, P., Bányai, L., Haug, H., Ortner, K., Becker, C.R., and Leitenstorfer, A. (2001) *Phys. Rev. Lett.*, **86**, 4684.
4 Petek, H. and Ogawa, S. (1997) *Prog. Surf. Sci.*, **56**, 239.
5 Kittel, C. (1963) *Quantum Theory of Solids*, John Wiley & Sons, New York.
6 Madelung, O. (1978) *Introduction to Solid State Theory*, Springer, Berlin.
7 Czycholl, G. (2000) *Theoretische Festkörperphysik*, Springer, Berlin.
8 Sandratskii, L. (1998) *Adv. Phys.*, **47**, 91.
9 Echenique, P., Berndt, R., Chulkov, E., Fauster, T., Goldmann, A., and Höfer, U. (2004) *Surf. Sci. Rep.*, **52**, 219.
10 Kira, M. and Koch, S.W. (2006) *Prog. Quantum Electron.*, **30**, 155.
11 Pitarke, J.M., Zhukov, V.P., Keyling, R., Chulkov, E.V., and Echenique, P.M. (2004) *ChemPhysChem*, **4**, 1284.
12 Pitarke, J., Silkin, V., Chulkov, E., and Echenique, P. (2007) *Rep. Prog. Phys.*, **70**, 1.
13 Hüfner, S. (1995) *Photoelectron Spectroscopy: Principles and Applications*, Springer Series in Solid-state Science, vol. 82, Springer.
14 Abrikosov, A.A., Gorkov, L.R., and Dzyaloshinski, I.E. (1963) *Methods of Quantum Field Theory in Statistical Physics*, Prentice-Hall, Englewood Cliffs.
15 Chulkov, E.V., Borisov, A.G., Gauyacq, J.P., Sanchez-Portal, D., Silkin, V.M., Zhukov, V.P., and Echenique, P.M. (2006) *Chem. Rev.*, **106**, 4160.
16 Hofmann, P., Sklyadneva, I., Rienks, E., and Chulkov, E. (2009) *New J. Phys.*, **11**, 125005.
17 Schmidt, A.B., Pickel, M., Donath, M., Buczek, P., Ernst, A., Zhukov, V.P., Echenique, P.M., Chulkov, E.V., and Weinelt, M. (2010) *Phys. Rev. Lett.*, **105**, 197401.
18 García-Lekue, A., Pitarke, J.M., Chulkov, E.V., Liebsch, A., and Echenique, P.M. (2002) *Phys. Rev. Lett.*, **89**, 96.
19 Grimvall, G. (1981) *The Electron–Phonon Interaction in Metals, Selected Topics in Solid State Physics* (ed. E. Wohlfarth), vol. 16, North-Holland, New York.
20 Eliashberg, G. (1960) *Sov. Phys. JETP*, **11**, 696.
21 Sklyadneva, I., Chulkov, E., and Echenique, P. (2008) *J. Phys. Condens. Matter*, **20**, 165203.
22 Haug, H. and Koch, S.W. (2009) *Quantum Theory of the Optical and Electronic Properties of Semiconductors*, 5th edn, World Scientific Publ., Singapore.
23 Klingshirn, C. (2005) *Semiconductor Optics*, 2nd edn, Springer, Berlin.

24 Schäfer, W. and Wegener, M. (2002) *Semiconductor Optics and Transport Phenomena*, 1st edn, Springer, Berlin.
25 Zimmermann, R. (1987) *Many-Particle Theory of Highly Excited Semiconductors*, Teubner, Leipzig.
26 Frenkel, J. (1931) *Phys. Rev.*, **37**, 1276.
27 Wannier, G.H. (1937) *Phys. Rev.*, **52**, 191.
28 Elliott, R. (1963) Theory of excitons, in *Polarons and Excitons* (eds C. Kuper and G. Whitefield), Oliver and Boyd, Edinburgh, pp. 269–293.
29 Knox, R.S. (1963) Theory of excitons, in *Solid State Physics Suppl. 5*, (eds F. Seitz and D. Turnbull), Academic Press, New York.
30 Dexter, D.I. and Knox, R.S. (1981) *Excitons*, John Wiley & Sons, Inc., New York.
31 Lindberg, M. and Koch, S.W. (1988) *Phys. Rev. B*, **38**, 3342.
32 Smith, R., Wahlstrand, J., Funk, A., Mirin, R., Cundiff, S., Steiner, J., Schäfer, M., Kira, M., and Koch, S. (2010) *Phys. Rev. Lett.*, **104**, 247401.
33 Groeneveld, R.M. and Grischkowsky, D. (1994) *J. Opt. Soc. Am. B*, **11**, 2502.
34 Kira, M., Hoyer, W., Stroucken, T., and Koch, S.W. (2001) *Phys. Rev. Lett.*, **87**, 176401.
35 Chatterjee, S., Ell, C., Mosor, S., Khitrova, G., Gibbs, H.M., Hoyer, W., Kira, M., Koch, S.W., Prineas, J.P., and Stolz, H. (2004) *Phys. Rev. Lett.*, **92**, 067402.
36 Chiaradia, P., Cricenti, A., Selci, S., and Chiarotti, G. (1984) *Phys. Rev. Lett.*, **52**, 1145–1147.
37 Reichelt, M., Meier, T., Koch, S.W., and Rohlfing, M. (2003) *Phys. Rev. B*, **68**, 045330.
38 Frohlich, H. (1954) *Adv. Phys.*, **3**, 325.
39 Mahan, G. (1990) *Many-Particle Physics*, Plenum.
40 Devreese, J. (2005) Polarons, in *Encyclopedia of Physics* (eds R. Lerner and G. Trigg), Wiley-VCH Verlag GmbH, Weinheim, pp. 2004–2027.
41 Sheng, P. and Dow, J. (1971) *Phys. Rev. B*, **4**, 1343.
42 Landau, L. and Pekar, S. (1948) *Z. Eksp. Teor. Fiz.*, **18**, 419.
43 Miyake, S. (1976) *J. Phys. Soc. Jpn.*, **41**, 747.
44 Feynman, R. (1955) *Phys. Rev.*, **97**, 660.
45 Ando, T., Fowler, A., and Stern, F. (1982) *Rev. Mod. Phys.*, **54**, 437.
46 Sak, J. (1972) *Phys. Rev. B*, **6**, 3981.
47 Xiaoguang, W., Peeters, F., and Devreese, J. (1985) *Phys. Rev. B*, **31**, 3420.
48 Peeters, F., Xiaoguang, W., and Devreese, J. (1988) *Phys. Rev. B*, **37**, 933.
49 Pines, D. and Bohm, D. (1952) *Phys. Rev.*, **85**, 338.
50 Tonks, L. and Langmuir, I. (1929) *Phys. Rev.*, **33**, 195.
51 Pines, D. and Noziéres, P. (1990) *The Theory of Quantum Liquids*, Addison-Wesley.
52 Raether, H. (1980) *Excitation of Plasmons and Interband Transitions by Electron*, Springer Tracts in Modern Physics, vol. 88, Springer.
53 Ritchie, R. (1957) *Phys. Rev.*, **106**, 874.
54 Raether, H. (1988) *Surface Plasmons on Smooth and Rough Surfaces and on Gratings*, Springer Tracts in Modern Physics, vol. 111, Springer.
55 Jacksonn, J. (1999) *Classical Electrodynamics*, John Wiley & Sons, Inc., New York.
56 Inglesfield, J. (1982) *Rep. Prog. Phys.*, **45**, 223.
57 Silkin, V., Lekue, A.G., Pitarke, J.M., Chulkov, E.V., Zaremba, E., and Echenique, P.M. (2004) *Europhys. Lett.*, **66**, 260.
58 Pitarke, J., Nazarov, V.U., Silkin, V.M., Chulkov, E., Zaremba, E., and Echenique, P. (2005) *Phys. Rev. B*, **70**, 205403.
59 Silkin, V., Pitarke, J.M., Chulkov, E., and Echenique, P.M. (2005) *Phys. Rev. B*, **72**, 115435.
60 Diaconescu, B., Pohl, K., Vattuone, L., Savio, L., Silkin, V.M., Hofmann, P.H., Pitarke, J.M., Chulkov, E., Echenique, P.M., Farias, D., and Rocca, M. (2007) *Nature*, **448**, 57.
61 Gross, E.K.U. and Kohn, W. (1985) *Phys. Rev. Lett.*, **55**, 2850.
62 Dederichs, P.H., Blügel, S., Zeller, R., and Akai, H. (1984) *Phys. Rev. Lett.*, **53**, 2512.

63 Halilov, S.V., Eschrig, H., Perlov, A.Y., and Oppeneer, P.M. (1998) *Phys. Rev. B*, **58**, 293.
64 Şaşığlu, E., Sandratskii, L.M., Bruno, P., and Galanakis, I. (2005) *Phys. Rev. B*, **72**, 184415.
65 Buczek, P., Ernst, A., Bruno, P., and Sandratskii, L.M. (2009) *Phys. Rev. Lett.*, **102**, 247206.
66 Şaşığlu, E., Schindlmayr, A., Friedrich, C., Freimuth, F., and Blügel, S. (2010) *Phys. Rev. B*, **81**, 054434.
67 Ishioka, K., Hase, M., Kitajima, M., and Petek, H. (2006) *Appl. Phys. Lett.*, **89**, 231916.
68 Dekorsy, T., Cho, G.C., and Kurz, H. (2000) Coherent phonons in condensed media, in *Light Scattering in Solids VIII* (eds M. Cardona and G. Güntherodt), Springer, Berlin, pp. 169–209.
69 Hase, M., Kitajima, M., Nakashima, S., and Mizoguchi, K. (2002) *Phys. Rev. Lett.*, **88**, 067401.
70 Misochko, O., Hase, M., Ishioka, K., and Kitajima, M. (2004) *Phys. Rev. Lett.*, **92**, 197401.
71 Ishioka, K., Hase, M., Kitajima, M., Wirtz, L., Rubio, A., and Petek, H. (2008) *Phys. Rev. B*, **77**, 121402R.
72 Scholz, R., Pfeifer, T., and Kurz, H. (1993) *Phys. Rev. B*, **47**, 16229.
73 Kuznetsov, A. and Stanton, C. (1994) *Phys. Rev. Lett.*, **73**, 3243.
74 Hu, X. and Nori, F. (1996) *Phys. Rev. B*, **53**, 2419.
75 Dhar, L., Rogers, J., and Nelson, K. (1994) *Chem. Rev.*, **94**, 157.
76 Först, M. and Dekorsy, T. (2007) Coherent phonons in bulk and low-dimensional semiconductors, in *Coherent Vibrational Dynamics* (eds S.D. Silvestri, G. Cerullo, and G. Lanzani), CRC press, Boca Raton, pp. 129–172.
77 Stevens, T., Kuhl, J., and Merlin, R. (2002) *Phys. Rev. B*, **65**, 144304.
78 Yee, K.J., Lim, Y.S., Dekorsy, T., and Kim, D.S. (2001) *Phys. Rev. Lett.*, **86**, 1630.
79 Garrett, G., Albrecht, T., Whitaker, J., and Merlin, R. (1996) *Phys. Rev. Lett.*, **77**, 3661.
80 Riffe, D.M. and Sabbah, A.J. (2007) *Phys. Rev. B*, **76**, 085207.
81 Zeiger, H., Vidal, J., Cheng, T., Ippen, E., Dresselhaus, G., and Dresselhaus, M. (1992) *Phys. Rev. B*, **45**, 768.
82 Sokolowski-Tinten, K., Blome, C., Blums, J., Cavalleri, A., Dietrich, C., Tarasevitch, A., Uschmann, I., Förster, E., Kammler, M., Horn von Hoegen, M., and von der Linde, D. (2003) *Nature*, **422**, 287.
83 Zijlstra, E., Tatarinova, L., and Garcia, M. (2006) *Phys. Rev. B*, **74**, 220301.
84 Ishioka, K., Basak, A.K., and Petek H. (2011) *Phys. Rev. B*, **84**, 235202.
85 Chang, Y.M., Xu, L., and Tom, H.W.K. (2000) *Chem. Phys.*, **251**, 283.
86 Matsumoto, Y. and Watanabe, K. (2006) *Chem. Rev.*, **106**, 4234.
87 Sokolowski-Tinten, K., and von der Linde, D. (2004) *J. Phys. Condens. Matter*, **16**, R1517.
88 Reis, D.A. and Lindenberg, A.M. (2007) Ultrafast X-ray scattering in solids, in *Light Scattering in Solids IX* (eds M. Cardona and R. Merlin), Springer, Berlin, pp. 371–422.
89 Kübler, C., Ehrke, H., Huber, R., Lopez, R., Halabica, A., Haglund, R.F., and Leitenstorfer, A. (2007) *Phys. Rev. Lett.*, **99**, 116401.
90 Perfetti, L., Loukakos, P., Lisowski, M., Bovensiepen, U., Berger, H., Biermann, S., Cornaglia, P., Georges, A., and Wolf, M. (2006) *Phys. Rev. Lett.*, **97**, 067402.
91 Schmitt, F., Kirchmann, P.S., Bovensiepen, U., Moore, R.G., Rettig, L., Krenz, M., Chu, J.H., Ru, N., Perfetti, L., Lu, D.H., Wolf, M., Fisher, I.R., and Shen, Z.X. (2008) *Science*, **321**, 1649.
92 Vallee, F. (1994) *Phys. Rev. B*, **49** (4), 2460.
93 Hase, M., Mizoguchi, K., Harima, H., Nakashima, S., and Sakai, K. (1998) *Phys. Rev. B*, **58**, 5448.
94 Zhang, J.M., Giehler, M., Göbel, A., Ruf, T., Cardona, M., Haller, E.E., and Itoh, K. (1998) *Phys. Rev. B*, **57**, 1348.
95 Ishioka, K., Hase, M., Ushida, K., and Kitajima, M. (2002) *Physica B*, **316–317**, 296.
96 Pisana, S., Lazzeri, M., Casiraghi, C., Novoselov, K., Geim, A., Ferrari, A., and Mauri, F. (2007) *Nat. Mater.*, **6**, 198.

97 Lee, J.D., Inoue, J., and Hase, M. (2006) *Phys. Rev. Lett.*, **97**, 157405.

98 Misochko, O., Ishioka, K., Hase, M., and Kitajima, M. (2007) *J. Phys.: Condens. Matter*, **19**, 156227.

99 Fukasawa, R. and Perkowitz, S. (1996) *Jpn. J. Appl. Phys.*, **35**, 132.

100 Cho, G., Dekorsy, T., Bakker, H., Hövel, R., and Kurz, H. (1996) *Phys. Rev. Lett.*, **77**, 4062.

101 Hase, M., Nakashima, S.i., Mizoguchi, K., Harima, H., and Sakai, K. (1999) *Phys. Rev. B*, **60**, 16526.

102 Chang, Y.M. (2002) *Appl. Phys. Lett.*, **80**, 2487.

103 Bakker, H., Hunsche, S., and Kurz, H. (1998) *Rev. Mod. Phys.*, **70**, 523.

104 Bovensiepen, U. (2007) *J. Phys. Condens.: Matter*, **19**, 083201.

105 Thomsen, C., Grahn, H.T., Maris, H.J., and Tauc, J. (1986) *Phys. Rev. B*, **34**, 4129.

106 Yamamoto, A., Mishina, T., Masumoto, Y., and Nakayama, M. (1994) *Phys. Rev. Lett.*, **73**, 740.

107 Nelson, K., Miller, R., Lutz, D., and Fayer, M. (1982) *J. Appl. Phys.*, **53** (2), 1144.

108 Wolfe, J.P. (1998) *Imaging Phonons*, Cambridge University Press, Cambridge.

109 Wright, O.B., Matsuda, O., and Sugawara, Y. (2005) *Jpn. J. Appl. Phys.*, **44**, 4292.

110 Cavalleri, A. and Schoenlein, R.W. (2004) Femtosecond X-rays and structural dynamics in condensed matter, in *Ultrafast Dynamical Processes in Semiconductors* (ed. K.T. Tsen), Springer, Berlin, pp. 309–337.

111 Nie, S., Wang, X., Park, H., Clinite, R., and Cao, J. (2006) *Phys. Rev. Lett*, **96**, 025901.

112 Tamura, S. (1983) *Phys. Rev. B*, **27** (2), 858.

113 Tamura, S. (1985) *Phys. Rev. B*, **31** (4), 2574.

114 Okubo, K. and Tamura, S. (1983) *Phys. Rev. B*, **28** (8), 4847.

115 Tas, G. and Maris, H. (1994) *Phys. Rev. B*, **49** (21), 15046.

116 Maris, H. (1990) *Phys. Rev. B*, **41** (14), 9736.

117 Monthoux, P. and Pines, D. (1992) *Phys. Rev. Lett.*, **69**, 961.

118 Hengsberger, M., Purdie, D., Segovia, P., Garnier, M., and Baer, Y. (1999) *Phys. Rev. Lett.*, **83**, 592.

119 Valla, T., Fedorov, A.V., Johnson, P.D., and Hulbert, S.L. (1999) *Phys. Rev. Lett.*, **83**, 2085.

120 Eschrig, M. and Norman, M.R. (2000) *Phys. Rev. Lett.*, **85**, 3261.

121 Lanzara, A., Bogdanov, P.V., Zhou, X.J., Kellar, S.A., Feng, D.L., Lu, E.D., Yoshida, T., Eisaki, H., Fujimori, A., Kishio, K., Shimoyama, J.I., Noda, T., Uchida, S., Hussain, Z., and Shen, Z.X. (2001) *Nature*, **412**, 510.

122 Kevan, S.D. (ed.) (1992) *Angle-Resolved Photoemission*, Elsevier, Amsterdam.

123 Knapp, J.A., Himpsel, F.J., and Eastman, D.E. (1979) *Phys. Rev. B*, **19**, 4952.

124 Hofmann, A., Cui, X.Y., Schäfer, J., Meyer, S., Höpfner, P., Blumenstein, C., Paul, M., Patthey, L., Rotenberg, E., Bünemann, J., Gebhard, F., Ohm, T., Weber, W., and Claessen, R. (2009) *Phys. Rev. Lett.*, **102**, 187204.

125 Hulpke, E. and Lüdecke, J. (1992) *Phys. Rev. Lett.*, **68**, 2846.

126 Rotenberg, E. and Kevan, S.D. (1998) *Phys. Rev. Lett.*, **80**, 2905.

127 Rotenberg, E., Schaefer, J., and Kevan, S.D. (2000) *Phys. Rev. Lett.*, **84**, 2925.

128 Mook, H.A. and Nicklow, R.M. (1973) *Phys. Rev. B*, **7**, 336.

129 Paul, D.M., Mitchell, P.W., Mook, H.A., and Steigenberger, U. (1988) *Phys. Rev. B*, **38**, 580.

130 Jarlborg, T. (2003) *Physica C*, **385**, 513.

131 Schäfer, J., Schrupp, D., Rotenberg, E., Rossnagel, K., Koh, H., Blaha, P., and Claessen, R. (2004) *Phys. Rev. Lett.*, **92**, 097205.

132 Dal Corso, A. and de Gironcoli, S. (2000) *Phys. Rev. B*, **62**, 273.

133 Benedek, G., Toennies, J.P., and Zhang, G. (1992) *Phys. Rev. Lett.*, **68**, 2644.

134 Karlsson, K. and Aryasetiawan, F. (2000) *Phys. Rev. B*, **62**, 3006.

135 Blackman, J.A., Morgan, T., and Cooke, J.F. (1985) *Phys. Rev. Lett.*, **55**, 2814.

136 Ododo, J.C. and Anyakoha, M.W. (1983) *J. Phys. F Metal Phys.*, **13**, 2335.
137 Plihal, M., Mills, D.L., and Kirschner, J. (1999) *Phys. Rev. Lett.*, **82**, 2579.
138 Edwards, D.M. and Hertz, J.A. (1973) *J. Phys. F*, **3**, 2191.
139 Vollmer, R., Etzkorn, M., Anil Kumar, P., Ibach, H., and Kirschner, J. (2003) *Phys. Rev. Lett.*, **91**, 147201.
140 Tang, W.X., Zhang, Y., Tudosa, I., Prokop, J., Etzkorn, M., and Kirschner, J. (2007) *Phys. Rev. Lett.*, **99**, 087202.
141 Mook, H.A. and Paul, D.M. (1985) *Phys. Rev. Lett.*, **54**, 227.
142 Magnuson, M., Nilsson, A., Weinelt, M., and Mårtensson, N. (1999) *Phys. Rev. B*, **60**, 2436.
143 Eberhardt, W. and Plummer, E.W. (1980) *Phys. Rev. B*, **21**, 3245.
144 Bünemann, J., Weber, W., and Gebhard, F. (1998) *Phys. Rev. B*, **57**, 6896.
145 Bünemann, J., Gebhard, F., Ohm, T., Weiser, S., and Weber, W. (2008) *Phys. Rev. Lett.*, **101**, 236404.
146 Bünemann, J., Gebhard, F., Ohm, T., Umstätter, R., Weiser, S., Weber, W., Claessen, R., Ehm, D., Harasawa, A., Kakizaki, A., Kimura, A., Nicolay, G., Shin, S., and Strocov, V.N. (2003) *Europhys. Lett.*, **61**, 667.
147 Higashiguchi, M., Shimada, K., Nishiura, K., Cui, X., Namatame, H., and Taniguchi, M. (2005) *Phys. Rev. B*, **72**, 214438.
148 Halilov, S.V., Eschrig, H., Perlov, A.Y., and Oppeneer, P.M. (1998) *Phys. Rev. B*, **58**, 293.
149 Brown, R.H., Nicholson, D.M.C., Wang, X., and Schulthess, T.C. (1999) *J. App. Phys.*, **85**, 4830.
150 Takumi Watanabe, Y.K., Hirano, M., and Hosono, H. (2008) *J. Am. Chem. Soc.*, **130**, 3296.
151 Mazin, I.I., Singh, D.J., Johannes, M.D., and Du, M.H. (2008) *Phys. Rev. Lett.*, **101**, 057003.
152 Macridin, A., Jarrell, M., Maier, T., and Scalapino, D.J. (2007) *Phys. Rev. Lett.*, **99**, 237001.
153 Tan, F., Wan, Y., and Wang, Q.H. (2007) *Phys. Rev. B*, **76**, 054505.
154 Kakehashi, Y. and Fulde, P. (2005) *J. Phys. Soc. Jpn.*, **74**, 2397.
155 Byczuk, K., Kollar, M., Held, K., Yang, Y.F., Nekrasov, I.A., Pruschke, T., and Vollhardt, D. (2007) *Nat. Phys.*, **3**, 168.
156 Raas, C., Grete, P., and Uhrig, G.S. (2009) *Phys. Rev. Lett.*, **102**, 076406.
157 Reed, M.K., Steiner-Shepard, M.K., Armas, M.S., and Negus, D.K. (1995) *J. Opt. Soc. Am. B*, **11**, 2229.
158 Mathias, S., Bauer, M., Aeschlimann, M., Miaja-Avila, L., Kapteyn, H.C., and Murnane, M.M. (2010) Time-resolved photoelectron spectroscopy at surfaces using femtosecond XUV pulses, in *Dynamics at Solid State Surfaces and Interfaces Volume 1: Current Developments* (eds U. Bovensiepen, H. Petek, and M. Wolf), Wiley-VCH Verlag GmbH, Berlin, pp. 501–536.
159 Kirchmann, P.S., Perfetti, L., Wolf, M., and Bovensiepen, U. (2010) Femtosecond time- and angle-resolved photoemission as a real-time probe of cooperative effects in correlated electron materials, in *Dynamics at Solid State Surfaces and Interfaces Volume 1: Current Developments* (eds U. Bovensiepen, H. Petek, and M. Wolf), Wiley-VCH Verlag GmbH, Berlin, pp. 475–498.
160 Gießel, T., Bröcker, D., Schmidt, P., and Widdra, W. (2003) *Rev. Sci. Instrum.*, **74**, 4620.
161 Cavalieri, A.L., Fritz, D.M., Lee, S.H., Bucksbaum, P.H., Reis, D.A., Rudati, J., Mills, D.M., Fuoss, P.H., Stephenson, G.B., Kao, C.C., Siddons, D.P., Lowney, D.P., MacPhee, A.G., Weinstein, D., Falcone, R.W., Pahl, R., Als-Nielsen, J., Blome, C., Düsterer, S., Ischebeck, R., Schlarb, H., Schulte-Schrepping, H., Tschentscher, T., Schneider, J., Hignette, O., Sette, F., Sokolowski-Tinten, K., Chapman, H.N., Lee, R.W., Hansen, T.N., Synnergren, O., Larsson, J., Techert, S., Sheppard, J., Wark, J.S., Bergh, M., Caleman, C., Huldt, G., van der Spoel, D., Timneanu, N., Hajdu, J., Akre, R.A., Bong, E., Emma, P., Krejcik, P., Arthur, J., Brennan, S., Gaffney, K.J., Lindenberg, A.M., Luening, K., and

Hastings, J.B. (2005) *Phys. Rev. Lett*, **94** 114801.

162 Gahl, C., Azima, A., Beye, M., Deppe, M., Döbrich, K., Hasslinger, U., Hennies, F., Melnikov, A., Nagasono, M., Pietzsch, A., Wolf, M., Wurth, W., and Föhlisch, A. (2008) *Nat. Photonics*, **2**, 165.

163 Haight, R. (1995) *Surf. Sci. Rep.*, **21**, 275.

164 Bauer, M. (2005) *J. Phys. D Appl. Phys.*, **38**, R253.

165 Cavalieri, A.L., Krausz, F., Ernstorfer, R., Kienberger, R., Feulner, P., Barth, J.V., and Menzel, D. (2010) Attosecond time-resolved spectroscopy at surfaces, in *Dynamics at Solid State Surfaces and Interfaces Volume 1: Current Developments* (eds U. Bovensiepen, H. Petek, and M. Wolf), Wiley-VCH Verlag GmbH, Berlin, pp. 537–554.

166 Passlack, S., Mathias, S., Andreyev, O., Mittnacht, D., Aeschlimann, M., and Bauer, M. (2006) *J. Appl. Phys.*, **100**, 024912.

167 Pietzsch, A., Föhlisch, A., Beye, M., Deppe, M., Hennies, F., Nagasono, M., Suljoti, E., Wurth, W., Gahl, C., Döbrich, K., and Melnikov, A. (2008) *New J. Phys.*, **10**, 033004.

168 Melnikov, A., Povolotskiy, A., and Bovensiepen, U. (2008) *Phys. Rev. Lett.*, **100**, 247401.

169 Perfetti, L., Loukakos, P.A., Lisowski, M., Bovensiepen, U., Wolf, M., Berger, H., Biermann, S., and Georges, A. (2008) *New J. Phys.*, **10**, 053019.

170 Melnikov, A., Prima-Garcia, H., Lisowski, M., Gießel, T., Weber, R., Schmidt, R., Gahl, C., Bulgakova, N.M., Bovensiepen, U., and Weinelt, M. (2008) *Phys. Rev. Lett.*, **100**, 107202.

171 Wietstruk, M., Melnikov, A., Stamm, C., Kachel, T., Pontius, N., Sultan, M., Gahl, C., Weinelt, M., Dürr, H.A., and Bovensiepen, U. (2011) *Phys. Rev. Lett.*, **106**, 127401.

172 Höfer, U., Shumay, I.L. Reuß, Ch., Thomann, U., Wallauer, W., and Fauster, T. (1997) *Science*, **277**, 1480.

173 Shumay, I.L., Höfer, U., Reuß, C., Thomann, U., Wallauer, W., and Fauster, T. (1998) *Phys. Rev. B*, **58**, 13974.

174 Knoesel, E., Hotzel, A., and Wolf, M. (1998) *J. Electron. Spectrosc. Relat. Phenom.*, **88–91**, 577.

175 Weinelt, M. (2002) *J. Phys. Condens. Matter*, **14**, R1099.

176 Passek, F., Donath, M., Ertl, K., and Dose, V. (1995) *Phys. Rev. Lett.*, **75**, 2746.

177 Aeschlimann, M., Bauer, M., Pawlik, S., Weber, W., Burgermeister, R., Oberli, D., and Siegmann, H. (1997) *Phys. Rev. Lett.*, **79**, 5158.

178 Schmidt, A., Pickel, M., Wiemhöfer, M., Donath, M., and Weinelt, M. (2005) *Phys. Rev. Lett.*, **95**, 107402.

179 Goris, A., Döbrich, K.M., Panzer, I., Schmidt, A.B., Donath, M., and Weinelt, M. (2011) *Phys. Rev. Lett.*, **107**, 026601.

180 Berthold, W., Höfer, U., Feulner, P., Chulkov, E., Silkin, V., and Echenique, P. (2002) *Phys. Rev. Lett.*, **88**, 056805.

181 Schmidt, A.B. (2008) Spin-dependent electron dynamics in front of ferromagnetic surfaces. PhD thesis, Freie Universität Berlin.

182 Zhukov, V.P. and Chulkov, E.V. (2009) *Physics Uspekhi*, **52**, 105.

183 Ogawa, S., Nagano, H., and Petek, H. (1997) *Phys. Rev. B*, **55**, 10869.

184 Pawlik, S., Bauer, M., and Aeschlimann, M. (1997) *Surf. Sci.*, **377–379**, 206.

185 Cao, J., Gao, Y., Miller, R.J.D., Elsayed-Ali, H.E., and Mantell, D.A. (1997) *Phys. Rev. B*, **56**, 1099.

186 Knoesel, E., Hotzel, A., and Wolf, M. (1998) *Phys. Rev. B*, **57**, 12812.

187 Gerlach, A., Berge, K., Goldmann, A., Campillo, I., Rubio, A., Pitarke, J.M., and Echenique, P.M. (2001) *Phys. Rev. B*, **64**, 085423.

188 Zhukov, V.P., Aryasetiawan, F., Chulkov, E.V., de Gurtubay, I.G., and Echenique, P.M. (2001) *Phys. Rev. B*, **64**, 195122.

189 Kittel, C. (1996) *Introduction to Solid State Physics*, John Wiley & Sons, Inc., New York.

190 Lengeler, B. (1978) Electronic structure of noble metals, and polariton-mediated light scattering, in *Springer Tracts in Modern Physics*, vol. 82 (ed. G. Höhler), Springer, Heidelberg, p. 1.

191 Schöne, W.D., Keyling, R., Bandić, M., and Ekardt, W. (1999) *Phys. Rev. B*, **60**, 8616.

192 Halas, N.J. and Bokor, J. (1989) *Phys. Rev. Lett.*, **62**, 1679.

193 Fann, W.S., Storz, R., Tom, H.W.K., and Bokor, J. (1992) *Phys. Rev. B*, **46**, 13592.

194 Lisowski, M., Loukakos, P.A., Bovensiepen, U., Stähler, J., Gahl, C., and Wolf, M. (2004) *Appl. Phys. A*, **78**, 165.

195 Jeong, S., Zacharias, H., and Bokor, J. (1996) *Phys. Rev. B*, **54**, R17300.

196 Jeong, S. and Bokor, J. (1999) *Phys. Rev. B*, **59**, 4943.

197 Ichibayashi, T. and Tanimura, K. (2009) *Phys. Rev. Lett.*, **102**, 087403.

198 Elsaesser, T., Shah, J., Rota, L., and Lugli, P. (1991) *Phys. Rev. Lett.*, **66**, 1757.

199 Anisimov, S.I., Kapeliovich, B.L., and Perel'man, T.L. (1974) *Sov. Phys. JETP*, **39**, 375.

200 Bonn, M., Denzler, D.N., Funk, S., Wolf, M., Wellershoff, S.S., and Hohlfeld, J. (2000) *Phys. Rev. B*, **61**, 1101.

201 Kaganov, M.I., Lifschits, I.M., and Tanatarov, L.V. (1957) *Sov. Phys. JETP*, **4**, 173.

202 Groeneveld, R.H.M., Sprik, R., and Lagendijk, A. (1995) *Phys. Rev. B*, **51**, 11433.

203 Petek, H., Weida, M.J., Nagano, H., and Ogawa, S. (2000) *Science*, **288**, 1402.

204 Bonn, M., Funk, S., Hess, C., Denzler, D.N., Stampfl, C., Scheffler, M., Wolf, M., and Ertl, G. (1999) *Science*, **285**, 1042.

205 Hagen, S., Kate, P., Leyssner, F., Nandi, D., Wolf, M., and Tegeder, P. (2008) *J. Chem. Phys.*, **129**, 164102.

206 Beaurepaire, E., Merle, J.C., Daunois, A., and Bigot, J.Y. (1996) *Phys. Rev. Lett.*, **76**, 4250.

207 Stamm, C. *et al.* (2007) *Nat. Mater.*, **6**, 740.

208 Stanciu, C.D., Hansteen, F., Kimel, A.V., Kirilyuk, A., Tsukamoto, A., Itoh, A., and Rasing, T. (2007) *Phys. Rev. Lett.*, **99**, 047601.

209 Downer, M.C. and Shank, C.V. (1986) *Phys. Rev. Lett.*, **56**, 761.

210 Dziewior, J. and Schmid, W. (1977) *Appl. Phys. Lett.*, **31**, 346.

211 Weinelt, M., Kutschera, M., Fauster, T., and Rohlfing, M. (2004) *Phys. Rev. Lett.*, **92**, 126801.

212 Sjodin, T., Petek, H., and Dai, H.L. (1998) *Phys. Rev. Lett.*, **81**, 5664.

213 Loukakos, P.A., Lisowski, M., Bihlmayer, G., Blügel, S., Wolf, M., and Bovensiepen, U. (2007) *Phys. Rev. Lett.*, **98**, 097401.

214 Weinelt, M., Kutschera, M., Schmidt, R., Orth, C., Fauster, T., and Rohlfing, M. (2005) *Appl. Phys. A*, **80**, 995.

215 Ramstad, A., Brocks, G., and Kelly, P.J. (1995) *Phys. Rev. B*, **51**, 14504.

216 Tabata, T., Aruga, T., and Murata, Y. (1987) *Surf. Sci.*, **179**, L63.

217 Wolkow, R.A. (1992) *Phys. Rev. Lett.*, **68**, 2636.

218 Krüger, P. and Pollmann, J. (1995) *Phys. Rev. Lett.*, **74**, 1155.

219 Pollmann, J., Krüger, P., Rohlfing, M., Sabisch, M., and Vogel, D. (1996) *Appl. Surf. Sci.*, **104**, 1.

220 Pollmann, J. and Krüger, P. (2000) Chapter 2: Electronic structure of semiconductor surfaces, in *Handbook of Surface Science* (eds K. Horn and M. Scheffler), Elsevier Science B.V.

221 Hagen F S. (2005) Adsorption molekularer schalter und elektronendynamik an Si(100)- und Si (100)-H-oberflächen. Diploma thesis, Freie Universität Berlin.

222 Eickhoff, C. (2010) Time-resolved two-photon photoemission at the Si(001)-surface. PhD thesis, Freie Universität Berlin.

223 Hamers, R.J. and Köhler, U.K. (1989) *J. Vac. Sci. Tech. A*, **7**, 2854.

224 Rohlfing, M., and Louie, S.G. (1999) *Phys. Rev. Lett.*, **83**, 856.

225 Vollmer, R., Etzkorn, M., Kumar, P.A., Ibach, H., and Kirschner, J. (2004) *Thin Solid Films*, **464–465**, 42.

226 Balashov, T., Takács, A.F., Wulfhekel, W., and Kirschner, J. (2006) *Phys. Rev. Lett.*, **97**, 187201.

227 Stöhr, J. and Siegmann, H. (2006) *Magnetism: From Fundamentals to Nanoscale Dynamics*, Springer, Berlin.

228 Weinelt, M., Schmidt, A., Pickel, M., and Donath, M. (2007) *Prog. Surf. Sci.*, **82**, 388.

229 Weinelt, M., Schmidt, A., Pickel, M., and Donath, M. (2010) Spin-dependent relaxation of hot electrons on ferromagnetic surfaces, in *Dynamics at Solid State Surfaces and Interfaces Volume 1: Current Developments* (eds U. Bovensiepen, H. Petek, and M. Wolf), Wiley-VCH Verlag GmbH, Berlin, pp. 115–144.

230 Thomassen, J., May, F., Feldmann, B., Wuttig, M., and Ibach, H. (1992) *Phys. Rev. Lett.*, **69**, 3831.

231 Hong, J. and Mills, D. (2000) *Phys. Rev. B*, **62**, 5589.

232 Nigh, H.E., Legvold, S., and Spedding, F.H. (1963) *Phys. Rev.*, **132**, 1092.

233 Vaterlaus, A., Beutler, T., and Meier, F. (1991) *Phys. Rev. Lett.*, **67**, 3314.

234 Prima-Garcia, H. (2007) Laser-induced structural changes at surfaces investigated with synchrotron radiation. PhD thesis, Freie Universität Berlin.

3
Surface States and Adsorbate-Induced Electronic Structure

Thomas Fauster, Hrvoje Petek, and Martin Wolf

The creation of a surface breaks the three-dimensional periodicity of a crystalline solid. The surface possesses two-dimensional periodicity, which is often further reduced from the truncated bulk structure by reconstruction due to energy minimization leading to a rearrangement of the atoms in the surface layer. The changed periodicity and the missing binding partners affect the electronic structure of the surface leading to the formation of surface states. The wave functions of surface states are localized mainly in the surface region. These states are accessible by many surface-sensitive techniques and their energetics and dynamics have been studied in great detail. In this chapter, we first concentrate on intrinsic surface states that constitute a genuine property of single crystal surfaces. Extrinsic surface states, on the other hand, are states that are localized on adsorbed atoms or molecules. These states maintain specific features of the adsorbed atom or molecule but are modified by the interaction with the substrate surface. The electronic structure of adsorbate layers will be discussed at the end of this chapter in section 3.5.

3.1
Intrinsic Surface States

3.1.1
Basic Concepts of Surface States

The occurrence of intrinsic surface states can be described in a simple one-dimensional picture considering the wave functions perpendicular to the surface. Various possibilities are sketched in Figure 3.1. For energies above the vacuum energy E_V, propagating wave functions exist in the vacuum region. The wave functions decay rapidly away from the surface for energies below E_V. In the bulk region, exponentially decaying standing waves are obtained for energies in a band gap of the projected bulk band structure. Matching the wave function and their derivatives at the surface may lead then to surface states with wave functions localized in the surface region as illustrated in the third row of Figure 3.1.

Dynamics at Solid State Surfaces and Interfaces: Volume 2: Fundamentals, First Edition.
Edited by Uwe Bovensiepen, Hrvoje Petek, and Martin Wolf.

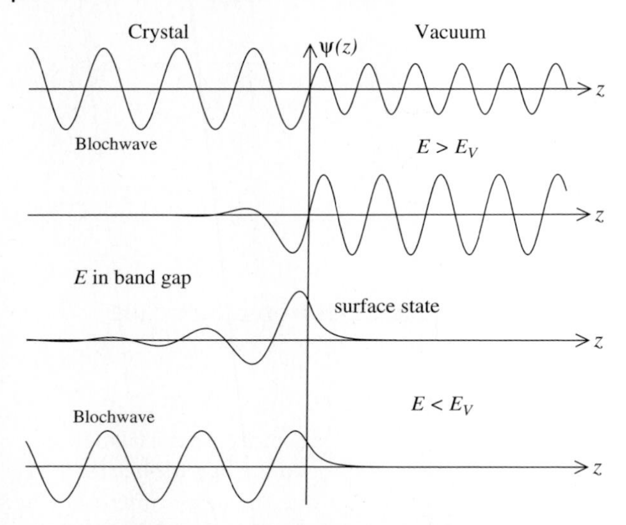

Figure 3.1 Matching wave functions at the surface can lead to surface states, when the energy is in a band gap of the projected bulk band structure and below the vacuum energy E_V.

Band gaps in the projected bulk band structure are quite common on many metal and semiconductor surfaces and depend on the parallel component $k_{||}$ of the wave vector relative to the surface normal. It should be noted that the electron outside the surface has the kinetic energy $\hbar^2 k_{||}^2/2m$ parallel to the surface. This energy has to be expended for the emission of the electron and the vacuum energy relative to the Fermi energy E_F is $E_V = \Phi + \hbar^2 k_{||}^2/2m$. The work function Φ is the minimum energy required for the emission of an electron.

The propagating wave functions sketched in Figure 3.1 correspond to free electron-like plane waves. The modified potential at the surface compared to the bulk may lead to solutions that have a large amplitude near the surface compared to the propagating waves. This situation is termed surface resonance. For further discussion, it is appropriate to distinguish between surface states with larger probability density on the crystal side versus the vacuum side. The former case is sketched in Figure 3.1 and represents a crystal-induced surface state discussed in more detail in Section 3.2. The latter situation of barrier-induced surface states occurs for image potential states that are the topic of Section 3.3.1. Relative to the substrate–overlayer interface, quantum well states (see Section 3.3.2) also fall into the class of barrier-induced surface states.

3.1.2
Scattering Model of Surface States

Surface states can be described and calculated in many ways from simple models to density functional theory (DFT). In this chapter, we choose a scattering approach that provides an intuitive insight into the nature of crystal- and barrier-induced surface states. Other methods such as the nearly free electron approximation or numerical integration of the Schrödinger equation in one dimension yield similar results.

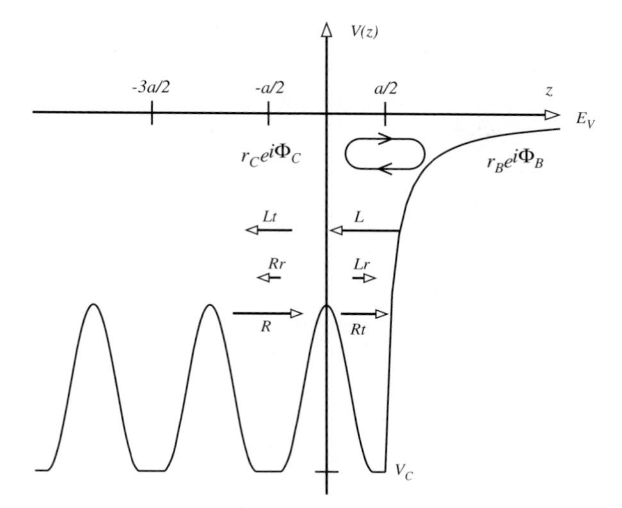

Figure 3.2 Scattering approach for surface states.

In Figure 3.2, the surface is assumed to be half a layer distance $a/2$ away from the topmost layers of surface atoms. At this boundary, the wave function traveling to the right toward the barrier returns after scattering at the barrier with complex amplitude $r_B \exp i\Phi_B$. Similarly, the wave function is reflected by the crystal with $r_C \exp i\Phi_C$. For a stationary condition, the reflection amplitudes r_B and r_C must be 1. The second requirement is that the phase shifts must add up coherently

$$\Phi_C + \Phi_B = 2\pi n. \tag{3.1}$$

The phase shifts are a function of energy and Eq. (3.1) may have several solutions characterized by the quantum number n. The distinction between crystal- and barrier-induced surface states depends on which phase shift Φ_C or Φ_B varies more rapidly with energy [1]. This criterion is equivalent to the intuitive argument about the probability density being more in the crystal or outside [1].

The calculation of the barrier phase Φ_B depends on the nature of the surface potential and will be deferred to Section 3.3. The crystal constitutes a semiinfinite solid and the existence of surface states depends largely on properties of its band structure. In the one-dimensional scattering approach [2], the potential is assumed to be a periodic arrangement of wells with separation a as depicted in Figure 3.2. At the minima exist arbitrarily small regions of constant potential V_C where the wave function can be expanded into plane waves with wave vector K

$$\psi(z) = L\,e^{-iKz} + Lr\,e^{iKz} + Rt\,e^{iKz}. \tag{3.2}$$

The first term describes a wave of amplitude L traveling to the left. It is reflected by the potential of the layer to the left with a reflection coefficient r. The last term is a wave of amplitude R that has been transmitted through the layer at the left with a transmission coefficient t. Without absorption, the transmission and reflection coefficients fulfill the condition $|t|^2 + |r|^2 = 1$. The wave vector K is given by the energy E relative to the potential V_C by $E - V_C = \hbar^2 K^2/2m$, where m denotes the (effective) mass of the electron. The solutions in the periodic potential with period a must obey Bloch's theorem

$$\psi(z+a) = e^{ika}\psi(z) \tag{3.3}$$

for all $z \leq a/2$ and in particular between the layers. The solutions for the Bloch wave vector k can be found by inserting Eq. (3.2) and the plane wave expansion at $z = -a/2$ from Figure 3.2 into Eq. (3.3) and by applying the same procedure to the derivatives of the wave functions. The following equations are obtained:

$$L/R = (e^{-i(K-k)a} - t)/r \quad \text{and} \quad R/L = (e^{-i(K+k)a} - t)/r. \tag{3.4}$$

Multiplication of these equations yields a second-order equation for e^{ika} and thus the Bloch wave vector k. The symmetry of Eq. (3.4) shows that a solution k implies the solution $-k$ with reversed amplitude ratio L/R. For complex wave vectors k, this ensures that there exists always a solution decaying into the bulk ($z \rightarrow -\infty$) with Im(k)>0

The complex band structure obtained from Eq. (3.4) is shown in Figure 3.3. The parameters V_C, m, and t have been adjusted to reproduce the nearly free electron band

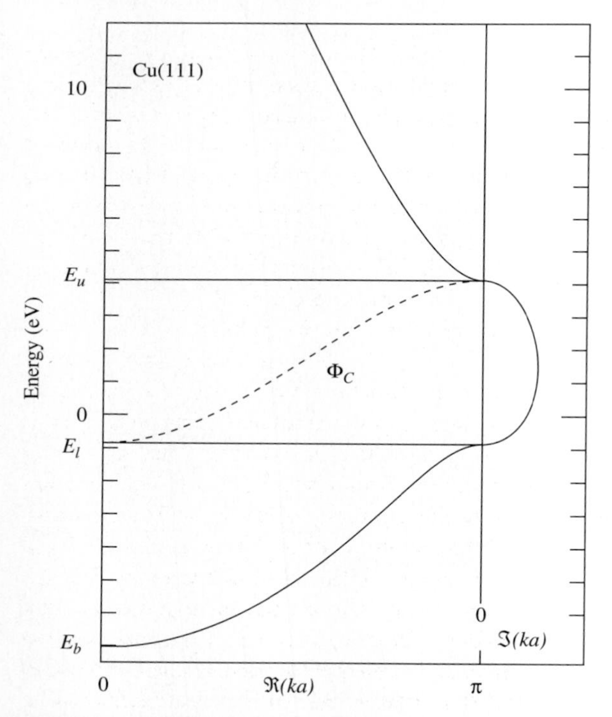

Figure 3.3 Complex band structure reproducing the nearly free electron band gap of Cu(111). The dashed line shows the crystal phase shift Φ_C.

gap of the Cu(111) surface. Note that even a high transmission coefficient $t = 0.93$ leads to sizeable band gap. The crystal phase shift can be obtained directly from Eq. (3.2) to

$$e^{-i\Phi_C} = (1 - e^{i(K-k)a} t)/r \tag{3.5}$$

and varies in a cosine-like curve from 0 to π across the band gap as illustrated by the dashed curve in Figure 3.3.

Within a nearly free electron approximation, the complex band structure and phase shift Φ_C can also be calculated analytically [3]. The results are almost indistinguishable from the scattering approach using a constant transmission coefficient t [2]. The scattering model has the advantage that it may be easily extended to overlayers and alloys [2].

3.2
Crystal-Induced Surface States

3.2.1
Tamm and Shockley Surface States

Surface states were first discussed by Tamm [4], Maue [5], and Shockley [6]. Surface states derived from *d*-bands are often called Tamm states and surface states with a free electron-like dispersion derived from *sp*-bands are called Shockley states. The common ingredient of these models is a rather abrupt transition of the potential from the crystal to the vacuum at the surface. This leads to an exponentially decaying wave function in vacuum for energies below the vacuum energy E_V as shown in the lower part of Figure 3.1. To match the bulk wave function $\psi(z)$ and its derivative $\psi'(z)$ at the surface, ψ'/ψ has to be negative [7]. This condition is often met when the potential has a minimum at the surface as shown in Figure 3.2.

For a step-like barrier, the phase shift Φ_B varies with energy from $-\pi$ at the bottom of the valence band to 0 at the vacuum-level E_V [3]. The crystal phase shift Φ_C varies across a band gap in the projected bulk band structure by π [8] (see also Figure 3.3). Note that only for energies inside a band gap the reflection coefficient r_C is 1. A variation from 0 at the bottom of the band gap to $+\pi$ at the top of the band gap is obtained if the potential at the surface has a minimum (see Figure 3.2). This situation is called Shockley-inverted gap because the symmetry of the wave functions with respect to the surface atoms is p_z at the bottom and s at the top of the band gap in contrast to the normal sequence of atomic orbitals. Note that the ranges of the phase shifts permit only the solution $n = 0$ in Eq. (3.1) and that this condition may not be fulfilled at any energy. The energies of the Shockley (or $n = 0$) surface states on the fcc(111) surfaces of noble metals can be calculated quite accurately assuming a step-like surface barrier [3, 9]. For the fcc(100) surfaces, only surface resonances exist at $k_{||} = 0$. Examples for the wave function and experimental observation of Shockley surface states are presented later in this chapter.

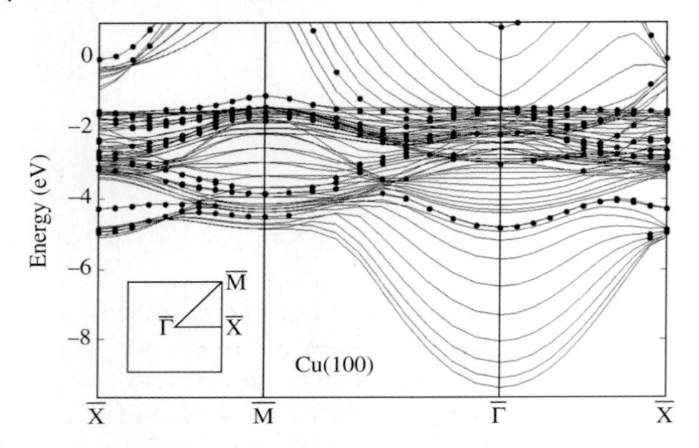

Figure 3.4 Surface band structure of Cu(100) (dotted lines) along high-symmetry lines. Reprinted with permission from Ref. [10]. Copyright (2003) by the American Physical Society.

For the step-like surface barrier, the phase is limited to the range $[-\pi, 0]$ and solutions of Eq. (3.1) are obtained only under certain conditions. This result can be generalized that the number of surface states on single-crystal surfaces containing a mirror plane is 0, 1, or 2 [8]. This statement is true anywhere in the projected bulk band gap. Two surface states can be found only at general points of the surface Brillouin zone (i.e. not at the center or zone boundary) [8].

Figure 3.4 shows the surface band structure of Cu(100) along the high-symmetry lines of the surface Brillouin zone (see inset) [10]. Surface states are marked by filled dots. The bulk bands appear as discrete lines because of the use of an 11-layer slab used in the calculation. The envelope indicates the projected bulk band structure, that is, the $E(k_{||})$ regions where bulk bands exits. At the center of the surface Brillouin zone $\bar{\Gamma}$ ($k_{||} = 0$), the surface resonance is visible around 1 eV. Among the many surface bands in the *d*-band region, the surface state at $\bar{\mathrm{M}}$ at -1.5 eV is a Tamm state with a weak dispersion and narrow linewidth (see also Figure 3.10).

3.2.2
Dangling Bond States

Surface states on semiconductor surfaces are often described as dangling bond states reflecting the cutting of the localized bonds between the atoms at the surface. They belong to the class of crystal-induced surface states because the probability density decays rapidly when going away from the surface. On the other hand, the electronic structure of the surface atoms is different from the bulk due to the cut bonds. In addition, strong reconstructions are commonly found at many semiconductor surfaces that reduce the number of dangling bonds compared to the truncated bulk.

The localized orbitals of the covalent bonds at semiconductors favor the use of tight binding models. Applied to surfaces, one obtains dangling bond states localized at certain surface atoms. However, the associated bands show considerable dispersion indicating the extended nature of the wave functions. An example is shown in Figure 3.5 for the Si(001)$c(4 \times 2)$ surface [11]. The shaded areas indicate the projected

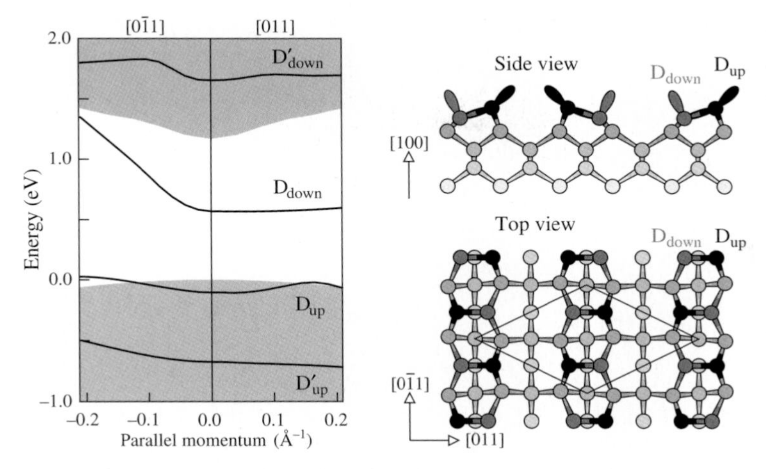

Figure 3.5 Dispersion of the dangling bond states for the Si(001)$c(4 \times 2)$ surface illustrated on the right. The shaded areas indicate the projected bulk bands. Reprinted with permission from Ref. [11]. Copyright (2005) by Springer.

bulk band structure and the energy scale is referenced to the valence band maximum. The occupied (unoccupied) dangling bonds are localized on atoms shifted up (down) in the asymmetric dimer reconstruction as indicated in the right part of Figure 3.5. The $c(4 \times 2)$ unit cell contains two dimers tilted in opposite orientation leading to two pairs of dangling bond bands. Figure 3.5 shows strong dispersion of the bands in the $[1\bar{1}0]$ direction of neighboring dimers. In the $[110]$ direction, the dispersion is considerably smaller due to the larger separation of the dimer rows compared to the distance between neighboring dimers.

3.3 Barrier-Induced Surface States

3.3.1 Image Potential States

When an electron is brought in front of a metal surface, the charge is screened in such a way that the field lines are perpendicular to the metal surface (see Figure 3.6). The polarization charge induced at the surface acts as if the metal would be replaced by an image charge of opposite sign at the opposite position from the metal surface. The force acting on the charge $-e$ at a distance z in front of the surface can be written as

$$F(z) = -\frac{e^2}{(2z)^2}. \tag{3.6}$$

Note that there is no force needed to move the charge in the direction parallel to the surface. The factor 2 in the denominator of Eq. (3.6) arises because charge and image charge are separated by distance $2z$.

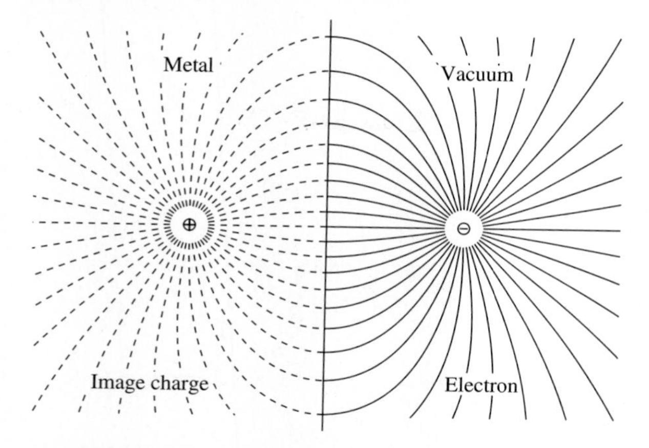

Figure 3.6 Electric field of a charge in front of a metal surface. The polarization charge induced at the surface has the same effect as if the metal would be replaced by an image charge (dashed lines).

The image force can be derived from the image potential

$$V(z) = E_V - \frac{e^2}{4(z - z_{im})}, \tag{3.7}$$

where z_{im} is the image plane position. The potential converges toward the vacuum-level E_V for large distances from the surface. The image plane is typically located about half an interlayer distance outside the outermost atomic layer.

The potential of Eq. (3.7) diverges and does not match the crystal potential at $z = z_{im}$. Chulkov *et al.* [12] suggested a simple one-dimensional model that reproduces energies of the observed surface states at many metal surfaces very well. Figure 3.7 shows the model potential for the Cu(111) surface. In the bulk region, it is described by a cosine function that opens the gap in the projected bulk band structure on the surface of interest [13]. In the surface region, the model potential is represented by a smooth cosine-like function that reproduces the energy of the Shockley surface state. In the vacuum region, this function merges into the long-range image potential in such a way that it describes the experimental energy of the first image potential state. The position of the image plane z_{im} is obtained consistently from the potential equations [12].

The one-dimensional potential is strictly valid only for electrons moving perpendicular to the surface, that is, with parallel momentum $k_{||} = 0$ at the center of the surface Brillouin zone. It can be extended assuming a free electron-like behavior for the motion parallel to the surface. The effect of the crystal potential might lead to an effective mass different from the free electron mass.

For a one-dimensional potential such as the one shown in Figure 3.7 for the Cu (111) surface, it is straightforward to obtain the solutions of the Schrödinger equation, for example, by numerical integration. The dashed lines in Figure 3.7 show the energies of two solutions together with the squares of the wave functions. The solution at the lower energy is the well-known Shockley surface state [6] that is

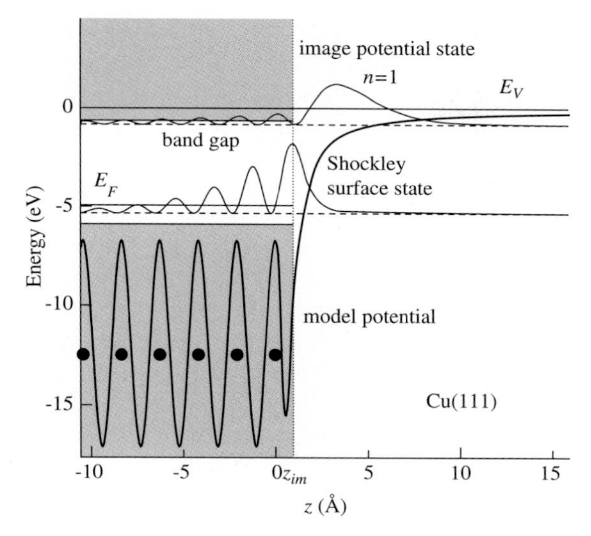

Figure 3.7 One-dimensional model potential and probability density for the Shockley surface state and the first image potential state ($n = 1$) at the Cu(111) surface. The shaded areas indicate the projected bulk bands bounded by the image plane. The dots give the positions of the atomic planes.

only slightly influenced by the image potential. The other solution is the $n = 1$ image potential state that has most of its probability density outside the surface [14].

The long-range nature of the image potential leads to an infinite series of image potential states. The energies show the same dependence as in the hydrogen atom

$$E_n = E_V - \frac{0.85\,\text{eV}}{(n+\alpha)^2}, \quad n = 1, 2, 3, \ldots \tag{3.8}$$

The energy is scaled down by a factor of 16 relative to the hydrogen atom due to the factor of 4 in the denominator of the potential in Eq. (3.7). Image potential states always have energies above the Fermi energy close to the vacuum level because the image potential and the states require an external charge that is absent in the ground state.

For the one-dimensional Coulomb potential of Eq. (3.7), the barrier phase shift may be written as [3, 15]

$$\Phi_B(E) = \pi\left[\sqrt{3.4\,\text{eV}/(E_V - E)} - 1\right]. \tag{3.9}$$

The divergence of Φ_B for $E \to E_V$ leads to an infinite series of barrier-induced surface states. Inserting the energy E_n into Eq. (3.9) yields in combination with the phase condition of Eq. (3.1) the following expression for the quantum defect α in Eq. (3.8) [3]:

$$\alpha = \frac{1}{2}(1 - \Phi_C/\pi). \tag{3.10}$$

The quantum defect varies from 1/2 to 0 across the band gap with increasing energy. At the upper edge of the band gap, the wave function has a node at the surface and is *s*-like with respect to the atom, whereas at the lower edge it has a maximum and p_z character. The different phase of the bulk wave function can be seen in Figure 3.7 for the image potential and Shockley surface state, respectively.

Image potential states at metal surfaces were predicted by Echenique and Pendry [14, 16] and first detected by inverse photoemission [17, 18]. The inverse photoemission results have been reviewed by Straub and Himpsel [19]. The limited resolution and sensitivity of inverse photoemission allowed only the $n = 1$ image potential state to be resolved clearly. With two-photon photoemission (see Section 3.4.2), the energy resolution could be improved and the lowest three image potential states could be identified [20]. The weak coupling of image potential states to surfaces makes them ideal observer states to monitor the geometric, electronic, and magnetic properties of metal surfaces [9, 21, 22]. The electron dynamics in image potential states at metal surfaces has been discussed in Volume I [23].

3.3.2 Quantum Well States

If a metal overlayer is deposited on top of a metal or semiconductor substrate, electronic states may exist for energies in the projected bulk band structure of the substrate. These states are confined to the overlayer by an effective potential well and are called quantum well states. It is straightforward to extend the phase accumulation model of Section 3.1.2 and add the phase shift of the Bloch wave of wavevector k after propagating back and forth through N layers of thickness a' [24]:

$$\Phi_C(E) + 2k(E)Na' + \Phi_B(E) = 2\pi n \tag{3.11}$$

For clarity, the energy dependence of the relevant terms is included, where $k(E)$ is taken from the band structure of the overlayer. The scattering model of Section 3.1.2 is a convenient way to calculate the dispersion $E(k)$. Equation (3.11) shows that the number of layers N can change the phase by large amounts. The crystal phase Φ_C may change only by π across the band gap. The barrier phase Φ_B varies significantly only close to the vacuum-level E_V (see Eq. (3.9)) and might otherwise be approximated by the phase for the step potential [3]. The quantum number n describing the total phase shift is therefore determined mainly by the number of layers N. It is convenient to sort the quantum well states by energy that is achieved by introducing the quantum number $\nu = N - n$ [24]. This numbering scheme permits the easy comparison of states for different layer thicknesses and fits the obvious sequence of states observed in experimental data. Note that the numbering scheme is not unique and might differ for energies at the bottom or top of a band in the overlayer band structure [24].

A special situation may occur for a quantum well straddling the vacuum energy. Here, the second and third terms of Eq. (3.11) may vary considerably with energy. The phase shift in the quantum well can be controlled by the layer thickness. One example are image potential states for Au on Pd(111) [25]. Figure 3.8 shows

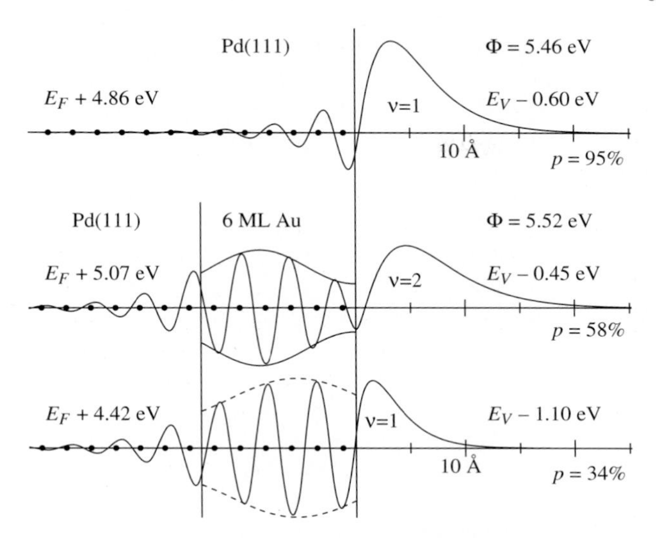

Figure 3.8 Wave functions for image-potential states coupled to a quantum well for Au on Pd(111). Reprinted with permission from Ref. [25]. Copyright (1995) by the American Physical Society.

calculated wave functions for six layers of Au on Pd(111). The lowest $\nu = 1$ state is located mainly in the overlayer and has an energy 1.1 eV below E_V, which is outside the range given in Eq. (3.10) for positive quantum defects. The $\nu = 2$ state corresponds more to an image potential state and has properties very similar to the first image potential state on the clean Pd(111) surface.

3.4
Experimental Methods

This section discusses briefly the main experimental techniques for the study of intrinsic surface states. Emphasis is put on the analysis of data regarding the dynamics at surfaces. For a comprehensive review of the results obtained by the various techniques and calculations, see Ref. [26].

3.4.1
Photoemission

Figure 3.9 shows on the left-hand side the energy diagram for the photoemission process. The incoming photon of energy $h\nu$ excites an electron from the initial state $|i\rangle$ with energy E_i to a final state $|f\rangle$. The final state energy E_f of the emitted electron is measured as the kinetic energy E_{kin} relative to the vacuum-level E_V. Energy conservation yields

$$E_f = E_i + h\nu = E_{kin} + \Phi. \tag{3.12}$$

The two-dimensional periodicity conserves the momentum parallel to the surface in a reduced Brillouin zone scheme

$$\vec{k}^{\,\mathrm{f}}_{\|} = \vec{k}^{\,\mathrm{i}}_{\|} \quad \text{with} \quad |\vec{k}^{\,\mathrm{f}}_{\|}| = \sqrt{2mE_{\mathrm{kin}}/\hbar^2}\,\sin\vartheta. \tag{3.13}$$

The momentum of the photon can be neglected in most cases. The second equation relates the momentum to the measured kinetic energy E_{kin} and emission angle ϑ. The power of angle-resolved photoelectron spectroscopy lies in a complete determination of the surface band structure $E_{\mathrm{i}}(\vec{k}_{\|})$.

A typical photoemission spectrum is sketched on the right-hand side of Figure 3.9. At the final state energy E_{f}, a peak in the spectrum appears. The low-energy cutoff arises from electrons at the vacuum-level E_{V} of the sample that have zero kinetic energy in the sample. In practice, a small bias voltage is often used to accelerate the emitted electrons toward the analyzer and to ensure that the low-energy cutoff is not set by the analyzer. The maximum energy E_{max} is gained by electrons at the Fermi-level E_{F} that obtain after the absorption of a photon with energy $h\nu$ the kinetic energy $E_{\mathrm{max}} - E_{\mathrm{V}} = h\nu - \Phi$. The width of the spectrum given by the low- and high-energy limits can be used to determine the work function Φ of the sample.

Figure 3.10 presents two examples for photoelectron spectra from surface states. Using a state-of-the art two-dimensional analyzer with an energy resolution of 3.5 meV and 0.15° angular resolution, the Shockley surface state for the Cu (111) surfaces has been measured [27]. It should be mentioned that such high-quality results require a careful sample preparation and are very sensitive to sample contamination.

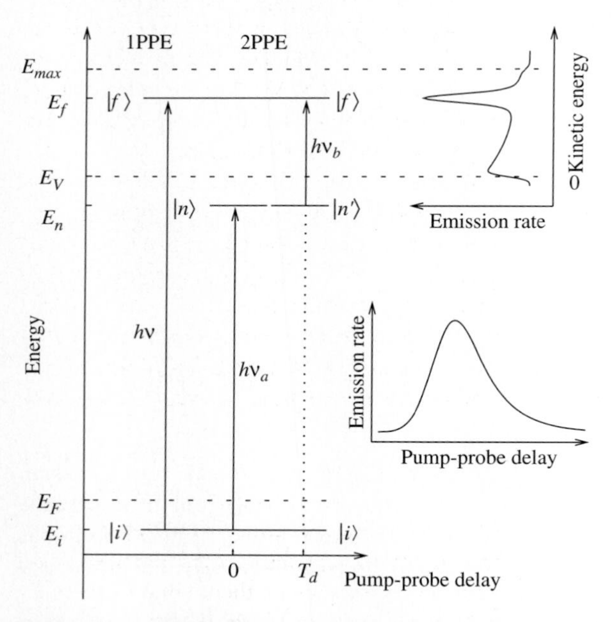

Figure 3.9 Energy diagram for one- and two-photon photoemission processes. Two-photon photoemission allows in addition time-resolved measurements.

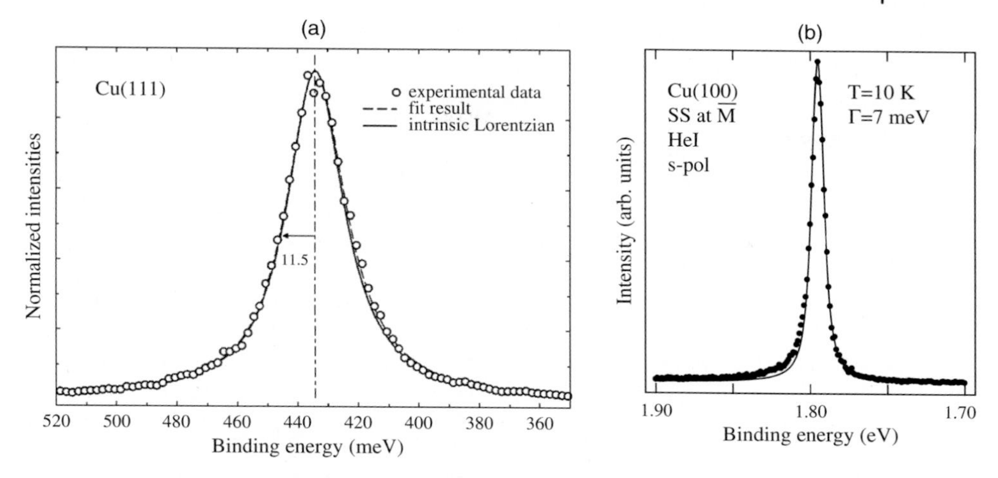

Figure 3.10 (a) Shockley surface state on Cu(111) at $\bar{\Gamma}$ measured at 30 K [27]. The Lorentzian peak width is 23 meV. (b) Tamm state at $\bar{M}$ measured at 10 K with an energy resolution of 5 meV. The fit to the data gives a Lorentzian peak width of 7 meV. Reprinted with permission from Ref. [28]. Copyright (1998) by Elsevier.

The lineshape can be fitted by a Lorentzian with a full-width at half-maximum (FWHM) of 23 meV. The experimental resolution is taken into account by convolution with a Gaussian.

The more localized Tamm surface states show less dispersion and the angular resolution is less important. As an example, a spectrum and the corresponding fit for the $\bar{M}$ Tamm state on Cu(100) is shown in Figure 3.10a [28]. The state shows a narrow intrinsic linewidth of $\Gamma = 7$ meV even at an energy of 1.8 eV below the Fermi energy. This is attributed to the localized character of the *d*-states and the small overlap with the *sp*-bands that provide the main decay channel for inelastic decay.

The photoemission lineshape is proportional to the hole spectral function of the sample times the Fermi distribution. The spectral function *A*, in turn, is used to describe the electronic structure of a solid in the presence of many-body effects. *A* can be viewed as the probability of finding an electron with energy E_i and momentum $\vec{k}$. The spectral function is determined by the unrenormalized dispersion $\varepsilon(\vec{k})$ and the complex self-energy $\Sigma = \Sigma' + i\Sigma''$. In general, Σ depends on the initial state $\varepsilon(\vec{k})$ including the momentum $\vec{k}$ and additional parameters of the many-body interactions. For electron–phonon coupling, this would be the temperature. Then, *A* has the form

$$A(E_\mathrm{i}, \vec{k}) = \frac{\pi^{-1}|\Sigma''|}{[E_\mathrm{i} - \varepsilon(\vec{k}) - \Sigma']^2 + \Sigma''^2}. \tag{3.14}$$

With a modern photoelectron spectrometer, the photoemission intensity can be measured for so many values $(E_\mathrm{f}, \vec{k})$ that any cut through the spectral function can be extracted. However, we briefly relate the spectral function to the traditional measuring modes of angle-resolved photoelectron spectroscopy, energy distribution curves (EDCs), and momentum distribution curves (MDCs).

An EDC is the photoemission intensity as a function of kinetic energy for a fixed photon energy and a fixed emission angle. The fact that the emission angle, not $\vec{k}$, is constant means that an EDC corresponds to a fairly complicated cut through the spectral function. Under certain conditions, for example, for normal emission or for a very small energy range, an EDC is taken at approximately constant $\vec{k}$. Even then, an EDC calculated from (3.14) may have a fairly complicated form. This is due to the energy dependence of Σ, which might be relevant in particular close to the Fermi level. A simple scenario arises when we assume that $\Sigma' = 0$ and that Σ'' is constant. Then, we get

$$A(E_{\rm i},\vec{k}) = \frac{\pi^{-1}|\Sigma''|}{[E_{\rm i}-\varepsilon(\vec{k})]^2+\Sigma''^2}, \tag{3.15}$$

which is a Lorentzian with the maximum at $\varepsilon(\vec{k})$ and an FWHM of $2|\Sigma''|$ in agreement with the experimentally observed lineshape (see Figure 3.10). However, care is necessary when an EDC linewidth is identified with $2|\Sigma''|$ because of the above-mentioned problem that an EDC is strictly measured at a constant emission angle, not at a constant $\vec{k}$ [29–31].

The situation is simpler in case of MDCs because they are readily represented by (3.14). The maximum of an MDC is reached when $E_{\rm i}-\varepsilon(\vec{k})-\Sigma' = 0$. On the basis of this, the renormalized dispersion is defined as the self-consistent solution of

$$E_{\rm i}(\vec{k}) = \varepsilon(\vec{k})+\Sigma'(E_{\rm i}(\vec{k})). \tag{3.16}$$

Equation (3.14) takes on a particularly simple form in the case of a linear dispersion. We consider only one direction in $\vec{k}$ space and write $\varepsilon(k) = vk$ such that the origin of the coordinates is at the Fermi-level crossing. Then, it is easy to show that (3.14) is a Lorentzian line in k for a given $E_{\rm i}$ with the maximum at

$$k_{\rm max} = (1/v)(E_{\rm i}-\Sigma') \quad \text{and} \quad \text{FWHM} = 2|\Sigma''/v|. \tag{3.17}$$

3.4.2
Two-Photon Photoemission

Two-photon photoemission can be viewed as regular (one-photon) photoemission from a state after excitation of the surface by another photon. This allows the spectroscopy of excited intermediate states with energies above the Fermi level, which are normally unoccupied. This energy range, in particular the part below the vacuum level, is otherwise accessible only by inverse photoemission.

The excitation of the surface is done by photons and the dipole selection rules apply for both steps in two-photon photoemission. Using two photons expands the parameter space compared to regular photoemission by the choice of energy and polarization of an additional photon. In order to reach the high intensities needed for the second-order process, femtosecond lasers are used. This adds the possibility to introduce a time delay between the two laser pulses that allows to follow the population in the excited state on short timescales. This last feature in particular is unique to two-photon photoemission and permits the detailed investigation of the electron dynamics at surfaces, which is the main topic of Volume I.

The energy diagram of two-photon photoemission is shown in Figure 3.9. The first photon with energy $h\nu_a$ excites an electron from an occupied initial state $|i\rangle$ below the Fermi energy E_F to an intermediate state $|n\rangle$. The second photon of energy $h\nu_b$ excites the electron into the final state $|f\rangle$ from where it can leave the surface if the energy E_f is above the vacuum energy E_{vac}. The kinetic energy E_{kin} of the emitted electron is measured with an electron energy analyzer as in regular photoelectron spectroscopy. The photon energies $h\nu_a$ and $h\nu_b$ should be below the work function $\Phi = E_V - E_F$ to avoid one-photon photoemission that is several orders of magnitude more intense than two-photon photoemission.

In two-photon photoemission spectra, the assignment of a peak to an initial or intermediate state is not obvious. In addition, the sequence of the photons $h\nu_a$ and $h\nu_b$ may be interchanged or two photons $h\nu_a$ and $h\nu_b$ may emit the electron. The latter possibility can be checked by an intensity variation (simply blocking the respective laser beams). The former possibilities can be checked by measuring spectra at photon energies varied by $\Delta h\nu_a$ and $\Delta h\nu_b$ and determining the change in the kinetic energy ΔE_{kin}. In most cases, only the fundamental photon energy $h\nu_0$ before frequency doubling or tripling is tuned. For surface states, ΔE_{kin} varies as an integer multiple of the variation of the fundamental photon energy $\Delta h\nu_0$. Bulk bands disperse with momentum perpendicular to the surface and the kinetic energy might vary in a different fashion with photon energies given by the dispersion of the bulk bands [32]. For parabolic bands, ΔE_{kin} is proportional to $\Delta h\nu_0$ with a factor given by the ratio of the effective masses [33]. A different approach to distinguish initial and intermediate states compares one- and two-photon photoemission spectra taken at the same total photon energy $h\nu = h\nu_a + h\nu_b$ [34]. Intermediate states are absent in one-photon photoemission spectra.

Two-photon photoemission spectra as a function of photon energy are shown in Figure 3.11 for the Cu(111) surface [35, 36]. On this surface, an occupied Shockley surface state and an unoccupied image potential state exist as shown in Figure 3.7. The spectra present two peaks that show a different variation of the kinetic energy with photon energy. The peak with the stronger variation with photon energy is assigned to an initial state, that is, the occupied surface state, because it requires the absorption of two photons of energies $h\nu_a = 3h\nu_b$ and $h\nu_b$. The image potential state peak shows as intermediate state a weaker variation because the electron is emitted after absorption of a photon with energy $h\nu_b$. The peak positions of the spectra in Figure 3.11 are projected to the baseline and marked with dots. Since the offset of the spectra is proportional to the photon energy, we obtain in this way a plot of photon energy versus kinetic energy. The straight lines show the expected linear behavior with different slopes for different states. At the photon energy $h\nu_b = 1.512$ eV, the two lines cross and a resonant excitation with photon energy $h\nu_a = 4.536$ eV from the surface to the image potential state occurs [37, 38]. The spectra in Figure 3.11 are normalized to the same height, so the intensity enhancement [39] at resonance is not visible.

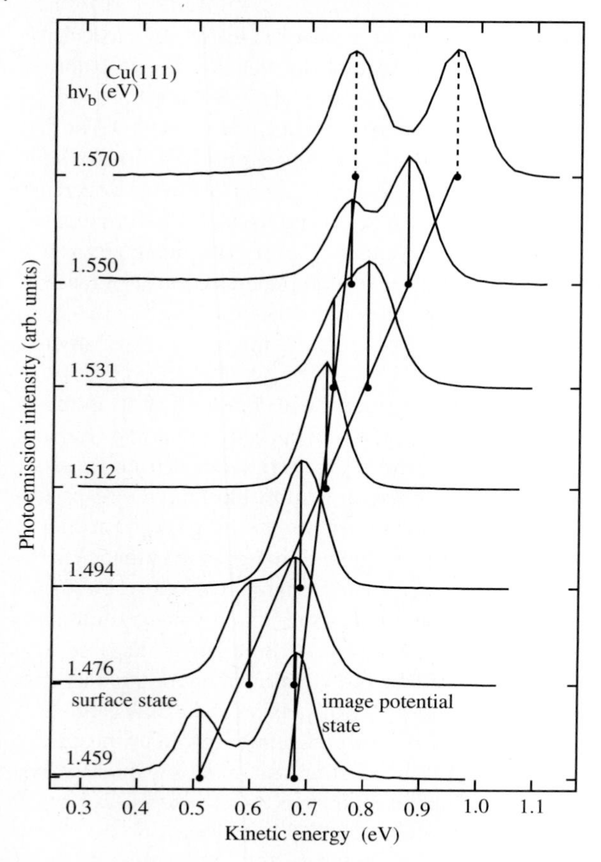

Figure 3.11 Series of two-photon photoemission spectra from Cu(111) for various photon energies with $h\nu_a = 3h\nu_b$. The spectra are normalized to same height and plotted with an offset proportional to $h\nu_b$.

Two-photon photoemission adds the additional parameter of a time delay between the two laser pulses. This allows to monitor the decay of the population of the intermediate state $|n\rangle$ as a function of time after the initial excitation. Note that scattering into other states $|n'\rangle$ may also be detected as indicated in Figure 3.9 [40].

An example for time-resolved two-photon photoemission is shown in Figure 3.12 for the Pd(111) surface that has an unoccupied Shockley surface state and an image potential state [41, 42]. The latter is excited as for Cu(111), but for the surface state $h\nu_b$ is absorbed before $h\nu_a$. The exponential decays (indicated by straight lines in the semilogarithmic plot of Figure 3.12) appear on different sides relative to the time zero. From this, the assignment of the excitation sequence is straightforward because the probe pulse must come after the pump pulse. The bottom traces of Figure 3.12 show measurements of the occupied surface Shockley state on Cu(111) that is excited by the simultaneous absorption of two photons. The symmetric shape

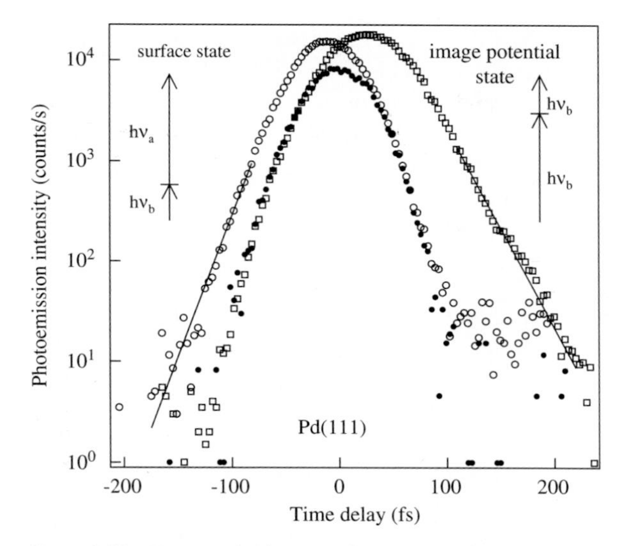

Figure 3.12 Time-resolved spectra for the surface (open circles, $n = 0$) and image-potential state (open squares, $n = 1$) on Pd(111). The solid dots show the cross correlation determined for the occupied surface state on Cu(111).

of the cross-correlation curves indicates a symmetric (Gaussian) pulse shape of the laser pulses.

Two-photon photoemission involves the absorption of two photons and the two optical transitions can usually be treated as independent consecutive processes. In the following paragraphs, we will discuss a few aspects of time-resolved two-photon photoemission that are fundamentally different from those of regular photoemission.

In a time-resolved two-photon photoemission experiment, the pulse duration can be comparable to the timescale of the temporal evolution of the involved states. In other words, the separation of the energy levels is comparable to the band width of the laser pulses. The laser fields excite the electronic states $|l\rangle$ into a superposition

$$|\Psi\rangle = \sum_l c_l(t)|l\rangle. \tag{3.18}$$

After the laser fields are turned off, the time dependence of the state l is given by

$$c_l(t) = c_l(0)\mathrm{e}^{-t/2\tau_l}\mathrm{e}^{\mathrm{i}\phi_l(t)}\mathrm{e}^{\mathrm{i}E_l t/\hbar}. \tag{3.19}$$

The exponential decay is described by a lifetime τ_l and the energy E_l determines the quantum mechanical phase. The term containing $\phi_l(t)$ describes changes in the phase by phase-breaking events, for example, (quasi-)elastic scattering of the electron with negligible change in energy or population $|c_l(t)|^2$ [43]. The population of the final state $|c_f(t)|^2$ is observed experimentally. It depends on the laser fields of both photons including the time delay between pump and probe pulse.

The time evolution of the population $|c_f(t)|^2$ can be calculated using the density matrix formalism [44]. The external laser fields couple the different states. The interaction with other states of the system is described by decay and dephasing rates that are treated as parameters. The resulting optical Bloch equations [45] may be solved numerically and analytically in certain limiting cases [38, 46].

The description of regular photoemission spectra by two-level optical Bloch equations yields a Lorentzian convoluted with the spectrum of the light source and the analyzer resolution function [46]. Note that an exponential decay in time leads to a Lorentzian in the energy domain. The Lorentzian width contains a sum of decay and dephasing rates of the hole, which cannot be separated. Therefore, the linewidth in photoemission cannot be identified directly with an inverse lifetime of the photohole.

Two-photon photoemission starts like regular photoemission from a constant initial state population, but the two-photon process samples a time-dependent intermediate state population with the second photon. This time dependence changes upon tuning the photon energy on or off resonance for a transition between discrete initial and intermediate states. In Figure 3.11, the linewidth at resonance ($h\nu_b = 1.512$ eV) is narrower than the linewidth observed for each state of resonance [36]. Using optical Bloch equations, the experimental lineshape can be described quite well by a Lorentzian and values for the decay and dephasing rates of the surface and image potential state can be obtained [38]. Decay and dephasing rates can also be separated by measuring the linewidth as a function of delay between pump and probe pulses [46]. For long delays, the Lorentzian is determined by the dephasing rate and the decay rate enters only when pump and probe pulse overlap. Note that for femtosecond laser pulses, the spectral band width may exceed the energy resolution of the analyzer in two-photon photoemission.

If the energy separation of intermediate states is comparable to the band width of the laser pulses, several states can be excited coherently and the temporal phase of the wave function enters directly. The optical Bloch equations have to be solved including several intermediate states [47, 48]. When the pump pulse is over, the intermediate state population is given by the square of the sum of all excited states $c_l(t)$ (see Eq. (3.19)):

$$\left|\sum_l c_l(t)\right|^2 = \sum_l |c_l(0)|^2 e^{-t/\tau_l} + 2\sum_{l \neq m} |c_l(0)||c_m(0)| e^{-t/2\tau_l - t/2\tau_m - \Gamma^*_{lm} t/\hbar} \cos((E_l - E_m)t/\hbar). \tag{3.20}$$

The oscillation frequencies in the last term depend on the energy differences of the coherently excited states. These oscillations are damped by the lifetimes of the states and the loss of phase coherence described by the dephasing rates Γ^*_{lm}. The first term of Eq. (3.20) describes the average population decay with time constants τ_l. An experimental illustration is given in Figure 3.13 for image potential states on Cu (100) [47, 49]. The oscillatory and exponential contributions can be separated and decay with different time constants. This allows a direct identification of different scattering processes by the analysis of decay and dephasing rates.

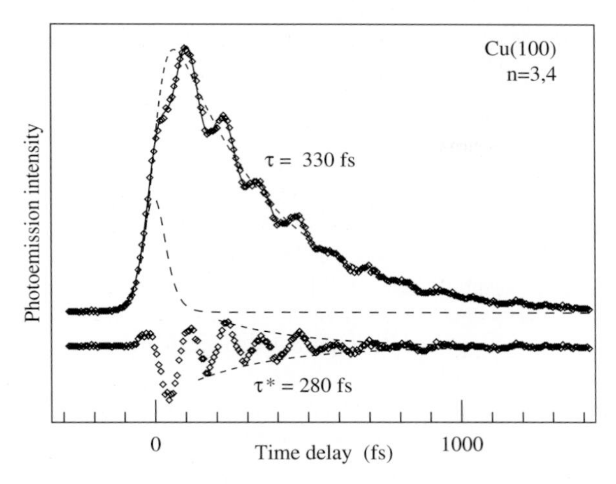

Figure 3.13 Coherent excitation of the $n = 3$ and $n = 4$ image potential states leads to beating patterns as a function of pump–probe delay.

3.4.3
Scanning Tunneling Methods

Scanning tunneling microscopy uses the exponential dependence of the tunneling current on distance and voltage to image the surface with resolution on the atomic scale. In the spectroscopic mode, the voltage is varied at fixed distance and the current is sensitive to the electronic structure of the surface (and tip). The high spatial resolution implies an integration over a range of parallel momenta $k_{||}$. Therefore, it seems impossible to determine energetic linewidths in scanning tunneling spectroscopy. However, the features in any energy spectrum have some intrinsic width and in suitable experiments the measured information may be related to the linewidth as measured with other techniques.

The first method simply records the onset of the tunneling into the Shockley surface states. The corresponding spectra are shown in Figure 3.14 for the (111) surfaces of Ag, Au, and Cu [50]. All spectra were taken at least 200 Å away from impurities and are averages of different single spectra from varying sample locations

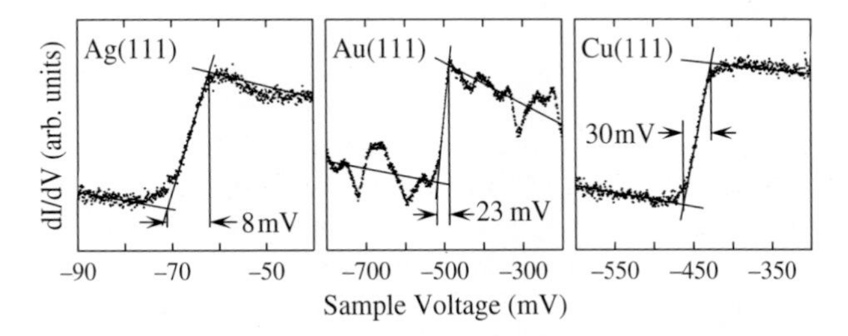

Figure 3.14 dI/dV spectra for the surface states on Ag(111), Au(111), and Cu(111). Reprinted with permission from Ref. [50]. Copyright (2000) by American Association for the Advancement of Science.

and tips. A step-like onset is observed with a material-dependent width Δ. This width is in good approximation proportional to the intrinsic linewidth Γ measured in photoelectron spectroscopy [51]:

$$\Delta = \frac{\pi}{2}\Gamma.$$

The previous method is limited to the onset, that is, to the bottom of the surface state band. In order to select states at higher energies, one can resort to select specific $k_{||}$ values by confining electrons by suitable barriers [52]. This method is closely related to the spatially resolved measurements discussed in the following paragraph and might be disturbed by the confining barriers in particular for small structures.

The most simple structures are steps that are ubiquitous on real surfaces. Figure 3.15 shows spectra at a descending step located at $x = 0$ on a Cu(111) surface. The current or the more pronounced differential conductivity $\mathrm{d}I/\mathrm{d}V$ shows an oscillatory pattern as a function of the distance x from the step. This can be explained by the interference of the electron wave reflected by the step with the original wave at the tip position [53]. The data can be fitted very well by the following function [54]:

$$\frac{\mathrm{d}I}{\mathrm{d}V} \propto 1 - |r| \exp\left(-\frac{x}{L_\phi}\right) J_0(2k_{||}x) \tag{3.21}$$

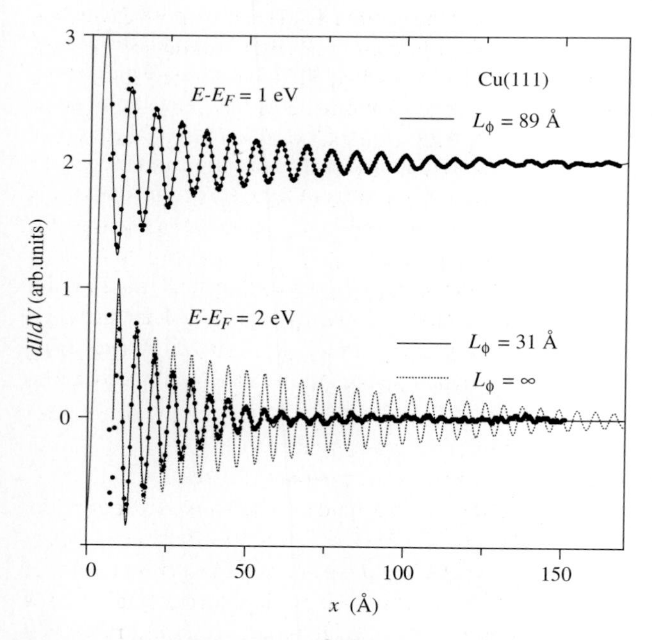

Figure 3.15 $\mathrm{d}I/\mathrm{d}V$ data perpendicular to a descending Cu(111) step obtained by averaging over several line scans of a $\mathrm{d}I/\mathrm{d}V$ image. Solid lines indicate fits. The significance of the deduced L is demonstrated by the dotted line: neglecting inelastic processes by setting $L_\phi \rightarrow \infty$ leads to a much slower decay rate than observed. Reprinted with permission from Ref. [53]. Copyright (1999) by the American Physical Society.

The Bessel function J_0 arises from the summation of all possible scattering paths and the exponential term includes additional damping described by the dephasing length L_ϕ. The arbitrary scaling and offset of the experimental data makes a fit of the reflectivity $|r|$ difficult. The dephasing length L_ϕ can be fitted reliably and shows a strong energy dependence as illustrated in Figure 3.15. The analysis of the data also yields the dispersion relation $E(k_{||})$ for the Shockley surface state in good agreement with results from photoemission or inverse photoemission. Scanning tunneling microscopy or spectroscopy can be performed at positive and negative voltages making both occupied and unoccupied states accessible. The obtained dephasing lengths can be converted to lifetimes $\tau = L_\phi / v_g$ using the group velocity $v_g = \partial E / \partial \hbar k_{||} = \hbar k_{||} / m^*$ for a parabolic band with effective mass m^*.

3.5 Adsorbate-Induced Electronic Structure

3.5.1 Bonding at Surfaces

The energetic location, charge density, and symmetry of adsorbate-induced electronic states determine many surface properties such as the chemical reactivity, the dynamics of electronic and vibrational excitations, the charge transport through interfaces between electronic materials, and so on [55]. For the understanding of the physical and chemical aspects of interfacial phenomena induced by electrons or photons as well as their dynamics, it is thus essential to elucidate the adsorbate–substrate interactions that define the interfacial electronic structure.

In this section, we briefly discuss several types of bonding at surfaces with a focus on the interactions that define the interfacial electronic structure. It is not our intention to provide a comprehensive review, as this can be found in various textbooks and review articles [56–58]. The bonding of atoms or molecules at surfaces can be loosely classified as physisorption, where the adsorption energy is typically less than 0.3 eV, and as chemisorption, with adsorption energies as high as several eV. Four types of bonding can be distinguished (namely, van der Waals, covalent, ionic, and metallic bonding), although these scenarios should be considered as limiting cases, and in reality often a mixture of these interactions contribute [57].

The van der Waals interaction arises from electron density fluctuations and is well defined only as long as the orbitals of the adsorbate and substrate do not overlap [57]. However, at the typical adsorption height of a physisorbed atom or molecule above the surface, this is not the case and the bonding to the surface arises from a balance between the Pauli repulsion and the attractive dispersion force, leading to a redistribution of the charge density and a (static) polarization of the adsorbate [57]. For physisorbed atoms or molecules with high polarizability (like Xenon) adsorbed on a metal surface, this polarization results in a dipole moment, which contributes significantly to the bond strength and also leads to a reduction in the work function (up to 1 eV in the case of xenon [59]). Figure 3.16a shows as an example the charge

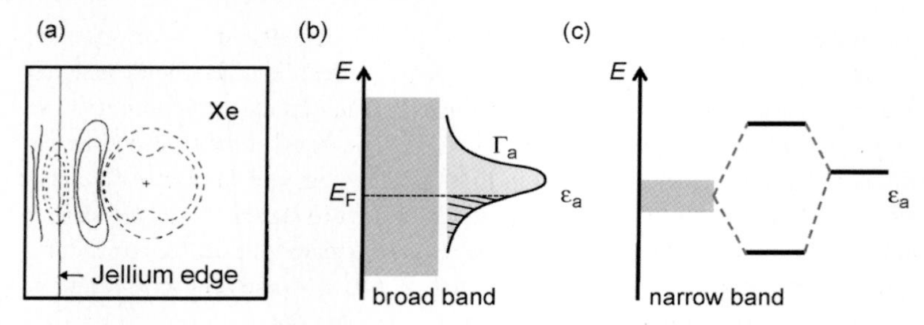

Figure 3.16 Bonding at surfaces. (a) Charge density difference plot for physisorbed xenon at a metal (jellium) surface calculated by DFT within the local density approximation. The resulting surface dipole reduces the work function. Reprinted with permission from Ref. [60]. Copyright (1981) by the American Physical Society. (b) Anderson–Newns model (weak coupling limit): the adsorbate-level ε_a broadens to a resonance due to the coupling with a broad band of substrate states. Depending on the position of the resonance with respect to the Fermi-level E_F and the resulting occupation of the resonance level, this situation corresponds to a case of weak chemisorption for partial occupation or to ionic bonding for complete transfer of one or more electrons. (c) Anderson–Newns model (strong coupling limit): splitting into bonding and anitbonding states for hybridization between a narrow substrate band and the adsorbate-level ε_a leads to strong chemisorption if the bonding (antibonding) state becomes completely occupied (unoccupied).

density difference plot for a Xe atom in front of metal surface calculated by density functional theory [60]. We conclude that in the case of physisorption, the electronic states of the adsorbate are perturbed in a relatively simple and predictable manner because their state occupation is close to the free atom or molecule and no hybridization with the states of the substrate occurs. Note that even with state-of-the-art *ab initio* methods a quantitative description of the van der Waals interaction is still challenging and subject of intense research [61, 62].

In the case of chemisorption, the occupation of the adsorbate electronic states is changed by charge transfer with the substrate and new electronic states are formed at the surface by hybridization with the electronic states of the substrate [58, 63]. Therefore, the bonding character depends strongly on the properties of the involved substrate wave functions and will be quite different for more localized electronic states (e.g., *d*-orbitals in metals or highly directional *sp*-hybrid orbitals in semiconductors) compared to the more delocalized *s*- or *p*-electrons (e.g., in noble metals) [64]. This is illustrated in Figure 3.16b and c depicting two cases for the so-called Anderson–Newns model [63, 64], which will be discussed in more detail in Section 4.1 in the context of charge transfer processes at interfaces. In this model, a single valence level of the adsorbate with energy ε_a interacts with the Bloch states of the substrate. The nature of the new electronic states, which are formed, depends on the interaction strength between the adsorbate and the substrate states, which is described by hopping matrix elements. If the adsorbate level interacts with a broad electronic band of the substrate (e.g., arising from delocalized *s*- or *p*-electrons), the hopping matrix elements will be typically much smaller than the band width. In this case, the hybridization of the adsorbate level with the quasi-continuum of Bloch states leads to the formation of a resonance in the projected density of states, which is centered around ε_a and has a finite width Γ_a (see Figure 3.16b). The width of this resonance determines the (elastic) lifetime $\tau = \hbar/\Gamma_a$ [63]. This scenario is usually referred to as weak chemisorptions case [63, 64]. Thereby, the bonding strength results from the gain of electronic energy by the partial occupation of the adsorbate resonance (i.e., depending on its position with respect to the Fermi energy E_F). This is illustrated in Figure 3.16b and will be explained further in Section 3.5.2 for unoccupied resonances of alkali atoms at metal surfaces, where charge transfer of the alkali s-electron to the metal occurs.

The Anderson–Newns model predicts in the case of strong coupling between a narrow substrate band (e.g., *d*-electrons) and the adsorbate-level ε_a a splitting into bonding and antibonding states (see Figure 3.16c). Strong chemisorption occurs if the lower lying bonding state is fully occupied, whereas the antibonding state becomes unoccupied and is located above the Fermi level. A similar interaction and splitting into bonding and antibonding states can also arise by the interaction with a low lying unoccupied state of the adsorbate and contribute to the chemical bond if the bonding state becomes (at least partially) occupied leading to an energy gain of the system. These new electronic states arising from hybridization between adsorbate and substrate orbitals will exhibit specific symmetries according to the irreducible representations of the surface point group (e.g., with respect to the reflection on a certain mirror plane of the surface). This leads to a classification of bonding and antibonding states according to their symmetry (e.g., π and σ orbitals) [65]. The symmetry character of such adsorbate-induced electronic states can be analyzed experimentally, for example, by polarization-dependent measurements in angle-resolved photoemission or X-ray absorption/emission spectroscopy [66, 67].

A straightforward extension of the Anderson–Newns model considers the interaction of the highest occupied molecular orbital (HOMO) and the lowest unoccupied molecular orbital (LUMO) of the adsorbate with the electronic states of the substrate near the Fermi level. The HOMO and LUMO levels are called the frontier orbitals and contribute substantially to the adsorbate–substrate bonding [63]. In the frequently studied example of carbon monoxide adsorption on transition metals, the frontier orbitals are the (occupied) 5σ and the (unoccupied) $2\pi^*$ states of the free CO molecule. As proposed by Blyholder, the CO metal bond is formed by a "donation"–"backdonation" mechanism, whereby the CO 5σ becomes partially unoccupied by donating charge to the substrate and the CO $2\pi^*$ orbital becomes partially filled [68]. It should be noted that the Blyholder model provides a rather simplified view and has been refined [58]. However, a quantitative treatment of CO metal bonding and, in particular, the prediction of the correct adsorption site has been a challenge for *ab initio* theory and has become feasible fairly recently [69, 70].

In summary, bonding at surfaces can be classified into physisorption and weak and strong chemisorption, whereby relatively simple models may be used if the state occupation of the adsorbate is close to a free atom, molecule, or ion. This is frequently justified for physisorption or weak chemisorptions systems. In contrast, a detailed and quantitative treatment of strong chemisorption is far beyond such intuitively simple pictures like the Anderson–Newns model and requires a description by state-of-art *ab initio* methods. For extended systems, DFT calculations have now reached a high level of accuracy to predict the ground-state electronic structure and the adsorption geometry of chemisorbed molecules on solid surfaces. With recent theoretical advances like time-dependent DFT and many-body Green's-function approaches, even excited state and spectroscopic properties become accessible to *ab initio* calculations. For further reading, the reader may refer to Refs [63, 71–73].

3.5.2
Energy-Level Alignment: Alkali–Metal Interfaces as a Model System

We will now consider the electronic structure at adsorbate–metal interfaces for the case where the charge density of the occupied and unoccupied states near the Fermi level is well localized on the adsorbate. This allows a description through simple concepts, of which alkali atoms adsorbed on noble metals are an illustrative example.

We start our discussion of interfacial electronic structure using the concept of the vacuum-level alignment between the substrate and the adsorbate electronic potentials, also known as the Schottky–Mott (SM) limit (Figure 3.17) [74–77]. The SM limit can correctly describe the physics of the electronic state alignment in certain limiting cases, but care has to be taken how it is applied. Therefore, we proceed cautiously with precise definitions of various interactions that define the electronic properties of surfaces. The SM limit tells us to combine the electronic potentials of substrate and adsorbate at their common vacuum-level E_v. As indicated in Figure 3.17, such alignment allows to relate the energy of the frontier HOMO and LUMO orbitals of the adsorbate to the Fermi-level E_F of the metal substrate based on the knowledge of the adsorbate ionization potential, I, and its electron affinity, A.

To establish the band alignment in the SM limit, we need to define the correct reference vacuum level. An electron at rest in vacuum free from any influence of external fields rigorously defines the vacuum level; this, however, is not our reference level E_v. As shown in Figure 3.18a, to every bare metal surface can be ascribed a surface dipole layer consisting of an evanescent tail of electron density spilling out into the vacuum from the geometric surface boundary, and on the metal, the corresponding region of positive charge. This charge separation at the metal–vacuum interface creates a dipole layer that depends on the crystal orientation.

As shown in Figure 3.18d, the work required to transport an electron from a metal to rest at a distance z from the surface that is much smaller than the lateral extent of the surface L can be separated into two components: the chemical potential, $\bar{\mu}$, which describes the screened Coulomb interaction between the conduction band electrons and the positive ion cores of the metal, and the contribution from $\Delta\varphi$, which describes the work required to transport an electron through the surface dipole layer. The crystal face-dependent work function of a metal, $\Phi = \Delta\varphi + \bar{\mu} = E_v - E_F$, which is measured in photoemission experiments as the zero kinetic energy

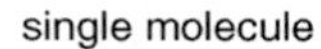

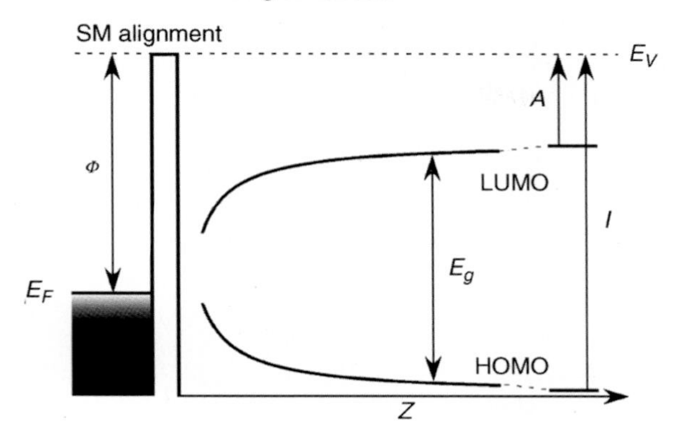

molecular multilayer

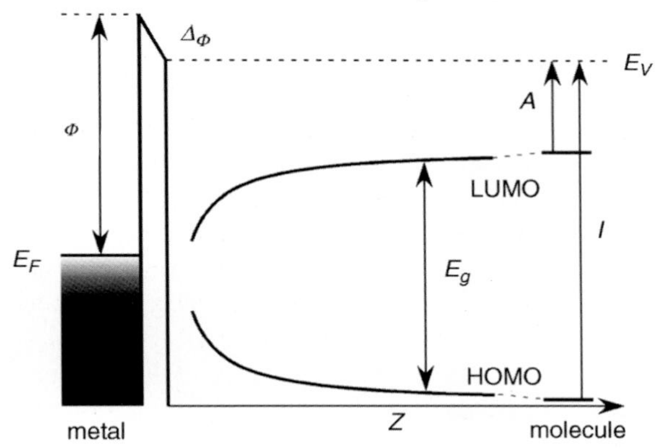

Figure 3.17 The band alignment at the adsorbate–metal interface in the single molecule and monolayer-to-multilayer limits. In the Schottky–Mott (SM) limit the alignment of the frontier orbitals with respect to the Fermi-level E_F is established by matching the vacuum levels E_v. The formation of adsorbate monolayer shifts the vacuum level of the interface by Δ_Φ. In addition, the image–charge interaction shifts the frontier orbitals by $\pm 1/4z$.

threshold for electrons to escape from the surface, combines these surface and bulk contributions. An electron at a molecule–surface distance $z < L$ is under the influence of the surface dipole potential and therefore the zero kinetic energy threshold that defines E_v does not represent the true vacuum level. For macroscopic distances from the surface, that is, $z \gg L$, the dipole potential decays as z^{-2} to the true vacuum potential, which is not generally accessible to photoemission experiments [78]. Because the molecular state alignment has chemical significance only close to a metal surface, in the presence of the surface dipole field, E_v of the bare metal surface (single molecule description in Figure 3.17) is the appropriate reference energy for the SM limit. As we will discuss later and is depicted in Figure 3.18c,

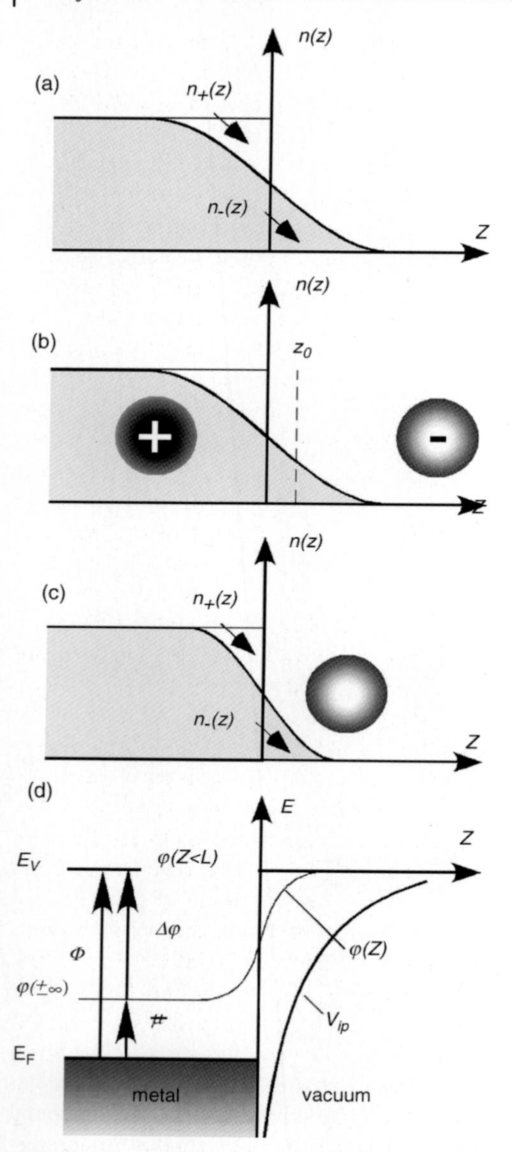

Figure 3.18 Factors that influence the band alignment at adsorbate–metal interfaces. (a) Charge spilling out from the geometric boundary of a metal surface into vacuum, and its depletion in the surface region, creates a crystal orientation-dependent surface dipole. (b) A charged particle (e.g., electron or cation) interacting with a metal creates an image charge; their mutual attraction lowers the energy of the particle by $V_{ip} = -1/4z$, where z is measured from a fictitious image plane, z_0. (c) The modification of the interfacial charge through the cushion effect decreases the surface dipole and, therefore, the work function Φ. (d) The contributions to the work function of a metal for $z < L$, where L is the lateral surface dimension, from the chemical potential $\bar{\mu}$ and the interfacial dipole $\Delta\phi$. The potential from the surface dipole layer $\phi(z)$ is significant in vacuum for $z < L$.

bringing adsorbates into direct contact with the substrate changes the surface dipole layer, thereby causing a change in the surface potential by an amount Δ_Φ; therefore, increasing the adsorbate coverage up to one monolayer defines a new adsorbate coverage-dependent reference energy E_v.

After defining the common reference energy, next we turn to molecule–surface interactions in order of their importance. Although it is too often omitted from discussion, the Coulomb potential established when an external charge polarizes the conduction band electrons in a metal constitutes the primary interaction that defines interfacial electronic structure on the length scales that precede the overlap between the adsorbate and the substrate charge clouds. The screening of the field of the external charge through the many-body response of conduction electrons creates a fictitious image charge of the opposite sign and equal distance from the reference image plane. The attraction between an external electron and its image charge is described by the image potential having the $V_{ip} = -1/4z$ form (measured from the image plane z_0 in atomic units), as is discussed in more detail in Section 3.1. Until the molecule and surface charge clouds begin to overlap, as shown in Figure 3.17, the V_{ip} potential describes the universal shift of an adsorbate ionization (HOMO) and affinity (LUMO) levels when electron is removed from the former and added to the latter. The shift is independent of the substrate and depends only on the adsorbate insofar as its size and shape determine the distance of the centroid of the charge from the image plane. The V_{ip} interaction stabilizes the hole created in the HOMO orbital and electron introduced into LUMO orbital, causing the former to shift upward and the later downward by equal amounts corresponding to $1/4z$, thereby reducing the HOMO–LUMO gap, E_g, by $1/2z$. As can be seen in Figure 3.17, the z^{-1} dependence of the Coulomb potential has a substantial effect even at distances as large as 1 nm, where the V_{ip} shift of the frontier orbitals is ± 0.36 eV. Moreover, at distances corresponding to equilibrium adsorption distance R_{ads} on a metal surface, the interaction can exceed 2 eV and the effect on E_g is doubled [79]. The actual shift of molecular levels at R_{ads} may be different from the value predicted by the V_{ip} potential due to additional interactions. In multilayer structures, however, where the second and subsequent layers do not form chemical bonds with the substrate, the V_{ip} potential defines the layer-to-layer energy shift of HOMO and LUMO. In this case, however, one has to correct for the dielectric screening by the surrounding adsorbate layers, which can be done with the dielectric continuum model [80]. One also has to consider whether or not creating a charged state introduces additional chemical interactions between adsorbates that can be a source of additional stabilization and whether the charge can delocalize into a dispersive band. The dominant effect of the image–charge interaction, and other effects that can influence the frontier orbital energy, have been dissected by Chiang *et al.* for layer-resolved core and valence-level shifts of rare gas overlayers on metal surfaces [80]. There are several other examples where the image–charge interaction has been shown to describe accurately the frontier orbital shifts, for instance, for C_{60} multilayers, by using methods such as photoemission, STM spectroscopy, and X-ray absorption spectroscopy [81, 82].

Next, we consider what happens when the adsorbate and substrate electron clouds overlap. The situation for the case of physisorption and weak chemisorption is depicted in Figure 3.18c. Even if there is no charge transfer, the exchange or Pauli repulsion between the adsorbate and the substrate will distort the interfacial charge distribution and, therefore, change the magnitude of the interfacial dipole. Because DFT calculations cannot describe accurately the van der Waals and polarization contributions to the molecule surface interaction, the "pillow" or "cushion" effect whereby weakly adsorbed molecules cause interfacial charge redistribution is best described by wave function-based approaches or the GW method [79, 83–85].

For pure physisorption, the dominant effect of adsorbate is to compress the evanescent electron density, thereby decreasing $n_{-}(z)$ in the vacuum and $n_{+}(z)$ in the substrate, as suggested by Figure 3.18c. The repulsion between the surface and the molecule charge distributions reduces the magnitude of the interfacial dipole, and consequently the work function. The cushion effect should be dominant for adsorbates such as rare gas atoms and saturated hydrocarbons where the energy separation of the frontier orbitals from E_F precludes substantial charge transfer. Weak chemisorption can either suppress the cushion effect by back donation of charge from the substrate to the adsorbate or enhance it if the charge transfer is in the opposite direction. In the case of weak chemisorption, the cushion effect is likely to be larger than the charge transfer, resulting in a reduced work function [83]. For strongly electronegative adsorbates, such as oxygen, halogens, or organic molecules with large values of A, however, the back donation can overcome the cushion effect to increase the size of the interfacial dipole, thereby raising the work function [2].

Assuming that the changes in the surface dipole moment are additive for each molecule comprising the dipole layer, the change in the work function with the coverage, represented by the adsorbate surface density σ, is given by $\Delta_{\Phi} = 2\pi\sigma\mu$ where μ is the induced dipole moment per adsorbate [86, 87]. At high coverage as the dipoles pack closer together, the relationship between Δ_{Φ} and μ is no longer linear due to dipole–dipole repulsion. The repulsive energy associated with the interaction between surface-aligned dipoles can be reduced, as described by the Topping model [88], through reverse charge transfer between the adsorbate layer and the substrate. The reverse charge transfer leads to the dipole depolarization, which reduces the magnitude of effective μ [86].

The shift in the work function is dominated by the first monolayer of adsorbates because the subsequent monolayers have no chemical interaction with the substrate. Nevertheless, the strong electric field gradient of the V_{ip} potential can polarize molecules that are not directly in contact with the substrate. The contribution of these induced dipoles to the work function change is much smaller than those of the first monolayer.

To summarize the discussion up to this point, the dominant contributions to the band alignment at a metal–molecule interface are the image charge and the surface dipole effects. The surface dipole layer modifies the reference energy for the SM alignment by the coverage and induced dipole-dependent change in the work function, Δ_{Φ}. The magnitude of Δ_{Φ} depends on the nature of the adsorbate- and substrate-dependent interactions such as the cushion effect and chemisorption-induced charge transfer. The change in the work function saturates at 1 monolayer (ML) coverage. By contrast, the image–charge interaction causes the occupied and unoccupied state energy shifts that are independent of the adsorbate and the substrate and depend only on the distance from the image plane, z. The shift of the frontier orbitals due to the image–charge interaction is with respect to E_v modified by the surface dipole layer. The surface dipole shift can be comparable in magnitude to the image–charge interaction, but its sign is determined by the direction of adsorption-induced charge transfer.

So far, we have restricted the discussion of the interfacial electronic structure to the case of physisorption or weak chemisorption; the same concepts, however, can be applied in the case of strong chemisorption when the substrate–adsorbate bonding is dominantly ionic. The chemisorption of alkali atoms on noble metals belongs to this special case, which has been studied systematically by two-photon photoemission for Li through Cs on Cu(111) and Ag(111) surfaces [87, 89, 90]. Representative two-photon photoemission (2PPE) spectra measured consecutively for increasing alkali atom coverage beginning with the bare Cu(111) surface and the scheme for 2PP excitation are shown in Figure 3.19.

For alkali atoms at their chemisorption distance R_{ads}, the image–charge repulsion is sufficiently strong to lift HOMO, that is, the ns valence level, above E_F, making it an unoccupied resonance in the surface density of states. Because of its high energy (2.8–3.0 eV above E_F) and narrow linewidth (<0.3 eV), ns electron is completely transferred to the conduction band of the substrate, leaving the ionic core at its corresponding R_{ads}. The ionic core of alkali atom and its image charge create an induced dipole corresponding to $\mu = 2R_{ads}$, which is much larger than the intrinsic dipole of a metal/vacuum interface (Figure 3.18c). Even though ns resonances of alkali atoms are completely depopulated, their binding energy with respect to E_v is nevertheless determined by the same interactions as in the physisorption case.

The 2PPE spectra in Figure 3.19 represent a photoexcitation process where in the first step absorption of a 3.1 eV photon transfers an electron from Cu(111) substrate to ns resonance of adsorbate, and this intermediate step is detected when absorption of an additional photon induces photoemission. Recording the unoccupied state spectra shows that, surprisingly, ns resonances of alkali atoms have a binding energy with respect to E_v of -2.0 and -1.85 eV for Cu(111) and Ag(111) surfaces, independent of the periodic trends in ionization potentials I and ionic sizes. Moreover, the ns resonances shift with the alkali atom coverage with a $-(\Delta_\Phi)^{3/2}$ dependence, rather than linearly as in the physisorption case, producing the coverage-dependent changes in Figure 3.19 [87].

The shift of the ns resonance to a common binding energy in the zero coverage limit can be fully explained by the image–charge interaction. In the case where HOMO is an unoccupied resonance, the energy of the neutral state corresponding to an electron transiently occupying HOMO and centered on the ionic core at R_{ads} has two contributions, as illustrated in Figure 3.20: the first one is the image–charge attraction between electron with its own hole, that is, $V_{ip} = -1/4R_{ads}$, and the second one is a repulsive interaction $V_\Delta = 1/2R_{ads}$ with the negative image charge of the

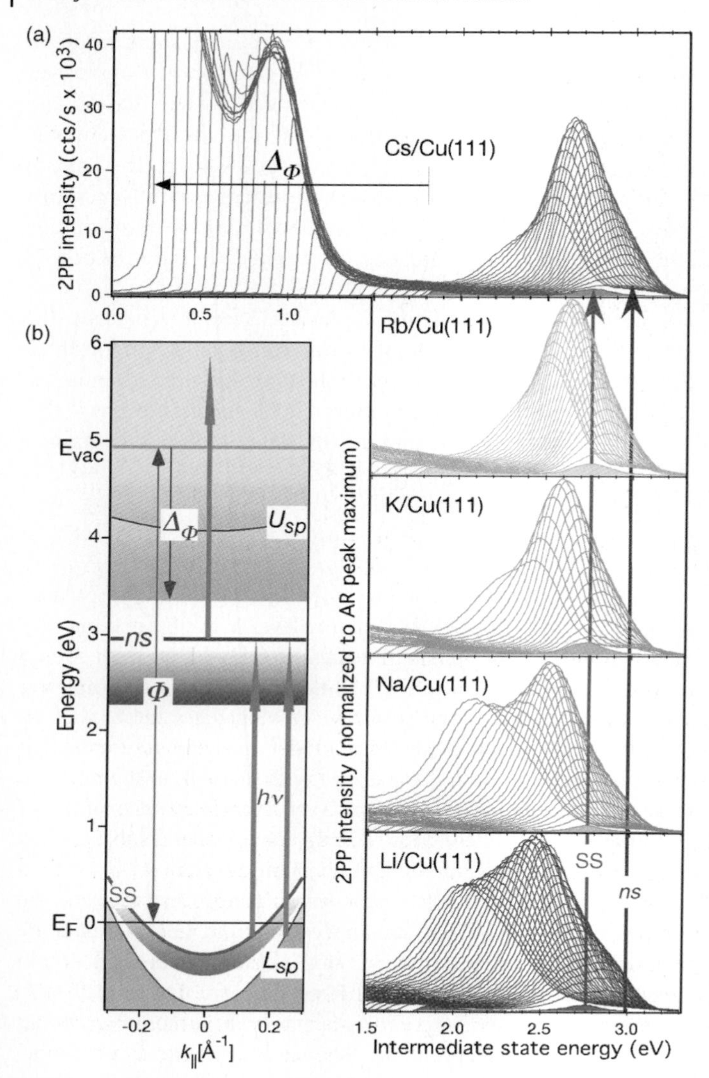

Figure 3.19 The (a) 2PPE spectra and (b) 2PP excitation scheme for Cu(111) surface during a continuous deposition of alkali atoms. The binding energy of the unoccupied *ns* resonance of alkali atoms is determined by the image–charge interaction. The formation of a surface dipole layer reduces the vacuum level by Δ_Φ and the binding energy of *ns* resonance by $-(\Delta_\Phi)^{3/2}$, as explained in Ref. [87]. Reprinted with permission from Ref. [87]. Copyright (2008) by the American Physical Society. (Please find a color version of this figure on the color plates.)

ionic core. The total image–charge interaction corresponds to a net repulsive potential $\Delta E = V_\Delta + V_{ip} = +1/4R_{ads}$. The binding energy of *ns* electron E_b with respect to E_v is then given with near quantitative accuracy by the joint contribution from the attractive potential of the ionic core I and the image–charge repulsion, that

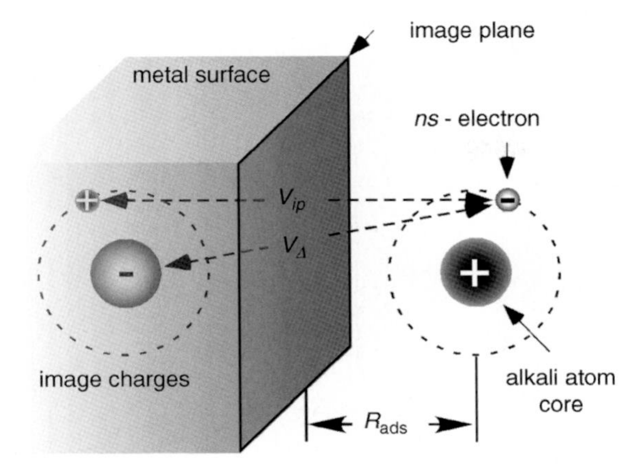

Figure 3.20 The image–charge interactions that define the binding energies of *ns* resonances of alkali atoms on noble metals. The repulsive interaction V_Δ between the image charge of the core and the valence electron destabilizes the *ns* state with respect to E_F. For charge introduced into the frontier orbitals of a neutral atom, only the attractive V_{ip} needs to be considered.

is, $E_b = 1/4R_{ads} - I$. As shown in Figure 3.21, the alkali period-independent E_b is a consequence of anticorrelation between the ionic size, which defines R_{ads}, and the ionization potential. How strongly an alkali ionic core can bind its valence electron and how close it can approach a metal surface are both determined by the interplay between the screened Coulomb attraction and the Pauli repulsion [87, 91]. As shown in Figure 3.21b, the attractive potential between the ionic core and *ns* valence electron and the net image–charge repulsion have nearly the same dependence on R_{ads}, but with an opposite sign, to give the period-independent binding energy. The total effective potential for electron at an alkali atom-covered Cu(111) surface can be obtained by combining the potential for the substrate, which can be represented by the 1D Chulkov potential [92], the pseudopotential for alkali atom, and the additional image–charge interactions of the combined system, as shown in Figure 3.22 [87]. Solving for the stationary states of the system using a wave packet propagation method gives good agreement with the binding energies obtained with the simple approach as described above [87]. It also explains the slightly larger alkali *ns* resonance binding energies for Cu(111) as compared to Ag(111), which arise from weaker repulsion by the conduction band electrons of the substrate within the *L*-projected band gaps of Cu and Ag [84, 86].

The final contribution to *ns* resonance energy to consider is from the dipole layer that forms at finite coverage. In the case of physisorption, the interaction between the dipole at the geometric boundary of metal substrate with a spatially separate charge on a physisorbed adsorbate is not appreciable [80]. In case of ionic chemisorption, the centroid of *ns* resonance electron distribution spatially overlaps with the positive end of the surface dipole formed by alkali atom and its image charge. The potential experienced by the electron interacting with the surface dipole layer, minus

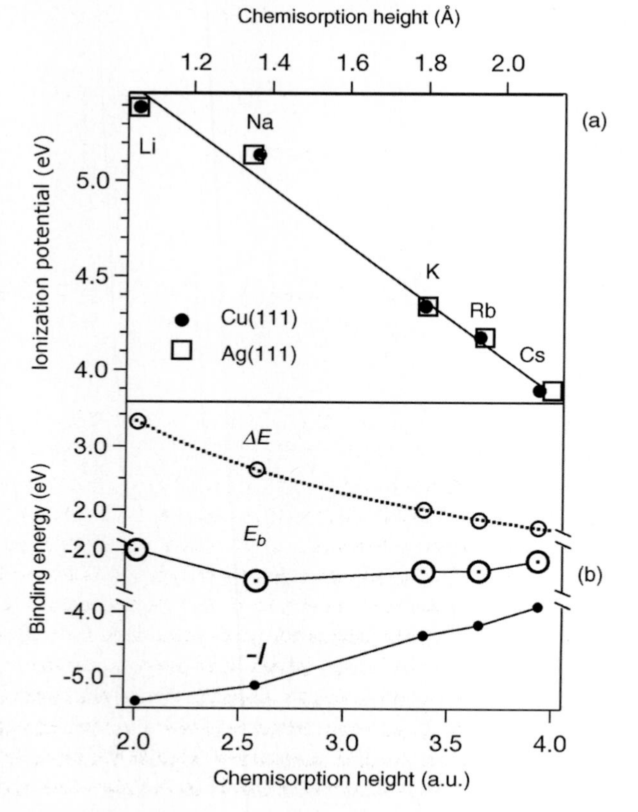

Figure 3.21 The factors that determine the period-independent electronic structure of alkali atoms on noble metals: (a) The anticorrelation between the ionization potential of free alkali atoms and the adsorption height R_{ads} of alkali atoms on Cu(111) and Ag(111) surfaces, which is a consequence of common Coulomb attraction and Pauli exclusion forces. (b) The alkali atom *ns* resonance binding energy E_b, which is given in the most rudimentary terms by the difference between the attractive potential between alkali atom and its *ns* electron, that is, the ionization potential, I, and the image charge repulsion $\Delta E = 1/4z$ (dotted line) evaluated at R_{ads} obtained from density functional theory calculations. Reprinted with permission from Ref. [87]. Copyright (2008) by the American Physical Society.

a disk encompassing a single adsorbate that avoids double counting, is given by $V_\mu = -(\Delta_\Phi)^{3/2}(2\mu)^{-1/2}z$ [93]. Indeed, in two-photon photoemission spectra of alkali atom-covered Cu(111) and Ag(111) surfaces, the *ns* resonance energy is seen to follow the $-(\Delta_\Phi)^{3/2}$ dependence with increasing coverage.

To summarize, for situations where the charge density of an occupied and unoccupied state is well localized on the adsorbate, that is, the electronic state lacks covalent character, the interfacial electronic structure can be described through simple concepts including the SM vacuum-level alignment, vacuum-level shift through the surface dipole formation, the screening of charge in the frontier orbitals,

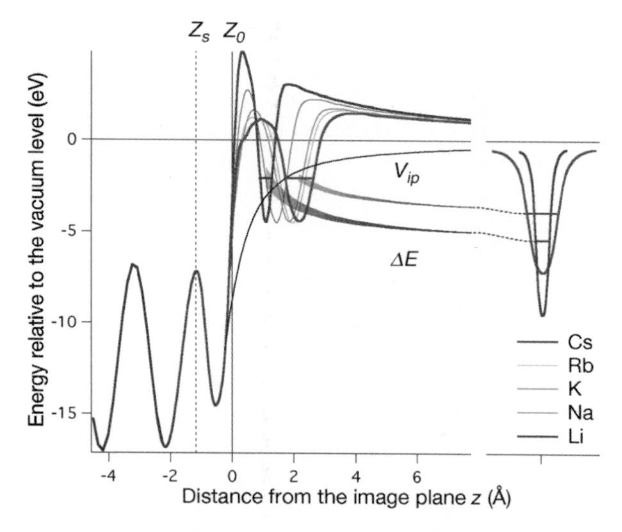

Figure 3.22 The effective electron potentials for single alkali atom (Li through Cs) on Cu(111) surface. The total potential combines the Chulkov potential for Cu(111), which includes the image potential V_{ip}, the pseudopotential for free alkali [92] atoms, and the main mode of interaction through the image–charge repulsion ΔE [87]. Both the V_{ip} and ΔE potential have the universal form expected for the Coulomb interaction. The adsorption height is measured from the image plane z_0, while z_s represents the geometrical boundary of the surface. Reprinted with permission from Ref. [87]. Copyright (2008) by the American Physical Society. (Please find a color version of this figure on the color plates.)

and the charge–dipole layer interaction. While the image–charge interaction has universal properties and can be calculated simply from the distance of the centroid of charge from the image plane, the full charge redistribution upon physisorption or chemisorption is much more difficult to obtain from first-principles calculations. Although the same physical interactions contribute to the electronic structure in the case of strongly covalent bonding, a full description requires the proper description of all interactions and correlation effects.

3.5.3
Electronic Band Structure: Chemisorbed and Physisorbed Adsorbates

Motivated by its relevance to catalysis, the electronic structure of adsorbates is frequently discussed in the context of surface reactions and the reactivity of the so-called "active sites" [55, 71]. Often, the *local* electronic structure of specific adsorption sites (like defects or steps) plays a key role, which can differ significantly from the electronic structure of extended, perfectly periodic systems [94]. Such local structures can be probed by dedicated microscopy or spectroscopic techniques (like X-ray spectroscopy [67]), which are sensitive to the local electronic environment. Adding ultrafast time-resolution to such techniques to study the dynamics of catalytic processes is very challenging and an active field of research in X-ray free electron laser science.

On the other hand, extended *periodic* structures, namely, well-ordered physisorbed or chemisorbed adsorbate overlayers on single-crystal metal surfaces, may serve as model systems for interfacial charge transfer and investigation of ultrafast dynamics at interfaces [76, 78]. In physisorbed and weakly chemisorbed systems, the adsorbate substrate interaction is weak and thus the adsorbate-induced band structure is derived predominantly from the superposition of the occupied and unoccupied molecular orbitals of the adsorbate layer. The width and dispersion of the bands is thereby determined by the adsorbate–adsorbate interactions (i.e., the overlap between nearest-neighbor orbitals in a tight binding picture). An excellent introduction to the formation of (adsorbate) band structures and bonding at surfaces can be found in Ref. [65].

In the following paragraphs, we briefly discuss the electronic band structure of adsorbates. For a comprehensive overview on band structure determination using angle-resolved photoemission spectroscopy (ARPES), the reader is referred to Ref. [66], where various examples for chemisorbed and physisorbed ordered overlayers can be found. A nice example for the formation of an adsorbate-induced band structure is the saturated benzene layer on Ni(110), which forms a well-ordered chemisorbed overlayer with a $c(4 \times 2)$ LEED structure [95]. ARPES measurements as function of light polarization and polar and azimuthal emission angle are consistent with an adsorption geometry parallel to the surface that maximizes the interaction of the π-electron system with the substrate [96]. Experimental evidence for strong lateral interactions within the densely packed benzene layer arises from the observation of a two-dimensional adsorbate band structure that exhibits a pronounced dispersion for the occupied $2a_{1g}$ state of benzene located ~11 eV below E_F (Figure 3.23b). The dispersion of this band can be followed by ARPES over several adsorbate-derived surface Brillouin zones nicely demonstrating the periodicity of the adlayer in real and reciprocal space (Figure 3.23a). The large band width of 0.8 eV, which is observed both in the [1–10] and in the [001] directions, appears first surprising as all other benzene-derived bands show no significant dispersion [95]. However, the intermolecular interactions between adsorbed planar hydrocarbon molecules are dominated by hydrogen–hydrogen interactions and will thus be largest for molecular orbitals with a maximum probability density at the hydrogen positions [66]. Inspection of the benzene orbitals shows that this is fulfilled in particular for the $2a_{1g}$ state of benzene (see inset in Figure 3.23b), whereas most other molecular orbitals exhibit an enhanced probability density at the carbon ring [66]. The band formation can be thus understood in a tight binding picture for a σ-type band with highest binding energy at the Γ point and upward dispersion toward the Brillouin zone boundary. Figure 3.23c displays schematically the Bloch wave function of the $2a_{1g}$ state for two characteristic cases: at the Brillouin zone center ($k_{||} = 0$), the Bloch wave function of the adsorbate band is a superposition with all $2a_{1g}$ orbitals in phase, whereas at the zone boundary ($k_{||} = \pi/b$), the wave function is a superposition with two neighboring $2a_{1g}$ orbitals in phase and four neighboring orbitals out of phase (see Figure 3.23c). Note that very similar observations have been made for the $(\sqrt{7} \times \sqrt{7})R19.1°$ structures of benzene on Ni(111) and Os(0001), where again the $2a_{1g}$ band exhibits a pronounced dispersion with a band width of 0.4–0.5 eV. The different magnitudes

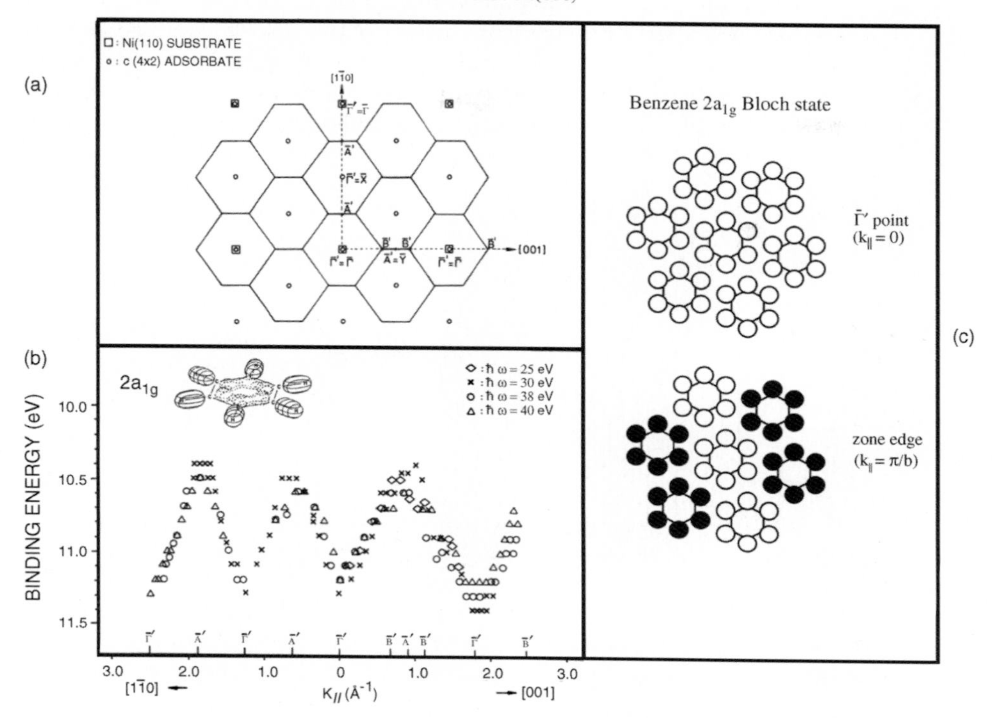

Figure 3.23 Adsorbate-induced band structure of the saturated $c(4 \times 2)$ layer of benzene on Ni(110): (a) Surface Brillouin zone of Ni(110) and adsorbate Brillouin zone for the c (4×2) structure of benzene/Ni(110). (b) Two-dimensional adsorbate bandstructure of the $2a_{1g}$ state obtained by angle-resolved photoemission spectroscopy. (c) Schematic representation of the Bloch wave function of the $2a_{1g}$ derived band at the Brillouin zone center (all $2a_{1g}$ orbitals are in phase) and at the zone boundary (two neighboring $2a_{1g}$ orbitals are in phase and four neighboring orbitals are out of phase). Reprinted with permission from Ref. [96]. Copyright (1999) by the Institute of Physics Publishing.

of the dispersion arises from the different sizes of the unit cell; the ~7% denser packing for the $c(4 \times 2)$ structure on Ni(110) increases the lateral interactions resulting in a stronger dispersion (0.8 eV). Another example for adsorbate band structures, which arise predominatly from lateral molecule–molecule interactions, are ordered ethylene overlayers on Ni(111) [96, 97], where the observed band structure can be nicely explained by negelecting the interaction with the substrate for all bands except for the $1b_{2u}$ HOMO state. The $1b_{2u}$ orbital acts as the frontier orbital in the chemical bonding with the surface and hybridizes with the Ni d-states. Note that in general, the simple tight binding picture for purely lateral interacting molecular orbitals will break down for states that are strongly hybridized on the substrate.

Finally, we discuss the influence of physisorbed rare gas overlayers on the electronic structure of surface and image potential states, which have been introduced in Section 3.3. Rare gas overlayers can be grown on noble metal surfaces with properties close to bulk rare gas crystals (band gap of 10–22 eV [98]) and they can act as atomically thin insulating layers, which decouple the wave function of image potential states from the metal [99]. The electron affinity of the rare gas layer (i.e., the energetic position of the conduction band minimum with respect to the vacuum

(a)

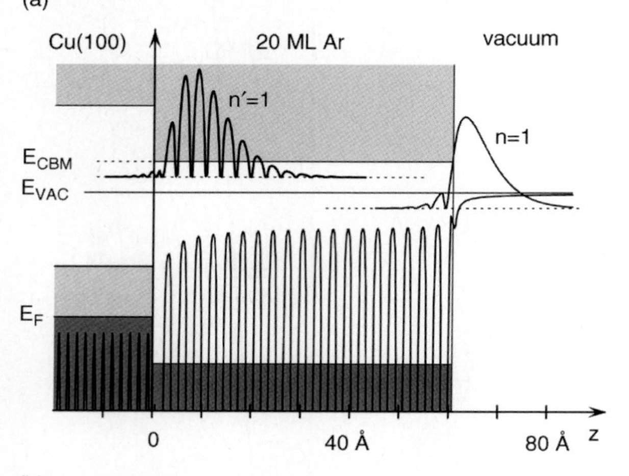

(b)

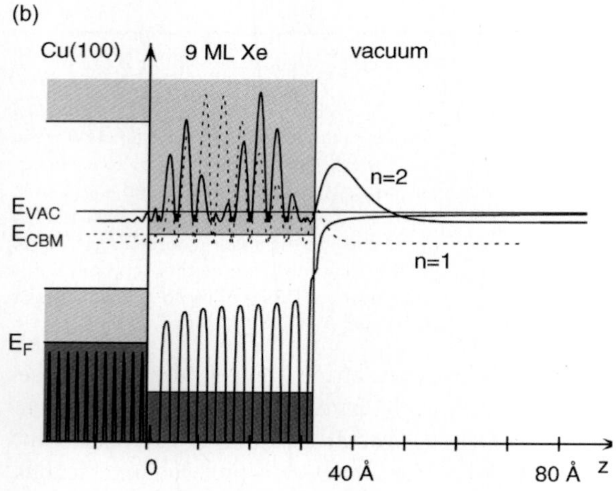

Figure 3.24 Schematic energy diagram for (a) a 20 ML Argon film and (b) a 9 ML Xenon film on Cu(100) with the corrugated effective potential for a single electron inside the Cu substrate and the Ar/Xe film, respectively, together with the image potential in the vacuum. Occupied and unoccupied electronic bands are indicated as dark- and light-shaded areas. (a) For Ar/Cu(100) the $n = 1$ image potential state is located at the rare gas/vacuum interface. In addition, an $n' = 1$ interface state is formed at the metal/argon interface due to the negative electron affinity of Ar. Reprinted with permission from Ref. [102]. Copyright (2005) by the American Physical Society. (b) The positive electron affinity of Xe/Cu(100) leads to the formation of a series of quantum well states located mainly inside the Xe adlayer. Reprinted with permission from [Güdde, J., private communication.].

level) plays a key role in defining the spacial distribution of probability density of surface and image states in the vicinity of the vaccum level (see below). To a first approximation, the modification of the image potential by the insulating layer can be described by treating the adlayer as a sheet of dielectric contiuum [100]. In this dielectric contiuum model, both the metal/dielectric and the dielectric/vacuum interface act as an image plane, whereby the affinity level serves as the energy reference inside the adlayer. A more realistic and quantitative description employs an atomically corrugated one-dimensional potential within the adlayer [101].

Figure 3.24a–c displays in a schematic energy diagram the electronic structure for rare gas multilayers on a Cu(100) surface together with such a corrugated effective potential for a single electron inside the metal, the rare gas layer, and in the vacuum, respectively. Also indicated is the energetic position of the projected band gap of the Cu(100) surface and the band edge of the respective rare gas layer. As argon has a negative electron affinity, the image potential states are located at the rare gas/vacuum interface and their wave functions decay strongly inside the band gap of the insulator. In addition, a series of quasi-bound interface states (or interface resonances) are formed at the metal/rare gas interface, which lie *above* the vacuum level and can therefore decay both elastically by electron transmission into the vacuum and inelastically by e–h pair excitation into the substrate [99]. Figure 3.24a shows for a 20 ML Argon film on Cu(100) the probability density of the $n=1$ image potential and the $n'=1$ interface state, located at the rare gas/vacuum and metal/rare gas interface, respectively [102]. As xenon has a positive electron affinity of 0.5 eV, the $n=2$ and higher image potential states are always located inside the Xe conduction band (whereas the $n=1$ state is located near band edge). This allows a large penetration of the image state wave function and leads to the formation of quantum well states inside the Xe adlayer. Figure 3.24b illustrates this for the $n=1$ and $n=2$ image potential states for 9 ML Xe/Cu(100).

In summary, for well-ordered physisorbed and weakly chemisorbed adsorbate overlayers, the adsorbate electronic band structure can be described by simple concepts leading to electronic bands derived from the occupied and unoccupied molecular orbitals of the adsorbate wereby the dispersion is determined predominantly by the adsorbate–adsorbate interactions. The electronic structure of surface states and image potential states is modified by rare gas overlyers (and other weakly bound adsorbates like organic molecules) whereby the position of the rare gas conduction band plays a decisive role in the binding energies and spatial extension of the surface state wave function.

References

1 Echenique, P.M. and Pendry, J.B. (1978) *J. Phys. C*, **11**, 2065.
2 Fauster, Th. (1994) *Appl. Phys. A*, **59**, 639.
3 Smith, N.V. (1985) *Phys. Rev. B*, **32**, 3549.
4 Tamm, I.E. (1932) *Z. Phys.*, **76**, 849;(1932) *Phys. Z. Sowjet*, **1**, 733.
5 Maue, A. (1935) *Z. Phys.*, **94**, 717.
6 Shockley, W. (1939) *Phys. Rev.*, **56**, 317.
7 Forstmann, F. (1970) *Z. Phys.*, **235**, 69.
8 Pendry, J. and Gurman, S. (1975) *Surf. Sci.*, **49**, 87.
9 Fauster, T. and Steinmann, W. (1995) in *Photonic Probes of Surfaces, Electromagnetic Waves: Recent*

Developments in Research (ed. P. Halevi), Vol. 2, North-Holland, Amsterdam, p. 347, Chap. 8.

10 Baldacchini, C., Chiodo, L., Allegretti, F., Mariani, C., Betti, M.G., Monachesi, P., and Del Sole, R. (2003) *Phys. Rev. B*, **68**, 195109.

11 Weinelt, M., Kutschera, M., Schmidt, R., Orth, Ch., Fauster, T., and Rohlfing, M. (2005) *Appl. Phys. A*, **80**, 995.

12 Chulkov, E.V., Silkin, V.M., and Echenique, P.M. (1997) *Surf. Sci.*, **391**, L1217;(1999) *Surf. Sci.*, **437**, 330.

13 Goodwin, E.T. (1939) *Proc. Cambridge Philos. Soc.*, **35**, 205.

14 Echenique, P.M. and Pendry, J.B. (1978) *J. Phys. C*, **11**, 2065.

15 McRae, E.G. (1979) *J. Vac. Sci. Technol.*, **16**, 654.

16 Echenique, P.M. and Pendry, J.B. (1990) *Prog. Surf. Sci.*, **32**, 111.

17 Dose, V., Altmann, W., Goldmann, A., Kolac, U., and Rogozik, J. (1984) *Phys. Rev. Lett.*, **52**, 1919.

18 Straub, D. and Himpsel, F.J. (1984) *Phys. Rev. Lett.*, **52**, 1922.

19 Straub, D. and Himpsel, F.J. (1986) *Phys. Rev. B*, **33**, 2256.

20 Giesen, K., Hage, F., Himpsel, F.J., Riess, H.J., and Steinmann, W. (1985) *Phys. Rev. Lett.*, **55**, 300.

21 Harris, C.B., Ge, N.-H., Lingle, R.L., Jr., McNeill, J.D., and Wong, C.M. (1997) *Annu. Rev. Phys. Chem.*, **48**, 711.

22 Schmidt, A.B., Pickel, M., Wiemhöfer, M., Donath, M., and Weinelt, M. (2005) *Phys. Rev. Lett.*, **95**, 107402.

23 Fauster, T. (2010) in *Dynamics at Solid State Surfaces and Interfaces* (eds U. Bovensiepen, H. Petek, and M. Wolf), vol. 1, Wiley-VCH Verlag GmbH, Weinheim, p. 53, Chap. 3.

24 Smith, N.V., Brookes, N.B., Chang, Y., and Johnson, P.D. (1994) *Phys. Rev. B*, **49**, 332.

25 Fischer, R. and Fauster, Th. (1995) *Phys. Rev. B*, **51**, 7112.

26 Echenique, P.M., Berndt, R., Chulkov, E.V., Fauster, T., Goldmann, A., and Höfer, U. (2004) *Surf. Sci. Rep.*, **52**, 219.

27 Reinert, F., Nicolay, G., Schmidt, S., Ehm, D., and Hüfner, S. (2001) *Phys. Rev. B*, **63**, 115415.

28 Purdie, D., Hengsberger, M., Garnier, M., and Baer, Y. (1998) *Surf. Sci.*, **407**, L671.

29 Smith, N.V., Thiry, P., and Petroff, Y. (1993) *Phys. Rev. B*, **47**, 15476.

30 Hansen, E.D., Miller, T., and Chiang, T.-C. (1998) *Phys. Rev. Lett.*, **80**, 1766.

31 LaShell, S., Jensen, E., and Balasubramanian, T. (2000) *Phys. Rev. B*, **61**, 2371.

32 Kentsch, C., Kutschera, M., Weinelt, M., and Fauster, Th., and Rohlfing, M. (2002) *Phys. Rev. B*, **65**, 035323.

33 Hao, Z., Dadap, J.I., Knox, K.R., Yilmaz, M.B., Zaki, N., Johnson, P.D., and Osgood, R.M. (2010) *Phys. Rev. Lett.*, **105**, 017602.

34 Pawlik, S., Burgermeister, R., Bauer, M., and Aeschlimann, M. (1998) *Surf. Sci.*, **402–404**, 556.

35 Schattke, W. and Van Hove, M.A. (2003) *Solid-State Photoemission and Related Methods*, Wiley-VCH Verlag GmbH, Weinheim, Berlin.

36 Wallauer, W. and Fauster, T. (1997) *Surf. Sci.*, **374**, 44.

37 Steinmann, W. and Fauster, T. (1995) in *Laser Spectroscopy and Photochemistry on Metal Surfaces* (eds H.L. Dai and W. Ho), World Scientific, Singapore, p. 184, Chap. 5.

38 Wolf, M., Hotzel, A., Knoesel, E., and Velic, D. (1999) *Phys. Rev. B*, **59**, 5926.

39 Giesen, K., Hage, F., Himpsel, F.J., Riess, H.J., and Steinmann, W. (1986) *Phys. Rev. B*, **33**, 5241.

40 Boger, K., Weinelt, M., and Fauster, Th. (2004) *Phys. Rev. Lett.*, **92**, 126803.

41 Fischer, R., Schuppler, S., Fischer, N., Fauster, T., and Steinmann, W. (1993) *Phys. Rev. Lett.*, **70**, 654.

42 Schäfer, A., Shumay, I.L., Wiets, M., Weinelt, M., Fauster, T., Chulkov, E.V., Silkin, V.M., and Echenique, P.M. (2000) *Phys. Rev. B*, **61**, 13159.

43 Fauster, T., Weinelt, M., and Höfer, U. (2007) *Prog. Surf. Sci.*, **82**, 224.

44 Blum, K. (1983) *Density Matrix Theory and Applications*, Plenum, New York.

45 Loudon, R. (1983) *The Quantum Theory of Light*, Oxford University Press, New York.

46 Boger, K., Roth, M., Weinelt, M., Fauster, T., and Reinhard, P.-G. (2002) *Phys. Rev. B*, **65**, 075104.

47 Höfer, U., Shumay, I.L., Reuß, C., Thomann, U., Wallauer, W., and Fauster, T. (1997) *Science*, **277**, 1480.

48 Klamroth, T., Saalfrank, P., and Höfer, U. (2001) *Phys. Rev. B*, **64**, 035420.

49 Reuß, C., Shumay, I.L., Thomann, U., Kutschera, M., Weinelt, M., Fauster, T., and Höfer, U. (1999) *Phys. Rev. Lett.*, **82**, 153.

50 Kliewer, J., Berndt, R., Chulkov, E.V., Silkin, V.M., Echenique, P.M., and Crampin, S. (2000) *Science*, **288**, 1399.

51 Li, J., Schneider, W.-D., Berndt, R., Bryant, O.R., and Crampin, S. (1998) *Phys. Rev. Lett.*, **81**, 4464.

52 Kliewer, J., Berndt, R., and Crampin, S. (2001) *New J. Phys.*, **3**, 22.

53 Bürgi, L., Jeandupeux, O., Brune, H., and Kern, K. (1999) *Phys. Rev. Lett.*, **82**, 4516.

54 Crampin, S., Kröger, J., Jensen, H., and Berndt, R. (2005) *Phys. Rev. Lett.*, **95**, 029701.

55 Ertl, G. (2009) *Reactions at Solid Surfaces*, John Wiley & Sons, Inc., New York.

56 Kolasinski, K.W. (2002) *Surface Science: Foundations of Catalysis and Nanoscience*, John Wiley & Sons, Inc., New York.

57 Scheffler, M. and Stampfl, C. (2000) in *Handbook of Surface Science: Vol. 2: Electronic Structure* (eds K. Horn and M. Scheffler), Elsevier, Amsterdam, p. 285, Chap. 5.

58 Nilsson, A. and Pettersson, L.G.M. (2008) in *Chemical Bonding at Surfaces and Interfaces* (eds A. Nilsson, L.G.M. Pettersson, and J.K. Norskov), Elsevier, Amsterdam, p. 57, Chap. 2.

59 Chen, Y.C., Cunningham, J.E., and Flynn, C.P. (1984) *Phys. Rev. B*, **30**, 7317.

60 Lang, N.D. (1981) *Phys. Rev. Lett.*, **46**, 842.

61 Tatchenko, A. and Scheffler, M. (2009) *Phys. Rev. Lett.*, **102**, 073005.

62 Klimes, J., Bowler, D.R., and Michaelides, A. (2011) *Phys. Rev. B*, **83**, 195131.

63 Gross, A. (2003) *Theoretical Surface Science: A Microscopic Perspective*, Springer, Berlin.

64 Hammer, B. and Norskov, B.K. (2000) *Adv. Catalysis*, **45**, 71.

65 Hoffmann, R. (1988) *Rev. Mod. Phys.*, **60**, 601.

66 Freund, H.J. and Kuhlenbeck, H. (1995) in *Applications of Synchrotron Radiation, Springer Series in Surface Science 35* (ed. W. Eberhardt), Springer, Berlin, p. 9–63.

67 Nilsson, A. and Pettersson, L.G.M. (2004) *Surf. Sci. Rep.*, **55**, 49.

68 Blyholder, G. (1964) *J. Chem. Phys.*, **68**, 2772.

69 Feibelman, P.J., Hammer, B., Norskov,, J.K., Wagner, F., Scheffler, M., Stumpf, R., Watwe, R., and Dumesic, J. (2001) *J. Phys. Chem. B*, **105**, 4018.

70 Qing-Miao, H., Reuter, K., and Scheffler, M. (2007) *Phys. Rev. Lett.*, **98**, 176103.

71 Greeley, J., Nørskov, J.K., and Mavrikakis, M. (2002) *Annu. Rev. Phys. Chem.*, **53**, 319.

72 Onida, G., Reining, L., and Rubio, A. (2002) *Rev. Mod. Phys.*, **74**, 601.

73 Marques, M.A.L. and Gross, E.K.U. (2004) *Annu. Rev. Phys. Chem.*, **55**, 427.

74 Ishii, H., Sugiyama, K., Ito, E., and Seki, K. (1999) *Adv. Mater.*, **11**, 605–625.

75 Braun, S., Salaneck, W.R., and Fahlman, M. (2009) *Adv. Mater.*, **21**, 1450–1472.

76 Koch, N. (2008) *J. Phys. Cond. Matter*, **20**, 184008.

77 Cahen, D., Kahn, A., and Umbach, E. (2005) *Mater. Today*, **8**, 32–41.

78 Zhu, X.Y. (2004) *Surf. Sci. Rep.*, **56**, 1–82.

79 Neaton, J.B., Hybertsen, M.S., and Louie, S.G. (2006) *Phys. Rev. Lett.*, **97**, 216405–216404.

80 Chiang, T.C., Kaindl, G., and Mandel, T. (1986) *Phys. Rev. B*, **33**, 695.

81 Maxwell, A.J., Brühwiler, P.A., Nilsson, A., Mårtensson, N., and Rudolf, P. (1994) *Phys. Rev. B*, **49**, 10717.

82 Fernandez Torrente, I., Franke, K., and Pascual, J.I. (2008) *J. Phys. Condens. Matter*, **20**, 184001.

83 Witte, G., Lukas, S., Bagus, P.S., and Wöll, C. (2005) *Appl. Phys. Lett.*, **87**, 263502–263503.

84 Bagus, P.S., Hermann, K., and Wöll, C. (2005) *J. Chem. Phys.*, **123**, 184109.

85 Thygesen, K.S. and Rubio, A. (2009) *Phys. Rev. Lett.*, **102**, 046802–046804.

86 Schmidt, L.D. and Gomer, R. (1966) *J. Chem. Phys.*, **45**, 1605–1623.

87 Zhao, J., Pontius, N., Winkelmann, A., Sametoglu, V., Kubo, A., Borisov, A.G., Sanchez-Portal, D., Silkin, V.M., Chulkov, E.V., Echenique, P.M., and Petek, H. (2008) *Phys. Rev. B*, **78**, 085419–085427.

88 Topping, J. (1927) *J. Proc. R. Soc. Lond. Ser. A*, **114**, 67–72.

89 Borisov, A.G., Sametoglu, V., Winkelmann, A., Kubo, A., Pontius, N., Zhao, J., Silkin, V.M., Gauyacq, J.P., Chulkov, E.V., Echenique, P.M., and Petek, H. (2008) *Phys. Rev. Lett.*, **101**, 266801–266804.

90 Wang, L.M., Sametoglu, V., Winkelmann, A., Zhao, J., and Petek, H. (2011) *J. Phys. Chem. A*, **115**, 9479.

91 Gadzuk, J.W. (2009) *Phys. Rev. B*, **79**, 073411–073413.

92 Chulkov, E.V., Silkin, V.M., and Echenique, P.M. (1999) *Surf. Sci.*, **437**, 330–352.

93 Borisov, A.G., Kazansky, A.K., and Gauyacq, J.P. (1999) *Surf. Sci.*, **430**, 165–175.

94 Freund, H.J. and Pacchionie, G. (2008) *Chem. Soc. Rev.*, **37**, 2224.

95 Huber, W., Weinelt, M., Zebisch, P., and Steinrück, H.P. (1991) *Surf. Sci.*, **253**, 72.

96 Steinrück, H.P. (1996) *J. Phys. Cond. Matter*, **8**, 6465.

97 Weinelt, M., Huber, W., Zebisch, P., Steinrück, H.-P., Reichert, B., Birkenheuer, U., and Rösch, N. (1992) *Phys. Rev. B*, **46**, 1675.

98 Schwentner, N., Koch, E.E., and Jortner, J. (1985) *Electronic Excitations in Condensed Rare Gases*, Springer, Berlin.

99 Güdde, J. and Höfer, U. (2005) *Prog. Surf. Sci.*, **80**, 49.

100 Harris, C.B., Ge, N.H., Lingle, R.L., McNeill, J.D., and Wong, C.M. (1997) *Annu. Rev. Phys. Chem.*, **48**, 711.

101 Machado, M., Berthold, W., Höfer, U., Chulkov, E.V., and Echenique, P.M. (2004) *Surf. Sci.*, **564**, 87.

102 Rohleder, M., Berthold, W., Güdde, J., and Höfer, U. (2005) *Phys. Rev. Lett.*, **94**, 017401.

Colour Plates

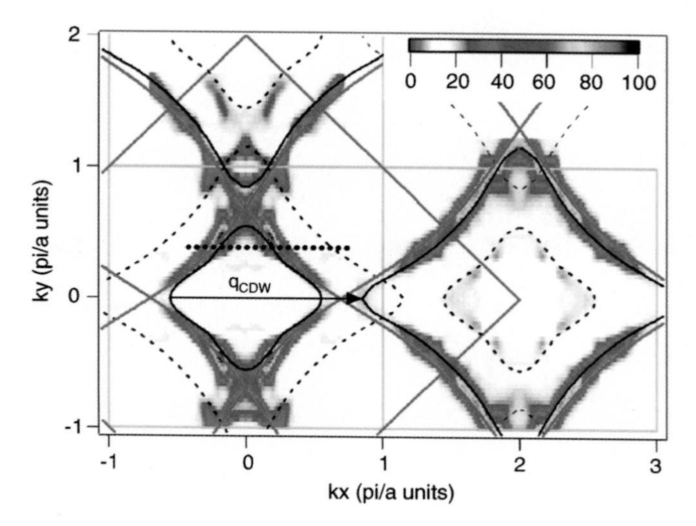

Figure 1.9 Fermi surface of $CeTe_3$ obtained by ARPES at the temperature of 25 K. In the horizontal direction, the Fermi surface is gapped due to charge density wave formation along this direction due to Fermi surface nesting with a momentum q_{CDW}. Reprinted with permission from Ref. [30]. Copyright (2004) by the American Physical Society.

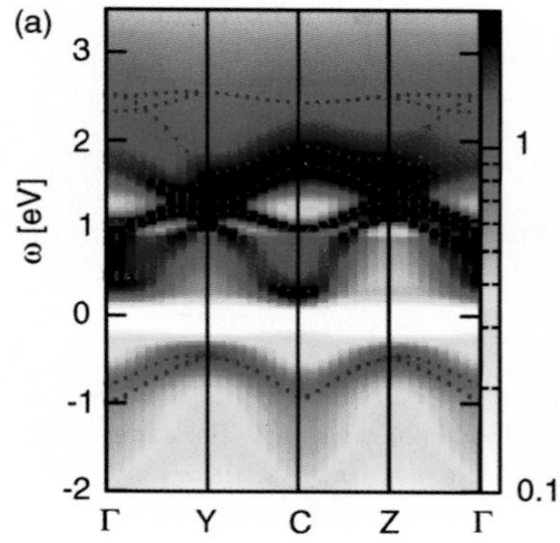

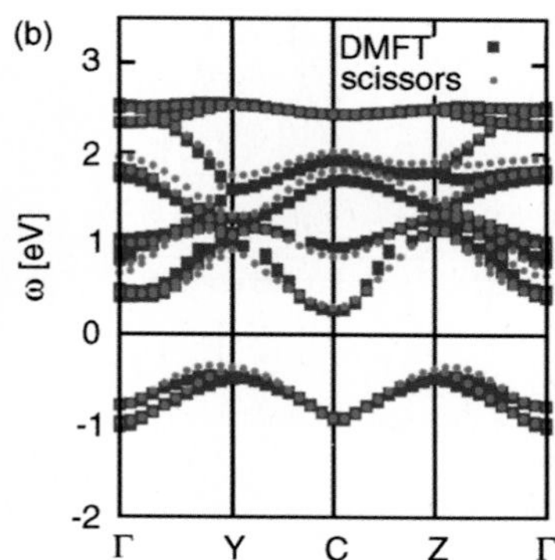

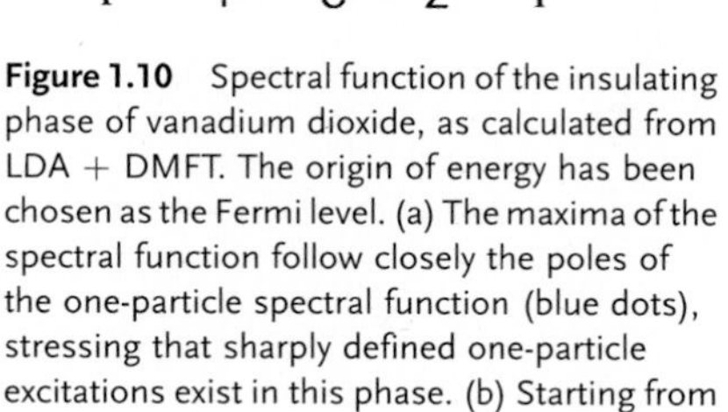

Figure 1.10 Spectral function of the insulating phase of vanadium dioxide, as calculated from LDA + DMFT. The origin of energy has been chosen as the Fermi level. (a) The maxima of the spectral function follow closely the poles of the one-particle spectral function (blue dots), stressing that sharply defined one-particle excitations exist in this phase. (b) Starting from the full LDA + DMFT solution, a static orbital-dependent potential has been constructed. The band structure corresponding to this potential is plotted in red dots and compared with the DMFT result of (a). Reprinted with permission from Ref. [37]. Copyright (2008) by the American Physical Society.

Dynamics at Solid State Surfaces and Interfaces: Volume 2: Fundamentals, First Edition.
Edited by Uwe Bovensiepen, Hrvoje Petek, and Martin Wolf.

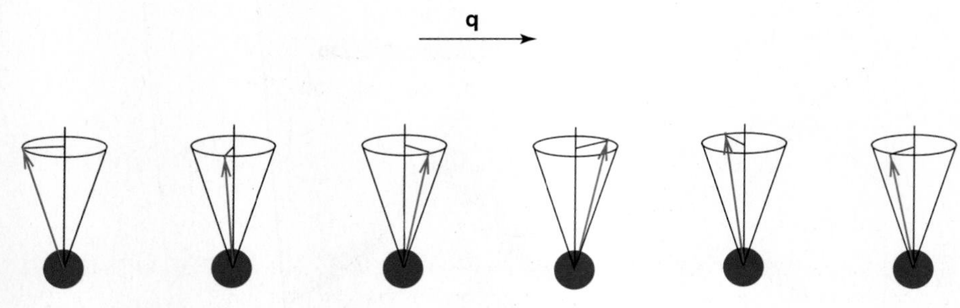

Figure 2.9 Schematic picture of the magnetic configuration corresponding to a spin wave with wave vector **q**.

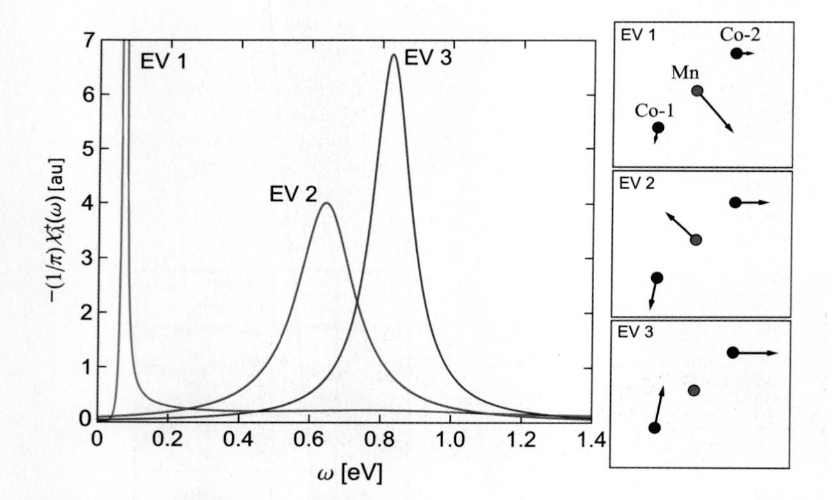

Figure 2.12 Enhanced susceptibility of Co_2MnSi, an example of the spectrum of the loss matrix for $\mathbf{q} = 0.28(1, 1, 0)2\pi/a$. a stands for the lattice constant. The three largest eigenvalues are shown, and other eigenvalues are of vanishing magnitude. Three clear peaks (EV 1, 2, and 3 corresponding to the labels in Figure 2.13) can be discerned. The panels show corresponding eigenvectors; arrows indicate the deviations of magnetic moments. Reprinted with permission from Ref. [65]. Copyright (2009) by the American Physical Society.

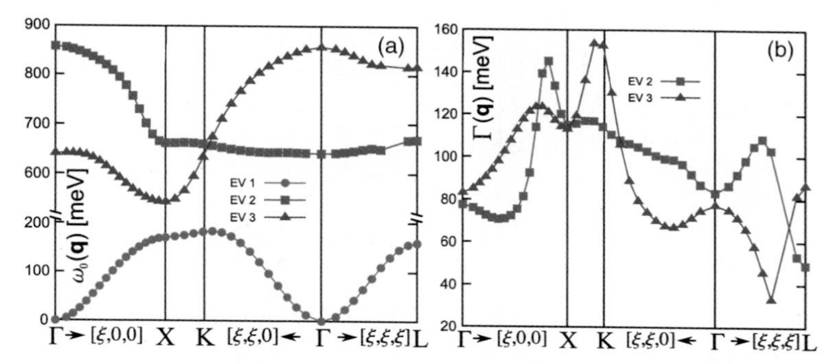

Figure 2.13 Energies (a) and inverse lifetimes Γ (b) of three SW modes in Co_2MnSi. Reprinted with permission from Ref. [65]. Copyright (2008) by the American Physical Society.

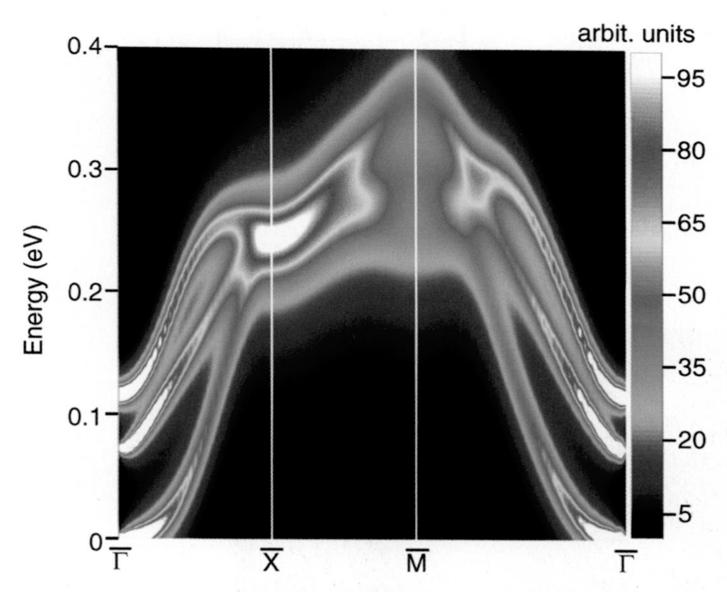

Figure 2.15 Spectral power $\mathrm{Im}\chi(q_{||}, \hbar\omega)$ of spin-flip excitations in three-monolayer iron film on Cu (100) obtained from LRDFT calculations. Reprinted with permission from Ref. [17]. Copyright (2010) by the American Physical Society.

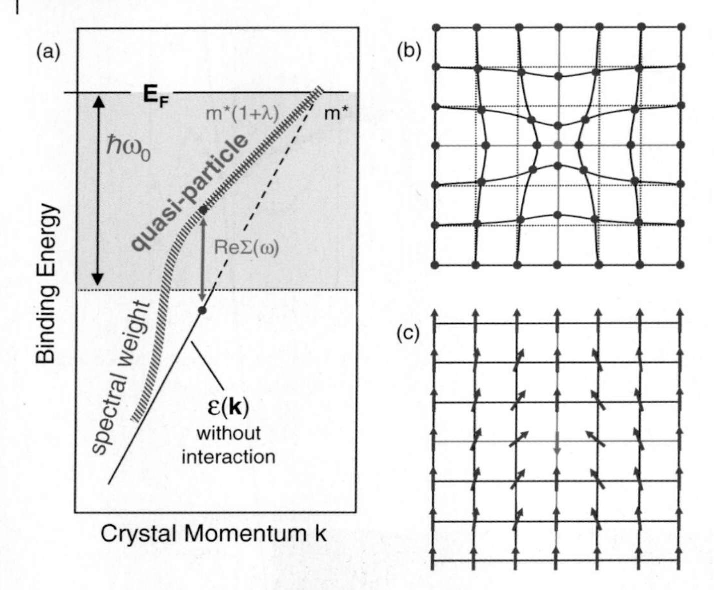

Figure 2.22 Schematic display of quasiparticle formation. (a) The electron is dressed with the excitation (phonon or spin excitation) up to an energy ω_0 below the Fermi energy, leading to a mass enhancement. (b) Electron–phonon coupling implies a distortion of the crystal lattice surrounding the electron. (c) Electron–magnon coupling implies spin scattering mediated by magnetic interactions.

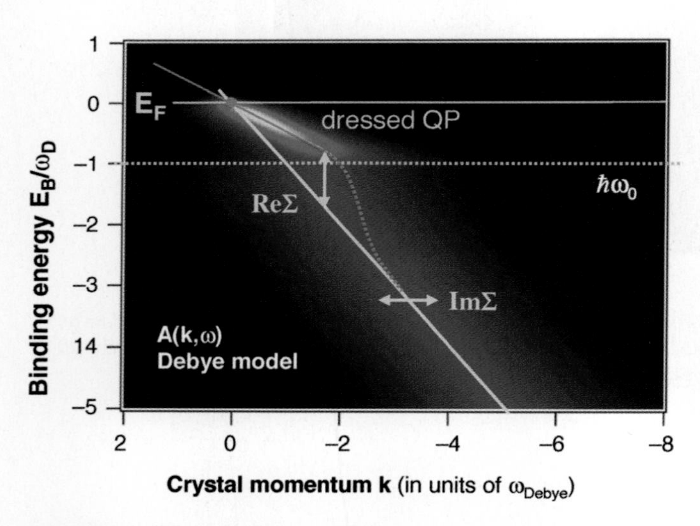

Figure 2.23 Spectral function generated numerically ($T = 20$ K, $T_{debye} = 340$ K, impurity scattering 10 meV). The shift of spectral weight near the Fermi level and a reduced slope become obvious. The self-energy Σ implies a broadening (given by ImΣ) and an energy shift (given by ReΣ) of the spectrum compared to the noncoupling case.

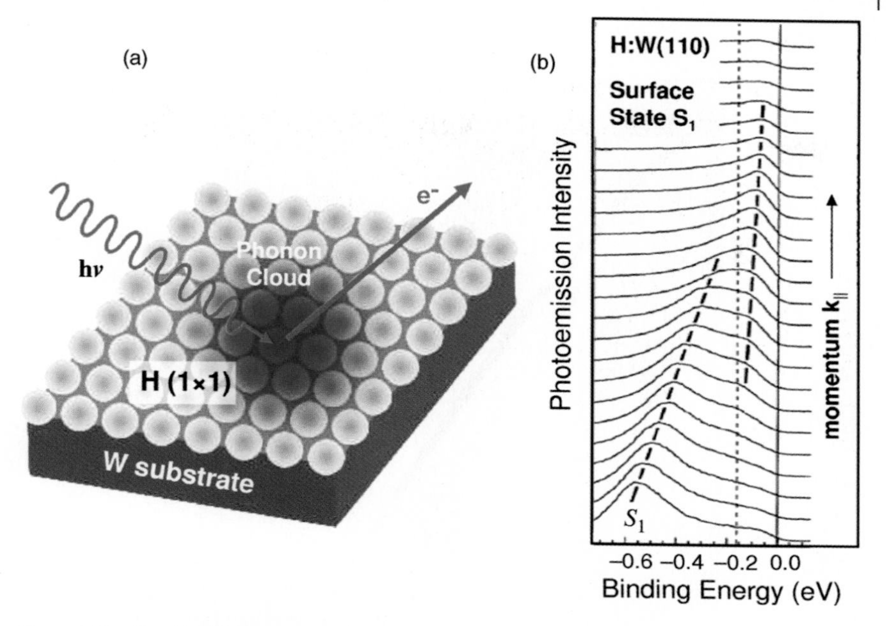

Figure 2.24 The H:W(110) system allows the study of adsorbate vibrations in the presence of interactions with a metallic electron bath. (b) Spectra of the surface state band ($T = 150$ K, $h\nu = 100$ eV). They reflect the electron–phonon coupling in an energy window of ≈ 160 meV that corresponds to the H stretching vibration.

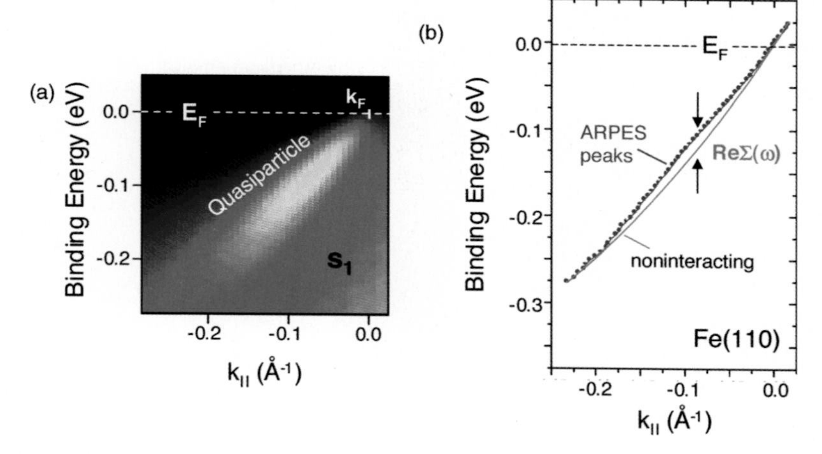

Figure 2.25 (a) ARPES data of the metallic Fe(110) surface state near $\bar{S}$ at $T = 85$ K. It is a surface resonance overlapping with the projection of bulk bands of opposite spin. (b) Peak position extracted from the band map using a fitting procedure (thick dots). The smooth line is a parabolic interpolation for the noninteracting band. A deviation from this band extends over a large energy scale below the Fermi level. Reprinted with permission from Ref. [131]. Copyright (2004) by the American Physical Society.

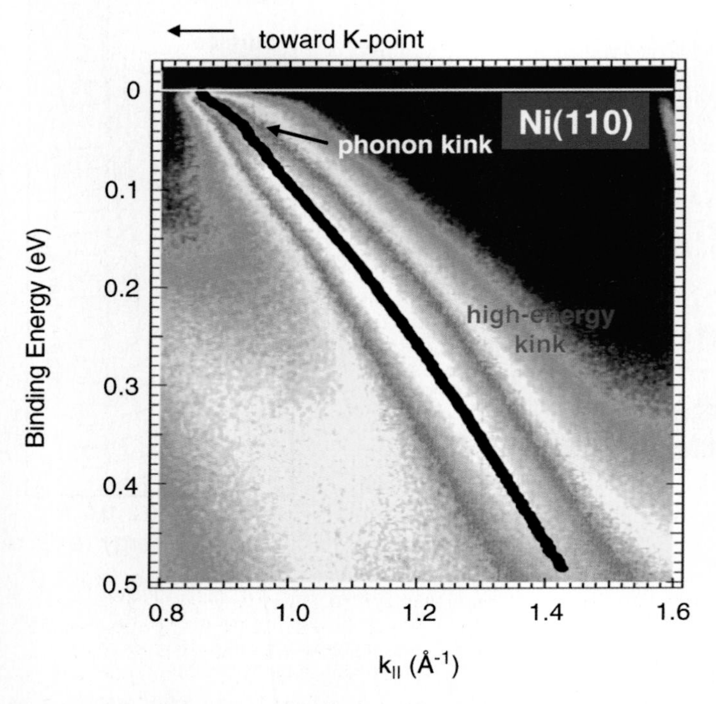

Figure 2.27 ARPES data of Ni(110) recorded at $h\nu = 100$ eV (corresponding to a high-symmetry plane). The band map shows the bulk band near the *K*-point. The momentum distribution curves are fitted to yield the peak positions. At low and high binding energy (as indicated), indications for kinks in the dispersion are discernible, corresponding to the phonon and magnon energy scale, respectively. Reprinted with permission from Ref. [124]. Copyright (2009) by the American Physical Society.

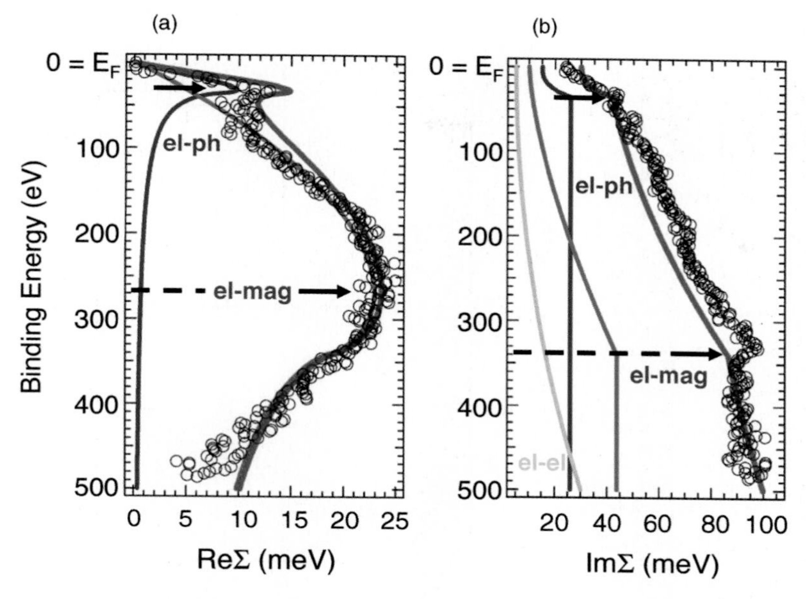

Figure 2.28 Self-energy analysis of the *K*-point band in Ni(110) derived from ARPES. (a) Real part Re Σ and (b) imaginary part Im Σ. In both panels, peaks in the self-energy are seen at energy scales that correspond to the phonons and magnons: in terms of Im Σ, this occurs at $\approx$ 30 meV and $\approx$ 340 meV, respectively. Note that in (a) a Gutzwiller reference band is used, which includes the electron correlation effects, so that they are effectively removed from Re Σ. In (b) for Im Σ, electron correlations have to be explicitly considered. The model curves for Re Σ and Im Σ are Kramers–Kronig transformable, consistent with the quasi-particle picture. Reprinted with permission from Ref. [124]. Copyright (2009) by the American Physical Society.

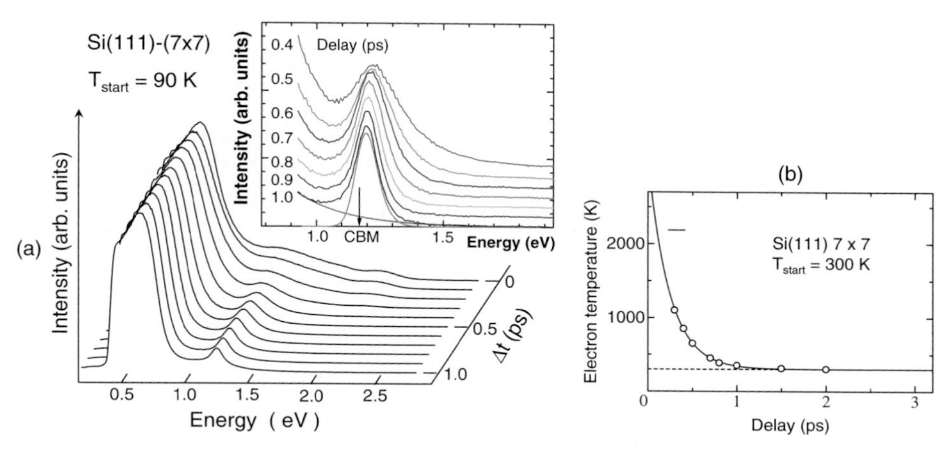

Figure 2.34 (a) Series of 2PPE spectra measured for Si(111) 7 × 7 for s-polarized 2.21 eV pump pulses (fluence $\approx$ 0.5 mJ cm^{-2}) and p-polarized 4.95 eV probe pulses. The inset shows the temporal evolution of the CB peak in an expanded scale from 0.4 to 1 ps delay. (b) The electron temperature at the CBM as a function of pump–probe delay. The solid curve is the fit of a single exponential decay component and a constant value of 296 K. The time constant is 240 fs. Reprinted with permission from Ref. [197]. Copyright (2009) by the American Physical Society.

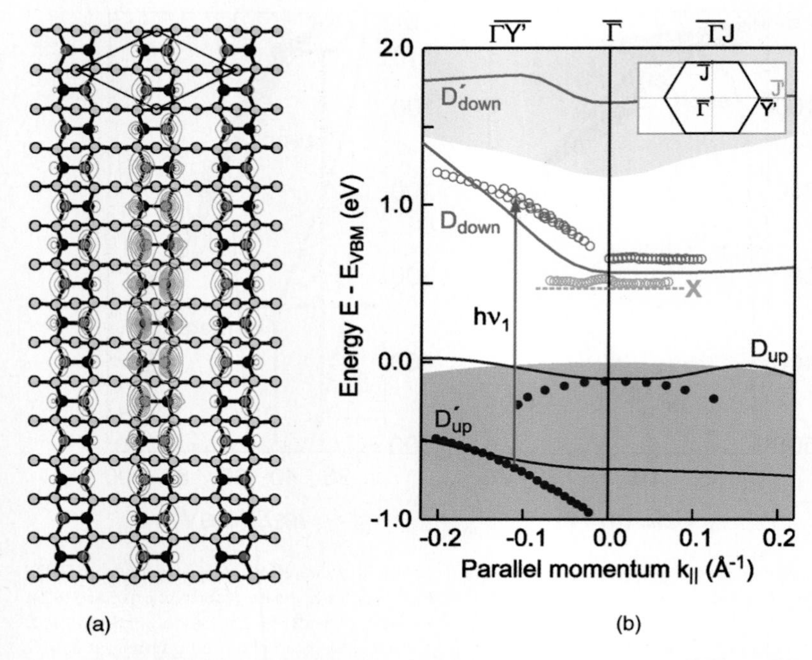

Figure 2.35 (a) The $c(4 \times 2)$ reconstruction of the Si(100) surface. Dimers are arranged in rows along the $[0\bar{1}1]$-direction and are alternately tilted along and perpendicular to the rows. The unit cell (solid rhombus) of the $c(4 \times 2)$ stacking comprises two dimers. The surface exciton is visualized by the contour plot of the probability density of the electron bound to the hole, located in the center of the depicted area Reprinted with permission from Ref. [214]. Copyright (2005) by Springer. (b) Measured (symbols) and calculated (solid lines, shaded areas) surface-projected band structure of the Si(100) $c(4 \times 2)$ surface. $\overline{\Gamma Y'}$ and $\overline{\Gamma J}$ correspond to the directions along and perpendicular to the dimer rows. The presence of two dimers per unit cell leads to two occupied (D_{up} and D'_{up}) and two unoccupied (D_{down} and D'_{down}) dangling bond bands. X is the signature of a surface exciton in the single-particle band structure. The silicon valence and conduction bands are indicated by the gray-filled areas. Reprinted with permission from Ref. [211]. Copyright (2004) by the American Physical Society.

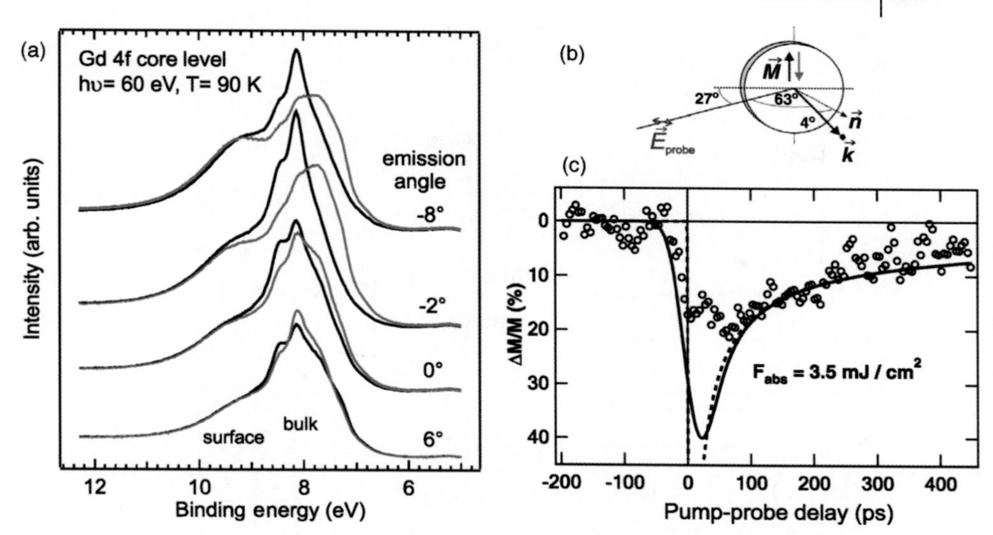

Figure 2.39 (a) Linear magnetic dichroism of the 4f photoemission line (surface and bulk components) of Gd(0001) (adapted from Ref. [234]). (b) Sketch of the measurement geometry. Spectra are recorded close to normal emission (0°) for opposite directions of the in-plane magnetization ($M\uparrow$ and $M\downarrow$). (c) Transient, relative change in the magnetization: measured 4f dichroism (open circles), calculated from lattice temperature and magnetization (dashed line), and convolved with a Gaussian of 50 ps FWHM duration (solid line). Reprinted with permission from Ref. [170]. Copyright (2008) by the American Physical Society.

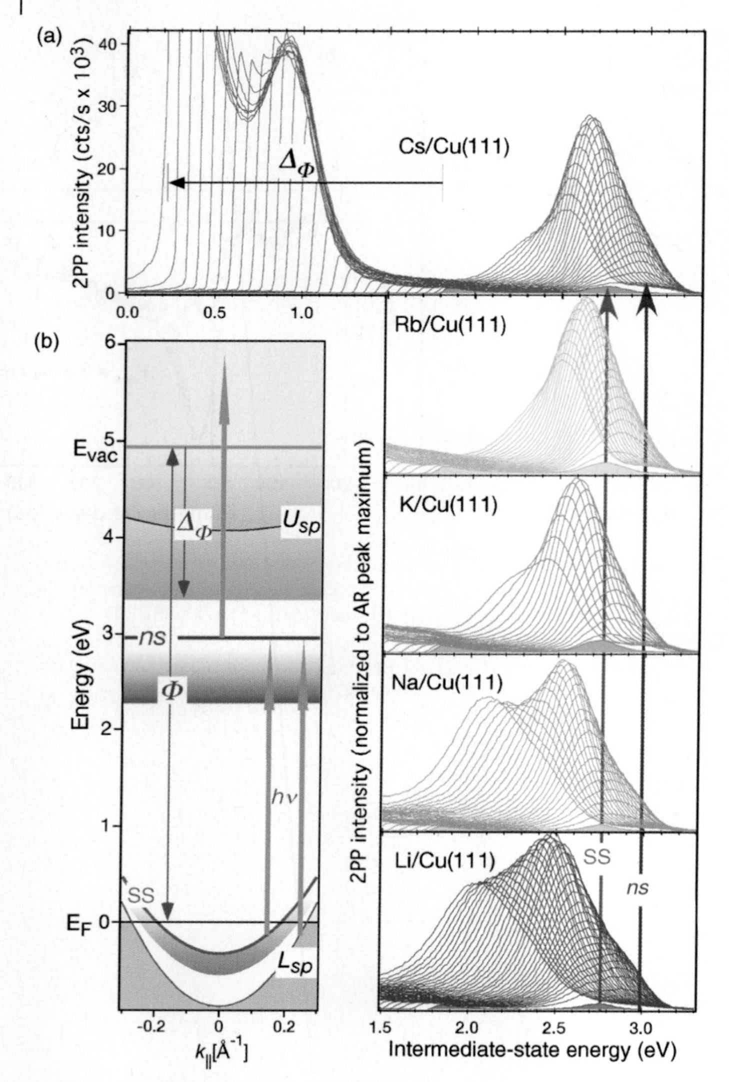

Figure 3.19 The (a) 2PP spectra and (b) 2PP excitation scheme for Cu(111) surface during a continuous deposition of alkali atoms. The binding energy of the unoccupied *n*s resonance of alkali atoms is determined by the image–charge interaction. The formation of a surface dipole layer reduces the vacuum level by Δ_Φ and the binding energy of *n*s resonance by $-(\Delta_\Phi)^{3/2}$, as explained in Ref. [87]. Reprinted with permission from Ref. [87]. Copyright (2008) by the American Physical Society.

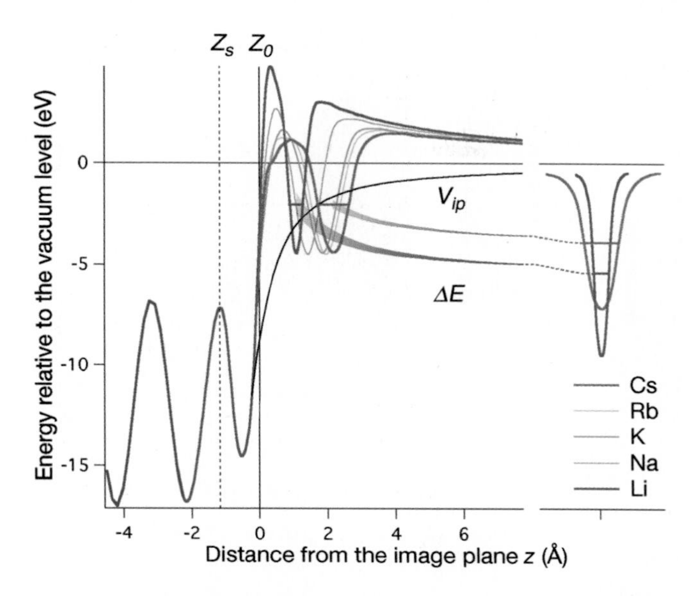

Figure 3.22 The effective electron potentials for single alkali atom (Li through Cs) on Cu(111) surface. The total potential combines the Chulkov potential for Cu(111), which includes the image potential V_{ip}, the pseudopotential for free alakli [92] atoms, and the main mode of interaction through the image–charge repulsion ΔE [87]. Both the V_{ip} and ΔE potential have the universal form expected for the Coulomb interaction. The adsorption height is measured from the image plane z_0, while z_s represents the geometrical boundary of the surface. Reprinted with permission from Ref. [87]. Copyright (2008) by the American Physical Society.

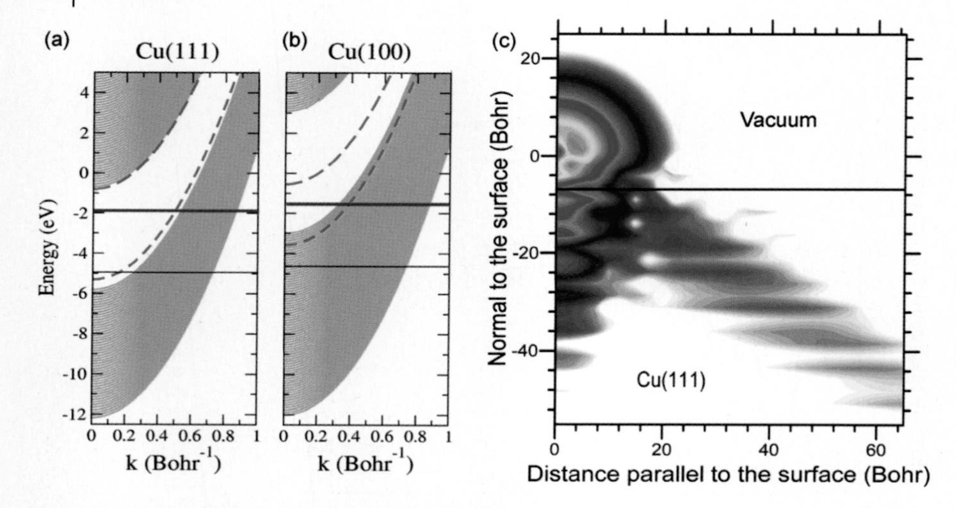

Figure 4.2 (a and b) Electronic structure of the Cu(111) and Cu(100) surfaces. Surface states, resonances, and the first image potential states are indicated by dashed lines. The thick horizontal lines indicate the energy of the resonances localized in Cs adsorbates at low coverage, while the thin horizontal line indicates the Fermi energy. (c) Logarithm of the electronic density associated with a Cs-localized resonance for Cs/Cu(111). The Cs atom is located at the origin of coordinates and the horizontal line indicates the position of the image plane of the metal. The electronic density decreases when going from red to violet (white corresponds to very small values of the density). Notice that the projected bandgap in Cu(111) prevents propagation along the normal to the surface. Reprinted with permission from Ref. [18]. Copyright (2001) by the American Physical Society.

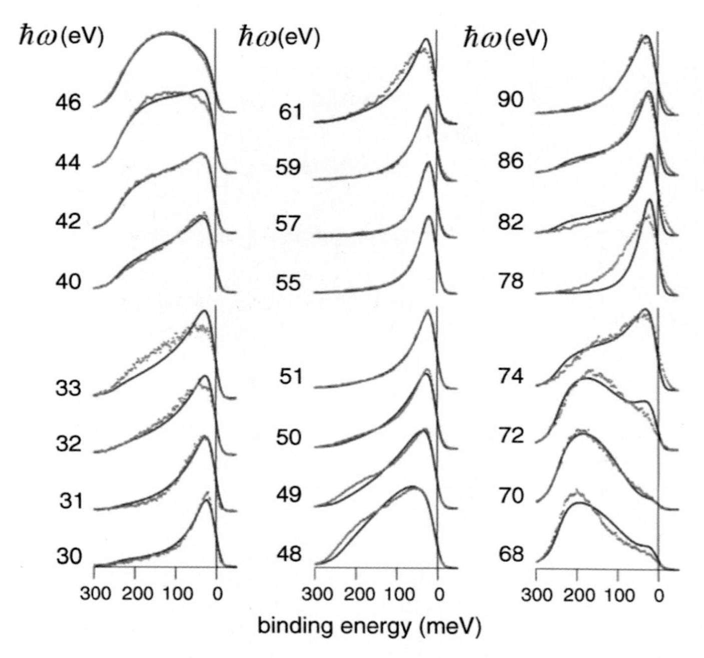

Figure 5.3 Experimental and theoretical energy distribution curves for Ti 3d photoemission in $TiTe_2$. Spectra are marked by photon energies. Reprinted with permission from Ref. [22]. Copyright (2007) by the American Physical Society.

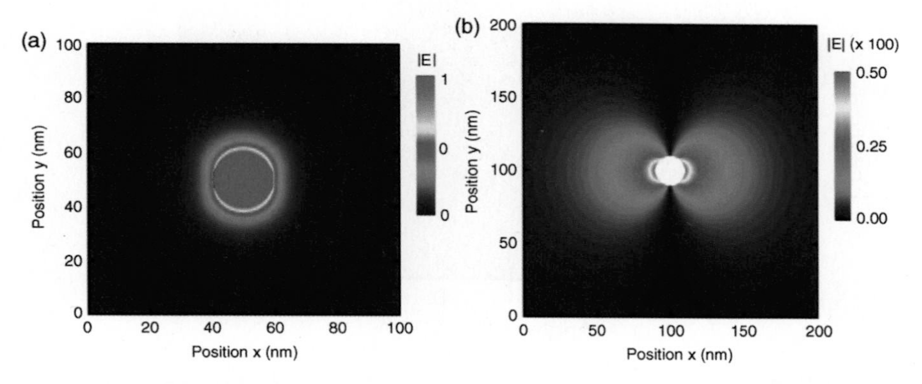

Figure 5.15 Magnitude of the electric field $|\vec{E}(\vec{r})|$ (arbitrary units) near a small nanoparticle with a radius of 10 nm. The field profile is calculated in the x–y plane using Eq. (2.40), assuming that it is given by that of a point-like oscillating dipole $\vec{p} = p_0 \cdot \vec{e}_y \cdot e^{i\omega t}$ at the center of the sphere. (b) Electric field magnitude $|\vec{E}(\vec{r})|$ given by the far-field term in Eq. (2.40). Close to the particle, the far-field amplitude is up to two orders of magnitude smaller than that of the corresponding near field.

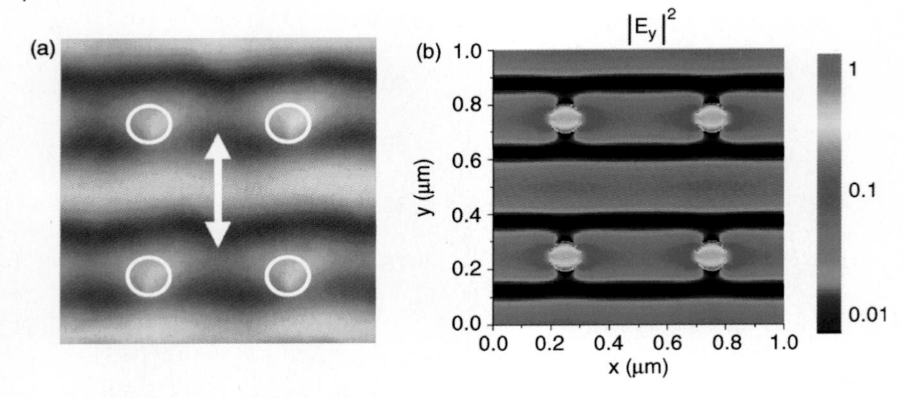

Figure 5.19 (a) NSOM image of the light transmitted through a subwavelength diffraction grating. The grating is fabricated by milling an 850 nm period array of 150 nm diameter nanoholes into a thin gold film deposited on a sapphire substrate. The grating is illuminated from the sapphire side with y-polarized light (see arrow) at 877 nm. The light at the air side of the grating is collected with a metal-coated NSOM fiber probe. Light regions correspond to high intensity, whereas the intensity drops to zero in the dark regions. Standing SPP waves at the grating interface are mapped. (b) Near-field intensity $|E_y|^2$ obtained from a three-dimensional finite difference time domain simulation. Reprinted with permission from Ref. [210]. Copyright (2002), American Institute of Physics.

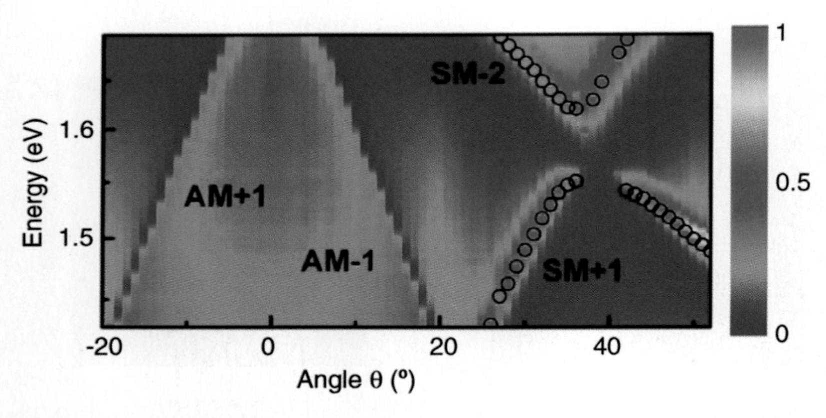

Figure 5.20 Experimentally measured angle-resolved transmission spectrum for a gold nanoslit array with a period of $a_0 = 650$ nm and a slit width of 50 nm. Open circles: Calculated SPP band structure near the crossing of SM[+ 1] and SM[−2] resonances. A bandgap splitting of 72 meV is revealed. Note the spectral narrowing of the transmission spectrum and the decrease in transmission intensity in the lower energy region of the SM[+ 1]/SM[−2] crossing.

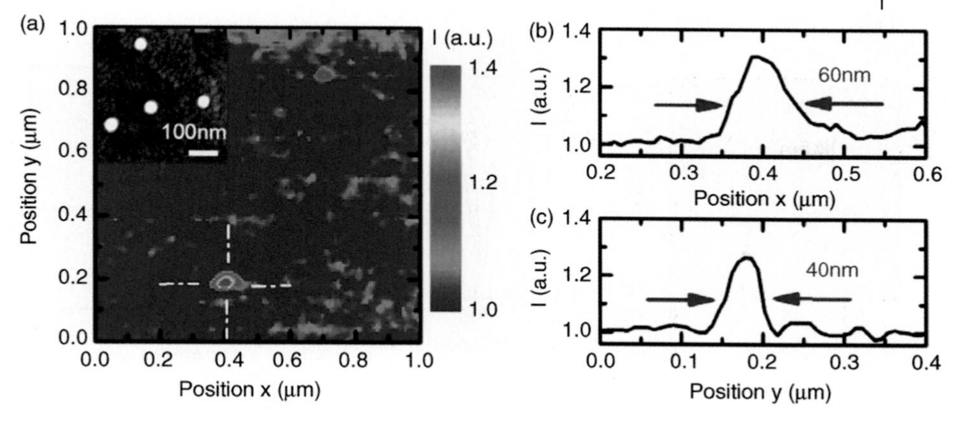

Figure 5.22 Two-dimensional optical images of individual gold nanoparticles on a glass substrate recorded by adiabatic nanofocusing scattering-type near-field scanning optical microscopy (s-NSOM). (a) Optical s-NSOM image of a single gold nanoparticle with <30 nm radius. In these experiments, SPP waves are launched onto a gold taper by grating coupling (Figure 5.21b) and the light scattering from the tip apex is recorded in the far field while scanning the tip across the surface of a dielectric substrate covered with a low concentration of gold nanoparticles. *Inset*: Scanning electron microscope image of the gold nanoparticles. (b) Cross sections of the optical intensity along the x and y directions (marked by dash-dotted lines in (a)). The optical resolution of about 40 nm and the large signal-to-background ratio confirm the efficient adiabatic nanofocusing of SPP waves at the tip apex. Reprinted with permission from Ref. [200]. Copyright (2011) American Chemical Society.

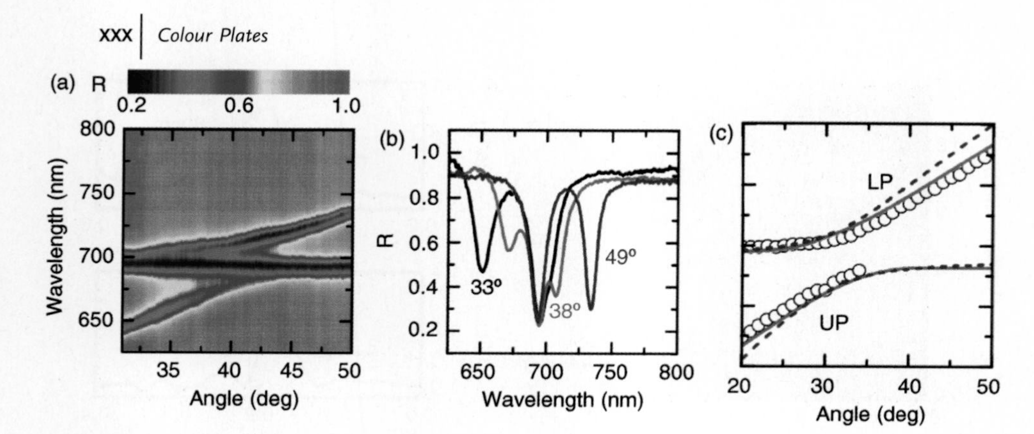

Figure 5.24 (a) Angle-resolved p-polarized linear reflectivity spectra ($T = 77$ K) of the J-aggregate dye deposited on a gold film perforated with a nanoslit grating of 430 nm period and 45 nm width. (b) Reflectivity spectra at different angles of 33°, 38°, and 49°. (c) Polariton dispersion relation obtained from the experimental spectra (open circles), a full vectorial solution of Maxwell's equations (solid line), and a coupled oscillator model (dashed line). Reprinted with permission from Ref. [256]. Copyright (2010) American Chemical Society.

4
Basic Theory of Heterogeneous Electron Transfer

Daniel Sanchez-Portal, Julia Stähler, and Xiaoyang Zhu

Heterogeneous electron transfer (HET), the redistribution of charge from a single donating electronic state to an accepting continuum, is a crucial step for many phenomena both of fundamental interest and with regard to application. For instance, desorption induced by electronic transitions or the field of photochemistry, and charge carrier separation in, for example, quantum dot-based solar cells. Depending on the electronic interaction of donor (D) and acceptor (A), two limiting cases, the strong and the weak coupling limit are defined and treated differently by theory. In the former, the electronic coupling is assumed to be sufficiently strong to describe the transfer reaction in the adiabatic limit where wave packet propagation occurs along a steady potential; this treatment is particularly appropriate when the characteristic charge transfer time is small compared to the nuclear motion of the donor (Franck–Condon approximation). This limit of HET will be discussed in Section 4.1. Approaching the nonadiabatic case for weak electronic interaction between donor and acceptor, Section 4.1.4 discusses the elastic electron transfer from a small adsorbate that impinges on the substrate on timescales on the order of the transfer event. Section 4.2 discusses the treatment of HET on the basis of Marcus theory where complex nuclear motions in polar or polarizable media strongly influence the charge transfer rate.

4.1
Resonant Charge Transfer in Chemisorbed Systems

In this first section, we focus in the resonant electron transfer between a "small" system (e.g., an atom or molecule), characterized by a discrete electronic spectrum, and a "large" system (e.g., a metallic substrate), characterized by a continuous electronic spectrum. We first neglect the influence of the atomic movements on the electron dynamics. This will be discussed in detail in the following sections, particularly in the context of the Marcus theory [1, 2] used to describe charge transfer processes in the presence of polarizable media such as solvents. The prototypical system here is a small adsorbate chemisorbed on a metallic substrate. In many of

Dynamics at Solid State Surfaces and Interfaces: Volume 2: Fundamentals, First Edition.
Edited by Uwe Bovensiepen, Hrvoje Petek, and Martin Wolf.

those systems, if an electron is excited from the metal into a bound, unoccupied resonance localized at the adsorbate, for example, using a laser pulse, it takes only a few femtoseconds for the electron to be transferred back into the substrate due to the strong adsorbate–surface interaction [2–4]. Thus, this transfer is much faster than the typical timescale associated with atomic movements, and therefore they have a limited influence on the observed electron dynamics. This approaches the limit of strong electronic coupling and adiabatic charge transfer of the Marcus theory.

4.1.1 Anderson–Grimley–Newns Hamiltonian

The study of chemisorbed adsorbates is commonly performed using the first-principles electronic structure calculations and, in particular, density functional theory (DFT) [5] that provides a reasonably accurate description of chemical interactions for a wide range of systems [6]. However, we will start our analysis using a simplified and physically transparent model, the Anderson–Grimley–Newns model of chemisorption [7–9]. We will later describe some of the methods based on more sophisticated electronic structure calculations. In this model, the description of the adsorbate is similar to that of a magnetic impurity in a metal as used in the Anderson model [10]. The adsorbate is described by a single electronic level of energy ε_a. Upon adsorption, nonzero Hamiltonian matrix elements V_{ak} couple the state localized at the adsorbate $|a\rangle$ with delocalized electronic levels in the metallic substrate $|k\rangle$. The total Hamiltonian reads in a second quantization form as

$$\hat{H} = \sum_{\sigma} \varepsilon_a \hat{n}_{a,\sigma} + \sum_{k,\sigma} \varepsilon_k \hat{n}_{k,\sigma} + \sum_{k} (V_{ak} \hat{c}^{+}_{a,\sigma} \hat{c}_{k,\sigma} + V^{*}_{ak} \hat{c}^{+}_{k,\sigma} \hat{c}_{a,\sigma}) + U \hat{n}_{a,\sigma} \hat{n}_{a,-\sigma}, \tag{4.1}$$

where σ denotes the electron spin. The last term in Eq. (4.1) describes the on-site charging energy of the adsorbate, that is, the Coulomb repulsion between two electrons of opposite spin in $|a\rangle$. The parameter U of such one-level system containing one electron could, in principle, be obtained as the difference between the electron affinity and the ionization potential in gas phase $U = A - I$. However, the value of U for the adsorbed system, along with the positions of A and I levels, is renormalized by the polarization (screening) of the underlying substrate. This is an important effect that cannot be disregarded [11].

Using the Hamiltonian in Eq. (4.1), Newns [9] was able to obtain closed formulas for the chemisorption energy and the charge state of the adsorbate within the unrestricted Hartree–Fock (mean field) approximation. However, we concentrate on a different aspect and try to answer the following question: How long does it take for an electron initially residing in the adsorbate, in state $|a\rangle$, to be transferred to the substrate? We focus only on the energy conserving electron transfer between the adsorbate level at energy ε_a and the electronic states of the substrate at the same energy. This is the so-called elastic or resonant charge transfer (RCT) time, and is a key parameter to understanding many physical and chemical phenomena at surfaces [2, 12]. Inelastic processes, in which the electron transfer is accompanied by the creation of electronic or vibration excitations, provide additional channels for electron transfer reducing the lifetime of the resonance and are discussed in detail in Chapter 2.[1)] The relative importance of elastic and inelastic charge transfer processes can strongly depend on the details of the band structure of the substrate as we will see later.

To study RCT processes, we disregard the last term in Eq. (4.1), since in many cases it can be taken as a renormalization of ε_a, and use a purely one-electron Hamiltonian. In this limit, the charge transfer corresponds to the dynamics of a wave packet initially localized at the adsorbate as it delocalizes in the continuum of states of the solid. To characterize this process, it is convenient to define the survival amplitude $A(t)$. It is the projection of the time-dependent electronic wave function onto the initial wave packet $A(t) = \langle a|\phi(t)\rangle$. The square of its modulus, $P(t) = |A(t)|^2$, gives the survival probability, that is, the probability to find the electron in $|a\rangle$ at a given time t. If we expand the electron wave function using the eigenfunctions $|\Phi_n\rangle$ and eigenvalues E_n of the Hamiltonian (4.1) and take its Fourier transform, we obtain

$$\tilde{A}(\omega) = \sum_n \int_0^\infty \mathrm{d}t\, |\langle a|\Phi_n\rangle|^2\, \mathrm{e}^{-(i/\hbar)\left(E_n - \hbar\omega - i\delta^+\right)t} = i\hbar G_{aa}(\hbar\omega). \tag{4.2}$$

Here, $G_{aa}(E) = \langle a|(z\hat{I} - \hat{H})^{-1}|a\rangle$, with $z = E + i\delta^+$ and δ^+ an infinitesimally small positive number, is the projection of the advanced Green's function [13] of the combined adsorbate/substrate system onto the initial state $|a\rangle$. It is quite simple to show [10, 14] that $G_{aa}(E) = (E - \varepsilon_a + f(E) + i\Delta(E))^{-1}$, where

$$f(E) = P \int_{\varepsilon_{\min}}^{\varepsilon_{\max}} \mathrm{d}\varepsilon\, \varrho(\varepsilon) \frac{|V(\varepsilon)|^2}{E - \varepsilon}, \tag{4.3}$$

$$\Delta(E) = \pi\varrho(E)|V(E)|^2 \tag{4.4}$$

with the density of states (DOS) of the clean substrate $\varrho(\varepsilon)$ and the energy-dependent electronic coupling $V(\varepsilon)$. In principle, the main effect of the interaction with the substrate is to shift the energy of the resonance and provide a finite width. In particular, if both ϱ and V are weakly energy dependent around ε_a, then $G_{aa}(E) \approx (E - \varepsilon_a + i\Delta(\varepsilon_a))^{-1}$ and the adsorbate level becomes a Lorentzian resonance of width $\Gamma_{\mathrm{RCT}} = 2\Delta(\varepsilon_a)$. In this case, the survival probability decays exponentially $P(t) = |A(t)|^2 = \mathrm{e}^{-t/\tau_{\mathrm{RCT}}}$ with a characteristic lifetime

$$\tau_{\mathrm{RCT}} = \frac{\hbar}{\Gamma_{\mathrm{RCT}}}. \tag{4.5}$$

In some cases, the changes in the electronic structure of the adsorbate/substrate system can be less intuitive than those described above, and localized levels may appear below the bottom of the substrate valence band [8, 9]. For example, if V is

1) This book contribution by M. Weinelt, E. Chulkov, R. Huber, J.M. Pitarke, K. Ishioka, and J. Schäfer.

roughly independent of energy and the adsorbate is coupled to a two-dimensional continuum, such as a surface state with parabolic dispersion, then it can be shown [15] that an additional resonance (peak) ε'_a always appears below the continuum. This is due to the constant density of states associated with a two-dimensional band that causes a logarithm divergence of $f(E)$ when E approaches the band bottom ε_{min} and guarantees that $\varepsilon'_a = \varepsilon_a + f(\varepsilon'_a)$ always has a solution with $\varepsilon'_a < \varepsilon_{min}$. This has been observed experimentally for Cu(111) and Ag(111): A clear spectroscopic signal appears localized in the neighborhood of adatoms of several species and below the minimum of the surface-state band [16, 17].

4.1.2
Main Factors that Determine RCT Decay Rates

According to Eq. (4.4) and the subsequent discussion, the RCT decay rate is determined by two factors: (i) the number of electronic states available in the substrate at the resonance energy and (ii) the strength of the adsorbate–substrate interaction. These effects are shown in Figure 4.1 for the case of a quantum dot (QD) in front of a semiconductor surface. The QD levels become broader and more delocalized as the interaction with the surface increases. However, those levels lying within the band gap are not broadened and remain fairly localized in the QD. This shows the importance of a detailed description of the substrate band structure when studying RCT processes [12]. An example of this is illustrated in Figure 4.2 for the case of a Cs-localized excited state on Cu(111) and Cu(100). Both surfaces present a projected band gap around $\bar{\Gamma}$ at the energy of the resonance. Hence, electron transfer has to involve states with a finite $k_{||}$ in the metal. The possibility to tunnel to the substrate along the surface normal direction, the most efficient direction due to

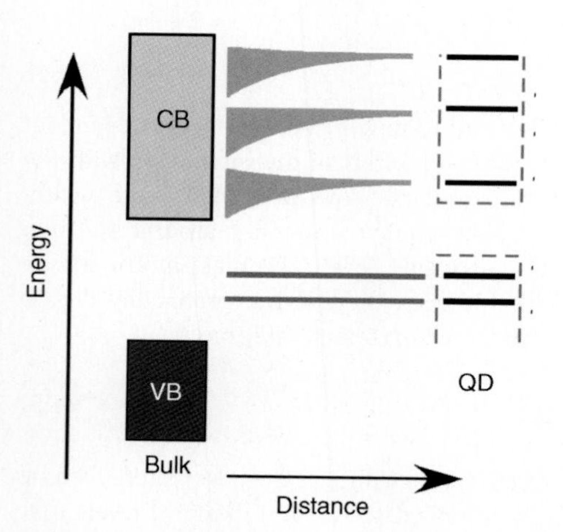

Figure 4.1 Electronic interaction between a QD and a bulk semiconductor surface. The lines between the QD and the bulk illustrate the broadening of discrete states of the QD with decreasing interfacial separation due to resonant interaction with bulk bands.

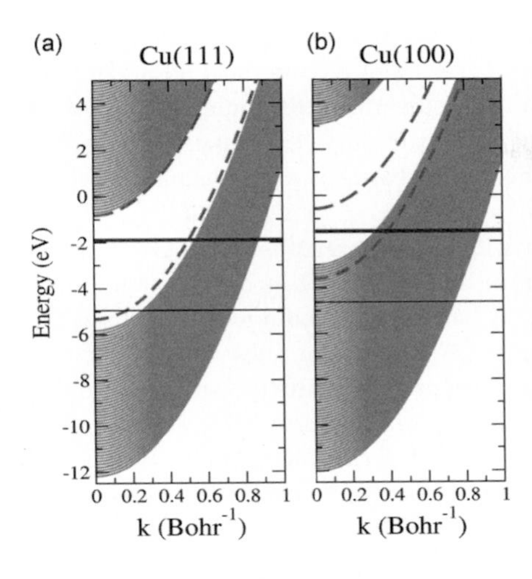

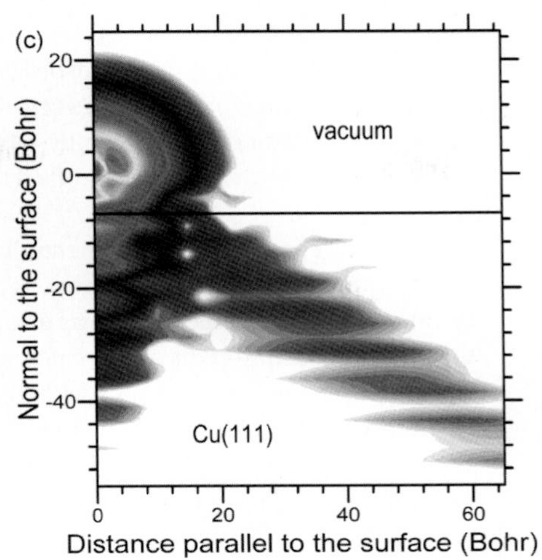

Figure 4.2 (a and b) Electronic structure of the Cu(111) and Cu(100) surfaces. Surface states, resonances, and the first image potential states are indicated by dashed lines. The thick horizontal lines indicate the energy of the resonances localized in Cs adsorbates at low coverage, while the thin horizontal line indicates the Fermi energy. (c) Logarithm of the electronic density associated with a Cs-localized resonance for Cs/Cu(111). The Cs atom is located at the origin of coordinates and the horizontal line indicates the position of the image plane of the metal. The electronic density decreases when going from red to violet (white corresponds to very small values of the density). Notice that the projected band gap in Cu(111) prevents propagation along the normal to the surface. (Please find a color version of this figure on the color plates.) . Reprinted with permission from Ref. [18]. Copyright (2001) by the American Physical Society.

the smaller effective barrier, is blocked and the effect of the projected band gap is the reduction in the RCT decay rates. Figure 4.2c clearly shows that for the Cs/Cu(111) system, the electronic flux into the metal coming from the adsorbate appears at a finite angle away from the surface normal. The effect is stronger for Cu(111) than for Cu(001) since the gap is considerably broader at the relevant energy. This is reflected in the much smaller RCT decay rates for Cu(111): The model calculations in Ref. [18] estimate Γ_{RCT} to be 7 and 112 meV, respectively, for Cs/Cu(111) and Cs/Cu(001). Interestingly, the inelastic electron–electron decay is similar in both systems [12, 18] ($\sim$20 meV), giving rise to very different lifetimes dominated by inelastic decay in the case of Cs/Cu(111) and by RCT for Cs/Cu(001). Experimental observations confirm widely different lifetimes for these two systems [4, 19]. The crucial role of the substrate band structure becomes even more clear if we compare with the typical values of $\Gamma_{RCT} \sim 1$ eV calculated for similar adsorbates at chemisorption distances for free-electron-like substrates [18, 20]. A thorough analysis of the influence of the substrate band structure in RCT was recently performed in the case of core-excited $Ar^*(2p_{3/2}^{-1}4s)$ on Ru(0001) using first-principles calculations [21].

The second factor determining the RCT rate is the size of the hopping matrix elements V_{ak} between the adsorbate level $|a\rangle$ and the metal states $|k\rangle$ at the energy of the resonance. These matrix elements contain information about (i) the height of the potential barrier separating the adsorbate potential well and the substrate and (ii) the spatial distribution and symmetry of the substrate electronic states. As a first approximation, for sufficiently large adsorbate–surface distances, the matrix elements V_{ak} can be assumed to reflect the overlap between substrate and adsorbate states.

Thus, RCT rates decay exponentially with distance in this situation. A more detailed analysis of this perturbative regime and an explicit formula for the calculation of these matrix elements was given by Bardeen [22] and provides the basis for the theory of scanning tunneling microscopy [23]. In general, substrate states with a slower decay into vacuum and with the appropriate symmetry can couple more efficiently to the adsorbate. Figure 4.3 shows an illustration of this effect at chemisorption distances for Cs/Fe(110). The RCT decay rate of the Cs resonance is estimated to be 50% larger for the majority spin channel than for minority spin [24].

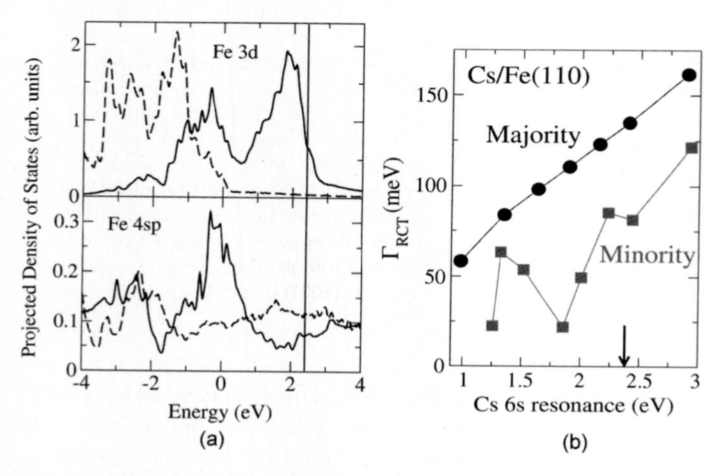

Figure 4.3 (a) Projected density of states (PDOS) onto the 3*d* and 4*sp* orbitals of a surface Fe atom in the Fe(110) substrate as a function of energy. Energies are referred to the Fermi level. Solid and dashed lines are for minority and majority spins, respectively. The vertical line indicates the calculated position (~2.3 eV above E_F) of the 6*s* Cs resonance of the Cs/Fe(110) system for the equilibrium Cs–surface distance at low Cs coverage. At the energy of the 6*s* Cs resonance, the total DOS in the Fe(110) surface is larger for the minority spin channel. As a consequence, one could expect a larger RCT rate for minority spin than for majority spin. However, the calculated result (see panel (b) and Ref. [24]) is just the opposite. This unexpected result is due to the larger extension of the 4*s* and 4*p* Fe orbitals toward the vacuum that translates into a much larger hopping matrix element between those orbitals and the 6*s* Cs state than that between the Fe 3*d* orbitals and Cs 6*s*, that is, $V_{Cs-Fe\ 4sp} \gg V_{Cs-Fe\ 3d}$. Since, at the relevant energy, the PDOS onto the 4*s* and 4*p* Fe orbitals is larger for majority spin, the RCT rate is also larger for majority spin. (b) Calculated RCT decay rates for a 6*s* Cs resonance on Fe(110) as a function of the energy position of the resonance [24]. The calculated energy of the 6*s* Cs resonance is indicated by the arrow. However, this position can be artificially shifted to check the robustness of the effect [24]. In spite of the larger total DOS for minority spin along the whole energy interval, the RCT decay is always faster for the majority spin channel reflecting the behavior of the 4*sp* PDOS.

This is surprising since DOS of Fe at the resonance energy is much larger for minority spin. However, the DOS is dominated by Fe $3d$ states that couple very inefficiently with the Cs adsorbate due to their strong localization around the Fe nuclei. The Fe $4s$ and $4p$ derived states extend further into vacuum and the hopping matrix element between these states and the 6s state of the Cs adsorbate is larger than that between the Fe $3d$ derived states and Cs $6s$, that is, $V_{\mathrm{Cs-Fe4}sp} \gg V_{\mathrm{Cs-Fe3}d}$. Therefore, the RCT rate is controlled by the tunneling from Cs to Fe $4s$ and $4p$ derived states, whose density is ~40% larger for majority spin than for minority spin at the relevant energy range. This is completely consistent with the known sign reversal of the magnetization of unoccupied states at the Fe(110) surface as we move away from the surface into vacuum [25].

4.1.3
Theoretical Approaches to Calculate RCT Rates in Realistic Systems

Most theoretical approaches for the calculation of the energies and widths of resonances associated with adsorbates on surfaces have been developed within the field of ion–surface interaction. Here, the RCT decay rate is a key ingredient to determine, for example, the relative abundance of different ion charges in the scattered beam. In spite of this central role, reliable nonperturbative calculations of RCT started to appear only during the past two decades of the last century. These approaches consist of two steps. First, an effective potential for the interaction of the electron with both the projectile and the surface must be constructed. In many of these works, the substrate is described using a jellium model, that is, the substrate electronic structure is free-electron-like, and the adsorbates considered are restricted to hydrogenic or alkali atoms. The interaction of the ion with the metal surface is described using classical image charges [20] or fitted to reproduce DFT results [26], and, to have the correct $-1/4z$ asymptotic behavior, an image tail is typically included in the electron–surface interaction [27]. Second, the energy and width of the adsorbate resonances are calculated. There are several methods to do so, and they produce similar results for a given one-electron effective potential. Among these methods, we can cite complex scaling [20, 28, 29], stabilization [28, 30], coupled angular momentum [31], wave packet propagation [12, 15], close coupling [32], Green's function [12, 33, 34], and recently, the embedding method [35–37] and deconvolution schemes [38]. In the following, we briefly describe some of these methodologies.

Complex scaling was one of the first methods used to obtain reliable results for hydrogen and the alkalis on metallic substrates [20, 29]. It consists of a rotation of the electron coordinates from the real axis into the complex plane that transforms the resonant wave functions into square-integrable ones [28]. The complex part of the corresponding energy eigenvalue can be related to the width of the resonance. Unfortunately, this method is difficult to apply to systems in which the substrate is described by a realistic potential, that is, beyond a jellium model.

The stabilization method uses the standard electronic solutions calculated in the spatial region where the resonance wave functions is mainly localized. From the dependence of these electronic eigenvalues on a particular parameter defining the basis set, the positions and widths of the resonances can be obtained [30]. For example, such a parameter can define the extension of the basis set elements, as far as it does not significantly affect the quality of the representation of the resonance itself. The main drawback of the method is that it requires a large number of calculations.

In the coupled angular momentum method [31], the electron scattering by the compound adsorbate–surface potential is studied using a basis set of spherical harmonics, that is, eigenstates of the angular momentum operator, centered in the adsorbate position. The approach is restricted to systems with cylindrical symmetry, so the projection of the angular momentum along the normal to the surface is a good quantum number. Looking at the scattering properties of the electron at negative energies (i.e., below the vacuum level) reveals the existence of resonances, which are associated with the atomic levels, perturbed by the surface. The analysis of this resonant scattering in terms of the time delay matrix [39] yields the energy position and the width of the atomic levels.

The wave packet propagation method starts with a wave packet that resembles in symmetry and localization the resonance wave function and explicitly calculates its time evolution using the time-dependent Schrödinger equation for a finite, but large, system. From this time evolution, the positions and widths of the adsorbate resonances can be obtained with high accuracy. One technical difficulty of the method is that in order to suppress the reflection of electrons at the boundaries of the finite system, it is necessary to introduce an absorbing potential. Thus, the Hamiltonian becomes non-Hermitian, which limits the number of time-dependent propagation techniques that can be used [12]. The wave packet propagation has been frequently used in combination with simplified models to describe the electron–substrate interaction, such as the one-dimensional Chulkov's potential for metals [40], in order to reduce the computational cost by reducing the dimensionality of the problem.

Equation (4.2) relates the dynamics of the resonance with the electronic Green's function projected onto the resonance wave packet. Therefore, it is necessary to compute only the Green's function in the surface region where the resonance wave packet is mainly localized period. Using recursive methods, a basis set of localized function (like atomic orbitals) or a tight-binding formalism, and combining the information of surface and bulk Hamiltonians, the surface Green's function can be easily calculated for the truly semi-infinite system. Recently, Sánchez-Portal [34] has utilized this methodology to study the RCT process in several systems including photoexcited argon [21] and alkalis [24] on transition metal surfaces and the case of the strongly bound sulfur in Ru(0001)–c(4×2)S for which the RCT time is in the attosecond scale [41]. In these calculations, the adsorbate–substrate interaction and surface Hamiltonian were obtained from DFT calculations with finite slabs. The use of *ab initio* methods is instrumental in these systems since there are no reliable simple models to describe the electronic structure of those substrates and their interactions with different adsorbates.

The embedding method [37] allows performing DFT calculations in which only a few atomic layers closer to the surface are explicitly taken into account. The influence of the semi-infinite substrate is then included by adding an "embedding potential" that forces the solutions to have the correct logarithmic derivatives at the surface–bulk interface. Although this method is very powerful, its implementation is relatively cumbersome and its applications to the problem of RCT are currently limited [35, 36].

Finally, Nordlander and Taylor [38] have recently developed a deconvolution scheme to try to extract the width of the adsorbate energy levels from the discrete spectrum obtained using DFT slab and cluster calculations. The surfaces are represented by finite clusters, and the widths and positions of the resonances are obtained by calculating the projected density of states (constituted by a collection of Dirac deltas due to the finite size of the substrate model) of the joint adsorbate/substrate system upon the atomic levels. Each of these deltas is broadened by an artificial width β such that $\beta \gg \Gamma_{\mathrm{RCT}}$, giving rise to a single broad peak. This broad peak is fitted to a Lorentzian, obtaining certain full width at half maximum (FWHM). The artificial broadening β is then subtracted from FWMH to obtain an estimation of the RCT rate, the physical width, that is, $\Gamma_{\mathrm{RCT}} \approx \mathrm{FWHM}-\beta$. Results seem to converge rapidly width (the size) of the studied models. This method is promising due to its simplicity. However, to date it has been applied only to a very limited number of relatively simple systems, such as Li on different substrates [38].

We close this section by making a final remark on the accuracy of DFT to describe the electronic spectrum of adsorbed systems. Although DFT calculations are known to severely underestimate the band gap of semiconductors, they usually provide a reasonable description of the electronic structure of metal surfaces. They are also quite successful in describing chemisorption.[2)] However, DFT presents some limitations to predict the alignment of the adsorbate levels with respect to those of the substrate. This is due to the renormalization of the adsorbate levels induced by the polarization of the substrate [42, 43]. The shifts are different for occupied and unoccupied molecular levels and, as a consequence, the adsorbate HOMO–LUMO gap shrinks with respect to the gas phase. This effect cannot be described by DFT calculations using standard functionals. The addition of a $-1/4z$ image tail to the effective one-electron potential can approximately reproduce only the stabilization of the affinity level. In general, since the renormalization depends on the energetic position of the level, this effect cannot be reproduced using a single one-electron potential and requires a many-body calculation of the energy-dependent self-energy. Simple methods have been recently proposed to construct approximate self-energies including some of these effects [11, 44].

4.1.4
Effect of the Adsorbate Motion

As mentioned in the introduction, typical RCT processes for chemisorbed molecules at metal surfaces take place in the range of few femtosecond or even faster [4, 41]. Within this short timescale, the influence of the atomic motion on the electron dynamics is typically very small. From the point of view of the electrons, the nuclear

2) DFT description of physisorbed systems is typically less reliable due to the lack of van der Waals-like interactions using standard local or semilocal functionals.

velocities can be disregarded and the analysis performed above remains valid. When the electrons remain in the same state (characterized by a set of quantum numbers) along the whole trajectory of the nuclei, this regime is referred to as adiabatic dynamics. The situation can be different in the case of fast projectiles impinging upon a surface or systems with very large RCT times. Then, the nuclear and electron dynamics can take place on comparable timescales, and the coupling between both of them cannot be neglected. This regime corresponds to the so-called nonadiabatic electron dynamics, where the motion of the nuclei induces transitions between different electronic states. Regarding its influence on the charge transfer processes, this nonadiabatic regime will be considered in detail within the framework of Marcus theory in Section 4.2.

The following rate equation approach is frequently used when the coupling between electron and nuclear dynamics is assumed to be weak. Here, the RCT rate is taken from that calculated for the static case at the instantaneous positions. Therefore, the coupling between the nuclear motions and the short-time electron dynamics is completely neglected. Then $\Gamma_{\mathrm{RCT}}(t) = \Gamma_{\mathrm{RCT}}^{\mathrm{static}}(z(t))$, where $z(t)$ is the distance from the projectile to the surface at time t, and

$$\frac{\mathrm{d}P(t)}{\mathrm{d}t} = -\Gamma_{\mathrm{RCT}}^{\mathrm{static}}(z(t))P(t) \tag{4.6}$$

gives the time evolution of the survival probability $P(t)$, that is, the probability that the electron still remains in the projectile/adsorbate at time t. In this case, the survival probability after the complete collision event, that is, the probability that the electron does not leave the projectile at all, depends only on the perpendicular velocity $v_{\perp}$ and the turning point z_{min} of the trajectory,

$$P(\infty) = \exp\left(-\frac{2}{v_{\perp}}\int_{z_{\mathrm{min}}}^{\infty} \mathrm{d}z\, \Gamma_{\mathrm{RCT}}^{\mathrm{static}}(z)\right). \tag{4.7}$$

Borisov *et al.* [15] have studied the validity of this approximation to describe collisions between H^- ions and a Cu(111) surface as a function of the projectile velocity. They compared the results of the rate equation with those obtained using the wave packet propagation method, considering explicitly the motion of the projectile. They showed that, in this particular system, the rate equation provides only a good description of the survival probability for velocities well below 0.1 a.u. This interesting result is linked to the presence of a wide projected bad gap in the Cu(111) substrate as shown in Figure 4.2. The static RCT rate is strongly influenced by the presence of this gap. However, in the case of a moving adsorbate, the gap can have an impact only on the RCT process if the electrons have enough time to explore the electron–surface interaction potential during the collision. At large velocities, the RCT becomes insensitive to these details of the electronic structure, and the results are identical to those obtained with a jellium model for velocities larger than 0.2 a.u. This insensitivity to the details of the electronic structure of the target at sufficiently large velocities has also been observed using metallic thin films and small spheres as targets [45, 46]. Usman *et al.* [46] have discussed in detail the limits of validity of the rate equation approach as a function of the different timescales in the problem.

Besides fast particles, the atomic motion becomes relevant in the case of resonant states with very long lifetimes. The 6*s* resonance of Cs on Cu(111) is a good example, and with a lifetime of 50 fs, is a good intermediate for surface reactions. In particular, the population of this resonance initiates the break of the Cu-Cs bond and, therefore, induces a displacement of the Cs adsorbate from its equilibrium position. Petek *et al.* [47] have found clear signatures of this bond-breaking process in their time-resolved two-photon photoemission (2PPE) spectra of this system. However, comparison with theoretical results [48] indicates that the atomic motion in this case is slow enough, so the 6s Cs resonance evolves adiabatically as the atom moves away from the surface.

To understand the difference between adiabatic and nonadiabatic regimes in more detail, it is quite convenient to expand the exact wave function of the system, including electronic and nuclear degrees of freedom, using a particular basis set [49]:

$$\Psi(\vec{R}_\mu, \vec{r}_i, t) = \sum_n a_n(t)\chi_n^{\text{nucl}}(\vec{R}_\mu, t)\Phi^{\text{el}}_{n,\{\vec{R}_\mu\}}(\vec{r}_i). \tag{4.8}$$

Here, $\Phi^{\text{el}}_{n,\{\vec{R}_\mu\}}(\vec{r}_i)$ are the solutions of the Schrödinger equation for the electrons at the instantaneous nuclear coordinates $\{\vec{R}_\mu\}$:

$$\left[\sum_i -\frac{\hbar^2}{2m_{\text{e}}}\nabla_i^2 + \sum_{i,j>i}\frac{1}{r_{ij}} + V^{\text{ext}}_{\{\vec{R}_\mu\}}(\vec{r}_i)\right]\Phi^{\text{el}}_{n,\{\vec{R}_\mu\}}(\vec{r}_i) = E_n^{\text{el}}(\vec{R}_\mu)\Phi^{\text{el}}_{n,\{\vec{R}_\mu\}}(\vec{r}_i). \tag{4.9}$$

Note that the parametric dependence of the electronic wave functions on the atomic positions comes from the external potential $V^{\text{ext}}_{\{\vec{R}_\mu\}}(\vec{r}_i)$, which, in our notation, includes the electron–nuclear attraction and the repulsion between nuclei. The nuclear wave packets $\chi_n^{\text{nucl}}(\vec{R}_\mu, t)$ are given by

$$\left[\sum_\mu -\frac{\hbar^2}{2M_\mu}\nabla_\mu^2 + E_n^{\text{el}}(\vec{R}_\mu)\right]\chi_n^{\text{nucl}}(\vec{R}_\mu, t) = i\hbar\frac{\partial}{\partial t}\chi_n^{\text{nucl}}(\vec{R}_\mu, t), \tag{4.10}$$

where M_μ is the nuclear mass and $E_n^{\text{el}}(\vec{R}_\mu)$ the potential energy surface (PES) corresponding to a particular electronic state *n*, for example, describing a negatively charged ion in front of a surface. The time evolution of the coefficients a_n is given by the following equation:

$$i\hbar\frac{\partial a_n}{\partial t} = -\sum_m a_m \left(\begin{array}{c}\langle\chi_n^{\text{nucl}}|\chi_m^{\text{nucl}}\rangle\langle\Phi_n^{\text{elec}}|\sum_\mu \frac{1}{2M_\mu}\nabla_\mu^2|\Phi_m^{\text{elec}}\rangle \\ +\sum_\mu \frac{1}{M_\mu}\langle\chi_n^{\text{nucl}}|\vec{\nabla}_\mu|\chi_m^{\text{nucl}}\rangle\langle\Phi_n^{\text{elec}}|\vec{\nabla}_\mu|\Phi_m^{\text{elec}}\rangle.\end{array}\right) \tag{4.11}$$

Owing to the large values of M_μ, the right-hand side of Eq. (4.11) is in many cases negligible, and the wave function of the system can be described with a single term of the expansion in Eq. (4.8). In this case, the nuclear motions do not induce transitions between different electronic states and their dynamics are described using a single PES that can be associated with a particular electronic state n. This corresponds to the adiabatic approximation, where the electronic properties of the system at each instant of time correspond to those of the static system with the same instantaneous nuclear positions. This adiabatic approximation breaks down in some cases, and the nonadiabatic effects associated with the cross-terms in Eq. (4.11) become relevant. As already mentioned, this happens at high velocities where the second term on the right-hand side of Eq. (4.11) becomes large or for very light nuclei such as hydrogen and helium. Nonadiabatic effects are also important when the system visits configurations $\{\vec{R}_\mu\}$ for which two or more electronic states are degenerate or quasi-degenerate. In this situation, $|E_n^{\rm el}(\vec{R}_\mu)-E_m^{\rm el}(\vec{R}_\mu)|$ is smaller or comparable to the matrix elements in Eq. (4.11) and significant transitions between these two electronic states can be induced by the nuclear movements.

Heavy nuclei are treated as classical particles in most theoretical approaches. Unfortunately, there is no unique way to go from the exact quantum formulation to a mixed quantum–classical dynamics in which electrons are treated quantum mechanically and nuclei classically [50, 51] and this involves quite severe approximations. The most straightforward approximation is based on the Ehrenfest theorem, where the classical limit can be reached by substituting the nuclear wave packets by δ functions [49]. The effective forces acting on the nuclei are calculated from a mean field potential averaged over the quantum (electronic) degrees of freedom. This is the so-called Ehrenfest dynamics that, if the electrons are forced to remain in the ground state, reduces to the Born–Oppenheimer adiabatic dynamics [49]. Ehrenfest dynamics, in combination with time-dependent DFT (TD-DFT) [52] used to obtain the electronic mean field, has become a popular approximation in recent years and has been applied to chemisorption problems [53]. Although Ehrenfest dynamics can describe some nonadiabatic effects, it is known to fail in some important situations such as current-induced heating or nonradiative electron–nuclear relaxation [54]. To overcome these limitations, several alternative methods have been proposed, many of them based on the intuitive idea of trajectory hopping between different PES corresponding to different electronic configurations [51]. The most popular of these methods is Tully's "fewest switchings" surface hopping method [55], which in recent years has been combined with TD-DFT methods within both linear response [50] and nonperturbative [56] regimes. This method has been applied to study many systems of interest, including the problem of charge separation in dye-sensitized semiconductor solar cells [56, 57]. These methodologies are paving the way for the first-principles treatment of the highly nonadiabatic regime of electron transfer that we consider in the next section. Here, however, we will limit our analysis to the more qualitative picture that arises from Marcus theory [1].

4.2
Electron Transfer in the Presence of Polar/Polarizable Media

Modern theory of nonadiabatic electron transfer (ET) is in large part due to the contributions of Marcus along with Hush, Levich, Jortner, and others [1, 58–60].

Marcus won the Nobel Prize in Chemistry in 1992 for his pioneering work in this field, so the nonadiabatic electron transfer theory is sometimes referred to as "Marcus theory" or the "Marcus picture." The term *nonadiabatic* (or *diabatic*) refers to the nuclear potential energy surfaces of the electron-donating and -accepting species that are considered to be weakly coupled. In contrast, an *adiabatic* transition is one in which coupling is sufficiently strong such that the donor and acceptor centers represent two local minima on the same continuous potential energy surface (cf. Figure 4.4).

In the following, the two limiting cases, nonadiabatic and adiabatic ET for weak and strong electronic coupling, respectively, are introduced for homogeneous electron transfer, that is, transfer from one distinct electronic state to another. These cases are then extended to the heterogeneous problem where the acceptor is a semiconductor or metal. Finally, Section 4.2.3 focuses on the regime of intermediate coupling, which becomes important when the transfer reaction occurs on comparable timescales as the solvent relaxation.

4.2.1
Nonadiabatic (Outer Sphere) Electron Transfer

To understand the generality of nonadiabatic electron transfer theory as well as the limitations and assumptions implicit to it, it is worthwhile to review the derivation of the Marcus rate expression. We begin our analysis with consideration of two localized states, a donor and an acceptor, in a solvent, as illustrated by the parabolas in Figure 4.4. Each of them describes one pathway through the very complex energy landscape of the solvent–solute system along the collective solvent (reaction) coordinate q. This picture and the following considerations are based on the assumption

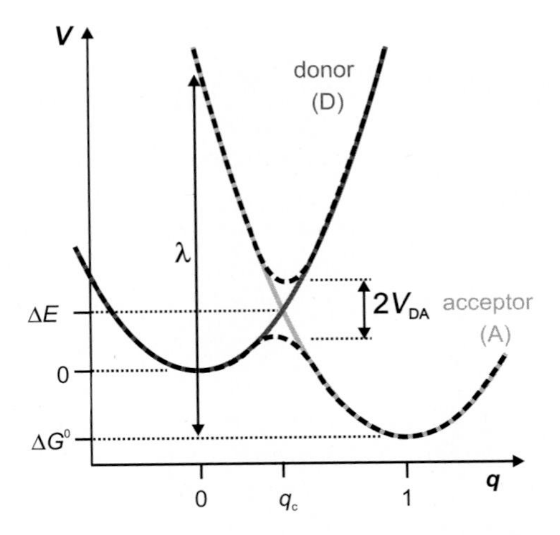

Figure 4.4 Diabatic (solid) and adiabatic (dashed) potential energy surfaces for homogeneous electron transfer between two separate states in a solution. For intermediate coupling (cf. Section 4.2.3), the avoided crossing is highly exaggerated.

of linear response of the solvent molecules and of a fixed distance of donor and acceptor states.

Before the transfer reaction occurs, the solvent has organized itself in such a way as to stabilize the excess electrons in the donor state. If the electrons were to make a transition to the acceptor state on a timescale much faster than the motion of the solvent molecules (vertical transition), then the solvent would no longer be in its most stable configuration. The free energy released in the subsequent relaxation of the solvent to its new most stable configuration (i.e., by reorientation of molecular dipoles) is called the solvent reorganization energy. The reorganization energy is typically referred to by the symbol λ and takes on values $\sim$0.1–1 eV in aqueous solutions [61]. If the electron transfer step were to proceed as described, it would generate a system with λ excess energy – considerably more than the available $k_B T$ at room temperature. What Marcus realized is that, because of the disparate timescales for electronic and nuclear motion, ET takes place only when the donor and the acceptor are both in such a configuration that the charge transfer step is energy conserving, specifically similar to the RCT processes described in previous sections. This means that the instantaneous nuclear arrangement during the charge transfer step is one in which the free energy of the system does not change whether the electron is found on the donor or the acceptor. The activation energy for such a charge transfer reaction is only the energy required for the deformation of the nuclear coordinates to the charge transfer configuration – considerably less energy than needed for the vertical transition from the stable donor configuration initially suggested. It is straightforward to show algebraically that, for parabolic potential energy surfaces, the activation energy required to reach the nuclear charge transfer arrangement is

$$\Delta E = \frac{(\Delta G^0 + \lambda)^2}{4\lambda}, \tag{4.12}$$

where ΔG^0 is the free energy difference between donor and acceptor states in equilibrium (cf. Figure 4.5). In general, the reorganization energy, λ, has contributions from both the surrounding solvent environment and intramolecular vibrational modes.

The likelihood of the system reaching the charge transfer configuration is proportional to the Boltzmann factor

$$\exp\left[-\frac{(\Delta G^0 + \lambda)^2}{4\lambda k_B T}\right].$$

The probability (per unit time) of the reaction occurring once this nuclear configuration has been reached is given by Fermi's golden rule for the thermally averaged transition rate between vibronic levels associated with the donor and acceptor electronic states:

$$k_{ET}(\Delta G^0) = \frac{2\pi}{\hbar}|V_{DA}|^2 F(\Delta G^0). \tag{4.13}$$

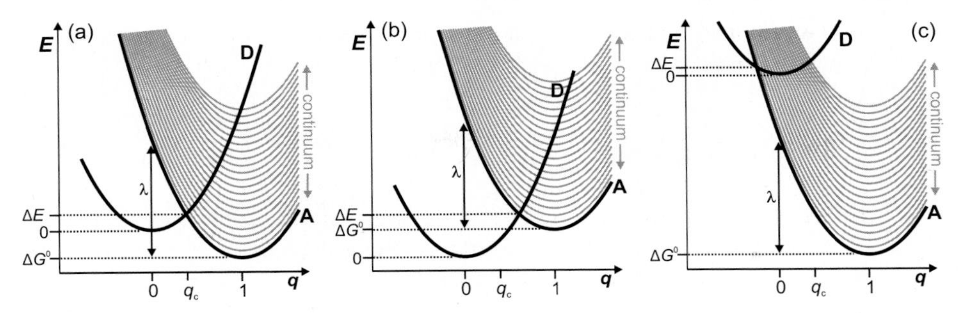

Figure 4.5 Free energy as a function of configuration coordinate q for electron transfer from a donor state to an acceptor state (thick curves). (a and b) Normal Marcus region for electron transfer with $\Delta G^0 < 0$ and $\Delta G^0 > 0$, respectively. (c) Case of the Marcus inverted region. The gray continuum represents a delocalized acceptor band (e.g., the conduction band of a semiconductor).

Here, V_{DA} is the electronic coupling matrix element and the function

$$F(\Delta G^0) = \frac{1}{Q_D} \sum_i e^{-E_{D,i}/k_B T} \sum_f \left| \langle \chi_{D,i} | \chi_{A,f} \rangle \right|^2 \delta(\Delta G^0 + E_{D,i} - E_{A,f}) \qquad (4.14)$$

is the thermally averaged Franck–Condon factor. $Q_D = \sum_i \exp(-E_{D,i}/k_B T)$ is the nuclear partition function in the donor electronic state, $\langle \chi_{D,i} | \chi_{A,f} \rangle$ is the overlap of initial and final state nuclear wave functions, and the summations are over vibronic levels associated with the initial (i) and final (f) electronic states. In the high-temperature (i.e., classical) limit, $F(\Delta G^0)$ is given by [62]

$$F(\Delta G^0) = \frac{1}{\sqrt{4\pi\lambda k_B T}} \exp\left[-\frac{(\Delta G^0 + \lambda)^2}{4\lambda k_B T} \right]. \qquad (4.15)$$

This expression assumes that $k_B T \gg \hbar\omega_Q$, where ω_Q is any vibrational mode involved in the transition, and that vibrational frequencies are not changed by charge transfer. The complete Marcus rate expression, in the high-temperature and weak coupling limit, is

$$k_{ET}(\Delta G^0) = \frac{2\pi}{\hbar} |V_{DA}|^2 \frac{1}{\sqrt{4\pi\lambda k_B T}} \exp\left[-\frac{(\Delta G^0 + \lambda)^2}{4\lambda k_B T} \right]. \qquad (4.16)$$

Equation (4.16) is generally valid at room temperature for most polar solvents – particularly in aqueous environments and for biological systems [59].

4.2.1.1 Continuum of Accepting States

When electron or hole transfer occurs between a localized donor species and a bulk semiconductor electrode, as illustrated by the gray continuum in Figure 4.5, one must consider a continuum of energy levels that may participate in the transition [63]. The Marcus rate expression in Eq. (4.16) must now be integrated over all possible electron (hole) accepting (donating) levels within the solid. For the case of molecule-to-semiconductor electron transfer, such as electron injection in dye-sensitized solar cells, the ET rate is given by

$$k_{\mathrm{ET}} = \frac{2\pi}{\hbar}\frac{1}{\sqrt{4\pi\lambda k_{\mathrm{B}}T}}\int |V_{\mathrm{DA}}(E)|^2(1-f(E))\varrho(E)\exp\left[-\frac{(E+\lambda)^2}{4\lambda k_{\mathrm{B}}T}\right]\mathrm{d}E, \qquad (4.17)$$

where $\varrho(E)$ is the density of states, $f(E)$ is the fractional occupation of states at energy E, and $V_{\mathrm{DA}}(E)$ is the k-space integrated and energy-dependent electronic coupling strength. The integration is from the conduction band minimum to maximum. In the normal Marcus regions, panels (a) and (b), the result of integration in Eq. (4.17) yields an electron transfer rate that depends on T, λ, and ΔG^0, qualitatively similar to Eq. (4.16) for the two localized states. However, under the broadband and constant DOS approximation, the two temperature-dependent terms in Eq. (4.17) cancel out after integration for the Marcus inverted region (Figure 4.5c) and k_{ET} becomes independent of T, λ, and ΔG^0 [64]. Thus, within these approximations, the Marcus inverted region does not exist or is insignificant for a localized state in resonance with a broadband of electronic states.

4.2.2
Adiabicity and the Effect of Strong Electronic Coupling

While Marcus theory has found broad applicability in a variety of systems, the preceding discussion of nonadiabatic ET may have limited validity when it comes to charge transfer between a localized state and a delocalized metal or semiconductor band. The central assumption underlying Eqs. (4.16) and (4.17) is that thermally activated nuclear rearrangement is the rate-limiting step in the charge transfer reaction. Furthermore, it is assumed that electronic coupling is weak enough to be treated perturbatively via the golden rule. This is not always the case. When electronic coupling between the donor and the acceptor is strong relative to the reorganization energy, ET occurs adiabatically. Instead of thinking of electron transfer as a discrete event occurring between two localized states, the adiabatic charge transfer reaction is best described as a time-dependent redistribution of charge on one continuous equilibrated potential energy surface, as shown by the adiabatic curve (dashed) in Figure 4.6. In the extreme case when the barrier between the donor and the acceptor becomes very small, the two states are no longer distinguishable, and electron "transfer" reduces to a quantum mechanical dephasing process. This clearly corresponds to the standard definition of adiabatic process presented in Section 4.1.4 in which, when the gap between different electronic states is large enough compared to the nonadiabatic terms appearing in Eq. (4.11), the electrons remain in the same state during the whole trajectory.

The treatment of strong electronic interaction at surfaces is best handled within the framework of chemisorption theory, which deals with the coupling of discrete atomic or molecular orbitals to electronic bands in solids and is presented in detail in Section 4.1. We have shown there that the interaction of a discrete adsorbate state with the substrate electronic band results in the formation of a broadened *adsorbate*

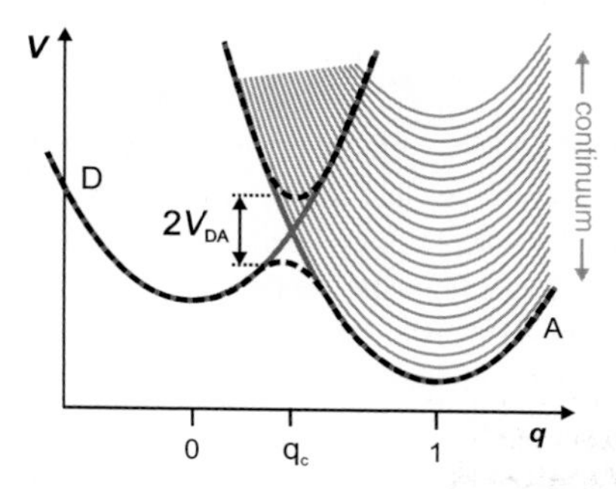

Figure 4.6 Free energy as a function of configuration coordinate q for electron transfer from a localized state (left) to a delocalized band (right). The gray (crossing) curves are parabolic diabatic free energy surfaces in the weak coupling limit. The dashed anticrossing curves are the adiabatic free energy surfaces of the coupled donor–acceptor system. The activation energy is lowered by the electronic coupling strength V_{DA}.

resonance of Lorentzian line shape with width 2Δ, with Δ given by Eq. (4.4) evaluated at the resonance energy. The hopping matrix element appearing in Eq. (4.4) is equivalent to the electronic exchange coupling energy (V_{DA}) in Eqs. (4.16) and (4.17). It is a measure of the electronic coupling strength between a discrete adsorbate orbital and a substrate electronic band. Interfacial ET processes are classified into three coupling regimes, depending on the relative magnitudes of Δ and λ [2].

In the case of $\Delta \ll 2\lambda/\pi$, Δ has little effect on the activation energy for charge transfer, which is determined almost entirely by λ and ΔG^0. This is the nonadiabatic limit presented in the previous section. In the second regime, or the intermediate coupling regime, $k_B T < \Delta \leq 2\lambda/\pi$. In this case, the transfer reaction occurs adiabatically and the activation energy is lowered by Δ (Figure 4.6). Electron transfer is still a thermally activated process and depends on nuclear rearrangement, but the constraints on the charge transfer configuration are lifted to some extent. This regime of ET will be discussed later. Finally, when $\Delta > 2\lambda/\pi$, we have reached the strong coupling regime and nuclear rearrangement plays little role in the charge transfer process. In the strong coupling limit, the donor and acceptor states are no longer distinguishable. The donor energy level is itself an eigenstate of the entire coupled system and has probability density throughout the solid. Charge transfer in this limit amounts to electronic dephasing between the adsorbate resonance and continuum states within the solid. This ultrafast process occurs on femtosecond timescales, with the electron lifetime in the adsorbate resonance governed by the uncertainty principle (Eq. (4.5)).

4.2.3
Intermediate Coupling and the Impact of Solvent Relaxation

As discussed previously, the treatment of electron transfer reactions depends crucially on the degree of electronic coupling between the donating and the accepting state. Accordingly, the preceding sections focused on the development of the rate constants for ET in the *weak* and the *strong* coupling limit. In the case of nonadiabatic transfer, the relaxation dynamics of the solvent are the rate-limiting step for the transfer reaction so that the ET rate is dominated by thermally activated transfer across the (sharp) nuclear barrier. If, however, the relaxation in the accepting well occurs on comparable timescales as the transfer reaction, that is, the electronic coupling leads to an adiabatic (cusped) potential energy surface (cf. Figure 4.4), interactions between the transfer and the solvation reaction have to be considered. This is particularly the case if the ET occurs in the presence of a *non-Debye* solvent, that is, a solvent that exhibits more than one characteristic relaxation time [65]. One approach to tackle this phenomenon is presented in the following. It is based on the assumption of a cusped potential energy surface allowing consideration of an adiabatic reaction *and* a potential barrier that is unaffected by solvent fluctuations [66].

Analogous to Section 4.2.1, the potential wells are defined by

$$F_D(q) = \lambda q^2, \tag{4.18}$$

$$F_A(q) = \lambda(q-1)^2 + \Delta G^0, \tag{4.19}$$

where $F_{D/A}$ denotes the nonequilibrium free energies, λ the reorganization energy that is associated with the force constant of the solvent, and ΔG^0 the equilibrium free energy of the transfer reaction. Linear combination of the solutions χ_D and χ_A of these diabatic harmonic potentials provides the adiabatic state (cf. Figure 4.7)

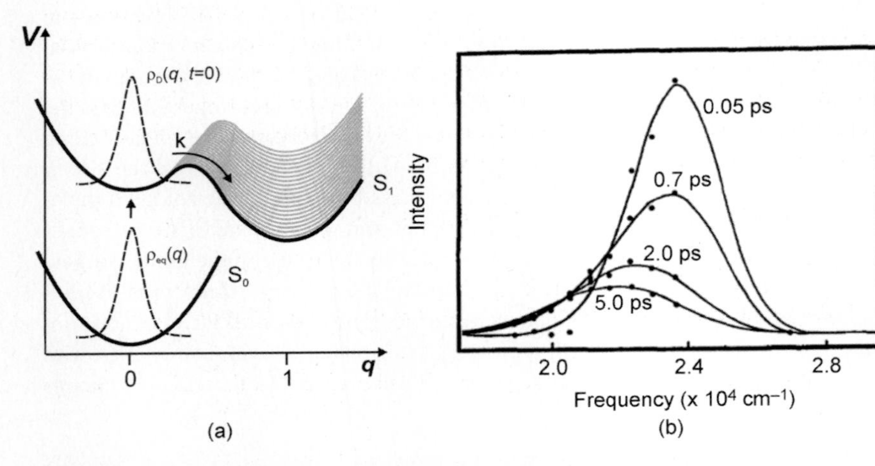

Figure 4.7 (a) Initial condition for excited-state electron transfer: the ground-state equilibrium distribution is projected to the excited state S_1. The electron transfer reaction is then determined by the time-dependent progression along the reaction coordinate q. The gray continuum illustrates a continuum of accepting states, comparable to nonadiabatic case discussed in Section 4.2.1. (b) Time-resolved emission spectra of 9,9′-bianthryl in acetone (markers) and empirical modeling on the basis of Eqs. (4.24a) and (4.24b) (solid curves). Reprinted from [69] with permission from Elsevier.

$$|\chi_{S_1}\rangle = d|\chi_D\rangle + a|\chi_A\rangle \tag{4.20}$$

considering the Hamiltonian matrix

$$\begin{pmatrix} H_{DD} & H_{DA} \\ H_{AD} & H_{AA} \end{pmatrix} \quad \text{with} \quad <\chi_D|\chi_A>=<\chi_A|\chi_D>= \text{const.} \tag{4.21}$$

To achieve the time-dependent evolution of the reaction coordinate, the generalized Langevin equation can be used [67]:

$$m \cdot \frac{\partial^2 q(t)}{\partial t^2} = -2\lambda \cdot q(t) - m \cdot \int_0^t \eta(t-\tau)\frac{\partial q(\tau)}{\partial \tau}\,d\tau + f(t), \tag{4.22}$$

where $\eta(t)$ is the longitudinal, time-dependent friction kernel and $f(t)$ a stochastic force related to η [65]. It is important to mention that the relaxation of the solvent is assumed to be overdamped, that is, $\ddot{q} = 0$ on the characteristic timescales of the solvent, and that the non-Debye character[3)] is accounted for by the introduction of two independent longitudinal relaxation rates τ_1 and τ_2 that contribute to $\eta(t)$.

Following Zusman [68], the probability distribution functions of donor and acceptor states, respectively, may then be described by

$$\dot{\varrho}_D(q,t) = -H_{DA}(q,t) \cdot \delta(q-q_c) + L_D(t)\varrho_D(q,t), \tag{4.23a}$$

$$\dot{\varrho}_A(q,t) = H_{DA}(q,t) \cdot \delta(q-q_c) + L_A(t)\varrho_A(q,t) \tag{4.23b}$$

with $H_{DA}(q,t) = (2\pi V_{DA}^2/\hbar) \cdot [\varrho_D(q,t) - \varrho_A(q,t)]$ so that the electronic coupling contributes only at the crossing point of the donor and acceptor potentials q_c, leading to a redistribution of the population between the donor and the acceptor. Even if the time-dependent Liouvillian operators $L_i(t)$ for the motion in the respective potential well are unknown, Eqs. (4.23a) and (4.23b) can be solved by Laplace transforming and replacing $L_i(t)$ with effective time-independent $L_i^{\text{eff}}(t)$:

$$s \cdot \hat{\varrho}_D(q,s) - \varrho_D(q,t=0) = -\hat{H}_{DA}(q,s) \cdot \delta(q-q_c) + L_D^{\text{eff}}(t)\hat{\varrho}_D(q,s), \tag{4.24a}$$

$$s \cdot \hat{\varrho}_A(q,s) - \varrho_D(q,t=0) = \hat{H}_{DA}(q,s) \cdot \delta(q-q_c) + L_A^{\text{eff}}(t)\hat{\varrho}_A(q,s) \tag{4.24b}$$

with the initial probability distribution $\varrho_D(q,t=0)$. As the solution

$$\Psi(q,q',t) = \left\{2\pi\langle q^2\rangle_{\text{eq}}\left[1-\Delta^2(t)\right]\right\}^{-1/2} \exp\left[-\frac{(q-q'\Delta(t))^2}{2\langle q^2\rangle_{\text{eq}}(1-\Delta^2(t))}\right] \tag{4.25}$$

is known for the limiting case of an overdamped oscillator, Eqs. (4.24a) and (4.24b) can be solved using the normalized time correlation function $\Delta(t)$ [65].

3) In the case of a Debye solvent where $\eta(t) \propto \tau_L\delta(t)$, which means it has only one characteristic longitudinal relaxation time, Eq. (4.14) reduces to an ordinary Langevin equation and the rate constant for ET becomes the one introduced in Section 4.1.

It is straightforward to project the equilibrium ground-state (S_0) distribution $\varrho_{eq}(q)$ (cf. Figure 4.7a) to the adiabatic excited-state potential S_1. This corresponds to (photo) excitation of the system under Born–Oppenheimer approximation. Using the ground-state equilibrium distribution $\varrho_{eq}(q)$, the initial distribution $\varrho(q, t=0)$ becomes a simple Gaussian. This approach has been successfully applied to model time-dependent emission spectra of large molecules in solution as shown in Figure 4.7b.

4.2.3.1 Classical Description and a Wide Band Acceptor

Returning to the classical description of electron transfer in the weak coupling limit of a Debye solvent using two diabatic potentials (Section 4.2.1), it is straightforward to introduce a coupling term V_{DA} that reduces the effective barrier in Eq. (4.15):

$$F(\Delta E) = \frac{1}{\sqrt{4\pi\lambda k_B T}} \exp\left[-\frac{(\Delta E - V_{DA})}{k_B T}\right]. \tag{4.26}$$

Here, ΔE is the nuclear barrier defined by the crossing of the two diabatic curves introduced earlier in Eq. (4.12) and the (intermediate) coupling term is treated as a perturbation of the nonadiabatic case. As for all considerations so far, this expression is based on the assumption of a thermal distribution in the donor diabatic potential as the initial condition. Analogous to Section 4.2.1.1, Eq. (4.26) can be extended to the heterogeneous case with a continuum of accepting states

$$k_{ET} \propto \int |V_{DA}|^2 (1 - f(E)) \varrho(E) \exp\left[-\frac{\Delta E(E) - V_{DA}}{k_B T}\right] dE \tag{4.27}$$

with energy-dependent diabatic barriers

$$\Delta E(E) = \frac{(E-\lambda)^2}{4\lambda}. \tag{4.28}$$

This situation is illustrated by the gray continuum in Figure 4.7a, although it should be noted that V_{DA} must not necessarily be constant for all accepting states, but may well depend on energy $V_{DA}(E)$ due to, for instance, varying degrees of localization of donor and acceptor states. The probability for a transfer reaction on the manifold of S_1 surfaces (gray) from $q=0$ to $q=1$ is thus determined by the reorganization energy λ of the solvent, the electronic properties of the accepting substrate $\varrho(E)$, and the coupling V_{DA} between them. This illustrates the complexity of heterogeneous electron transfer in general and the challenge of a description *beyond* the classical limit discussed here.

4.3 Transient Electronic Coupling: Crossover between Limiting Cases

While Section 4.1 dealt with heterogeneous electron transfer in the limit of frozen nuclear coordinates, and Section 4.2 discussed how HET is influenced or even dominated by the nuclear response, the remaining part of this chapter will focus on an example of experimental evidence for a crossover between these limiting cases. When searching for an appropriate system to investigate the interplay of electron transfer and nuclear motion, it is crucial to match the critical timescales of these processes. Typical excited-state lifetimes in front of metal surfaces or wide band semiconductors range from few femtoseconds up to hundreds of picoseconds, while characteristic times of nuclear motion[4)] usually occur on pico- or nanosecond timescales. Adsorption of atoms or molecules on single crystal surfaces is capable of increasing excited electron lifetimes, as, demonstrated for image potential states at rare gas/metal interfaces. It could be shown experimentally [70, 71] and theoretically [70–73] that both the polarizability of the rare gas medium and the relative position to the substrate band gap determine the changes in the excited-state lifetimes. The HET process itself occurs in these systems elastically and nuclear motion does not need to be considered. The dynamic screening of the medium is caused by (electronic) polarization of the adsorbate and therefore occurs much faster than the electron transfer from the image potential state to the substrate. This is, for example, different when introducing a *polar* medium as a sudden creation of charge density in such an environment can cause a reorientation of molecular dipoles or even translational motion of the molecules. It has been shown experimentally for various polar adsorbates on metal surfaces [74–78] that this dipolar screening sufficiently enhances the excited-state lifetime to enable the observation of nuclear reorientation on comparable timescales as the electron transfer process itself: The rearrangement of the molecular dipoles is reflected in an increase of the binding energy of the excited electron state due to the so-called *electron solvation*.

Figure 4.8a shows the time-dependent evolution of the population of the solvated electron state in an amorphous ice film on two different substrates, Cu(111) and Ru(001), to illustrate the competition of electron transfer and screening by the polar molecules. At early times ($t < 300$ fs), the population decays much faster in the case of the ruthenium substrate. This is a result of the four times narrower surface band gap of Ru(001) compared to Cu(111) around 3 eV above E_F [79, 80]. As discussed in Section 4.1.2, the width of the band gap strongly affects elastic electron transfer. At later times ($t > 300$ fs), however, the electron transfer rate (related to the slope of the traces) assimilates for both samples. The dipoles of the ice started reorienting and thereby dynamically screening the excited electron from the continuum of accepting states in the metal. This reduces the wave function overlap of the electron with the substrate and therefore the transfer rate. However, as shown in Ref. [64], nuclear motion does not rate limit the transfer process itself; the HET occurs in the strong coupling limit and is fast compared to any response of the solvent molecules.

This observation is illustrated by the phenomenological potential energy landscape in Figure 4.9. It merges the two relevant coordinates, the real space coordinate z, along which the actual transfer process occurs, and the nuclear (solvent) coordinate q. The two limiting cases of strong and weak coupling, discussed in Sections 4.1 and 4.2, are indicated by panels (b) and (c), respectively, that display horizontal (vertical) cuts

4) These timescales refer to a relaxation of the ion cores in quasi-equilibrium with the electronic system and not to coherent phonons.

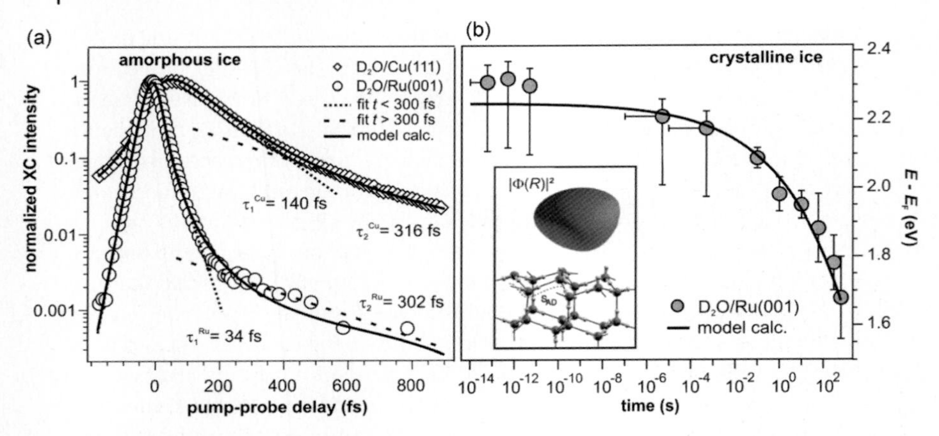

Figure 4.8 (a) Population dynamics of the solvated electron state in amorphous ice layers on two single crystal surfaces probed by 2PPE. For $t < 300$ fs, the orientational band gap of the substrate affects the electron transfer significantly, while dynamical screening of the polar molecules leads to a slowing down of charge transfer at larger delays. Reprinted with permission from [81]. Copyright (2006) American Chemical Society. (b) Energetic relaxation of trapped electrons at the vacuum interface of supported ice crystallites at 30K. The dynamics span a wide range of timescales up to minutes due to efficient decoupling of the excess electrons by orientational defects at the ice surface (inset). Reprinted with permission from [82]. Copyright (2009) American Chemical Society.

through the two-dimensional potential energy surface. In the experiment, the photoexcitation creates an excess electron in the ice at (z_D, q_A) and initially electron transfer occurs only along z and does not involve nuclear motion (process (1) in Figure 4.9a). As electron solvation proceeds (dashed arrows), the binding energy of the electrons increases. Simultaneously, the electrons are screened from the continuum of accepting states in the metal (shaded area) by the molecular dipoles. This is illustrated by the increasing barrier in Figure 4.9b. This *decoupling* of the excited electron from the continuum of acceptor states eventually leads to a situation where nuclear rearrangement, that is, motion along q, can *promote* the transfer reaction. At this point, the HET involves both electron and nuclear coordinates (process (2) in Figure 4.9a).

As previously discussed whether and when this point is reached strongly depends on whether the excited electrons are quickly decoupled from the metal or semiconductor acceptor adequately to allow nuclear motion to play a role. Although this is not the case for the solvated electrons in amorphous ice layers (Figure 4.8), crystallization of the ice can create very deep traps for electrons at the vacuum interface. Density functional theory calculations [82] suggest that these result from orientational defects at the ice surface, as illustrated by the inset in Figure 4.8b. For this system, the electrons stay in the excited state up to several minutes (note the logarithmic time axis), enabling the observation of their energetic relaxation over 17 orders of magnitude in time. This long residence time illustrates that resonant elastic electron transfer is basically suppressed and a treatment along the lines of Section 4.1 inappropriate. The rate-limiting step for HET is thermally activated fluctuation of

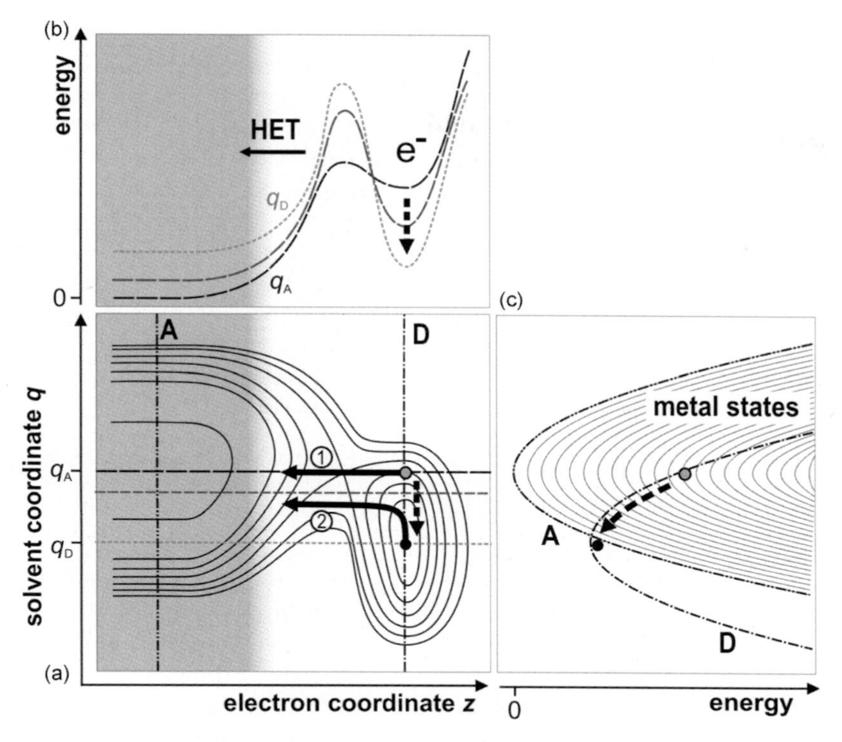

Figure 4.9 Energy potential landscape governing the crossover of HET between the limiting cases of the strong (1) and weak (2) electronic coupling. Horizontal cuts (dashed lines) (a) along the real space (electron) coordinate are displayed in (b) and vertical cuts (dash-dotted) are shown in (c). The energetic relaxation of the excited carriers due to solvation is indicated by the dashed arrows. reprinted with permission from Ref. [64]. Copyright (2007) by Institute of Physics Publishing.

the solvent along q (Figure 4.9) that eventually leads to the charge transfer from the adsorbate to the metal along the real space coordinate z.

4.4 Conclusions

Heterogeneous electron transfer is a fundamental problem that cuts across traditional boundaries of scientific research. Because of the ubiquitous nature of heterogeneous electron transfer in a wide range of physical and chemical processes, we have seen the nearly independent development of theories and languages that on the surface seem orthogonal but in reality describe the same problem at different limits. In this chapter, we attempt to provide a unified view of heterogeneous electron transfer at three limits as determined by the relative magnitude of electronic coupling energy versus nuclear reorganization energy. When electronic coupling is much larger than reorganization energy, we are at the strong coupling limit and heterogeneous electron transfer is an ultrafast process described by resonant charge transfer or chemisorption theories. This picture is most common in surface physics. When electronic coupling is much smaller than reorganization energy, we are in the weak coupling limit and heterogeneous electron transfer is well described by the Marcus theory commonly used in chemistry. Interestingly, the small polaron theory developed by Holstein [83] is essentially equivalent to the Marcus theory. When the electronic coupling energy is of the same order as nuclear reorganization energy, we are in the intermediate coupling region, which is perhaps most difficult to describe theoretically. We present the semiclassical approach of Zusman and others, which treats this coupling region. More sophisticated treatments are not discussed here, but may be found in the work on transition state theory in describing chemical dynamics and rate processes. Our discussion on heterogeneous electron transfer is by no means exhaustive or rigorous. We hope this chapter will serve as an introduction to readers interested in heterogeneous electron transfer and provide a conceptual starting point for further reading or research.

References

1 Marcus, R.A. (1956) *J. Chem. Phys.*, **24**, 966; Marcus, R.A. (1957) *J. Chem. Phys.*, **26**, 867; Marcus, R.A. (1957) *J. Chem. Phys.*, **26**, 872; Marcus, R.A. (1960) *Disc. Faraday Soc.*, **29**, 21.

2 Zhu, X.-Y. (2004) *Surf. Sci. Rep.*, **56**, 1–83.

3 Borisov, A.G., Gauyacq, J.P., Chulkov, E.V., Silkin, V.M., and Echenique, P.M. (2002) *Phys. Rev. B*, **65**, 235434.

4 Petek, H., Nagano, H., Weida, M.J., and Ogawa, S. (2001) *J. Phys. Chem.*, **105**, 6767.

5 Parr, R.G. and Yang, W. (1989) *Density Functional Theory of Atoms and Molecules, International Series of Monographs on Chemistry 16*, Oxford University Press, New York.

6 Horn, K. and Scheffler, M. (eds) (2000) *Handbook of Surface Science*, vol. 2, Electronic Structure (series editors N.V. Richardson and S. Holloway), Elsevier, Amsterdam.

7 Davison, S.G. and Sulston, K.W. (2006) *Green's Function Theory of Chemisorption*, Springer.

8 Grimley, T.B. (1967) *Proc. Phys. Soc. London*, **90**, 751; Grimley, T.B. (1967) *Proc. Phys. Soc. London*, **92**, 776.

9 Newns, D.M. (1969) *Phys. Rev.*, **178**, 1123.

10 Anderson, P.W. (1961) *Phys. Rev.*, **124**, 41.

11 Sau, J.D., Neaton, J.B., Choi, H.J., Louie, S.G., and Cohen, M.L. (2008) *Phys. Rev. Lett.*, **101**, 026804.

12 Chulkov, E.V., Borisov, A.G., Gauyacq, J.P., Sánchez-Portal, D., Silkin, V.M., Zhukov, V.P., and Echenique, P.M. (2006) *Chem. Rev.*, **106**, 4160.

13 Economou, E.N. (2006) *Green's Functions in Quantum Physics*, Springer, Berlin.

14 Fano, U. (1961) *Phys. Rev.*, **124**, 1866.

15 Borisov, A.G., Kazansky, A.K., and Gauyacq, J.-P. (1999) *Phys. Rev. B*, **59**, 10935.

16 Limot, L., Pehlke, E., Kröger, J., and Berndt, R. (2005) *Phys. Rev. Lett.*, **94**, 036805.

17 Olsson, F.E., Persson, M., Borisov, A.G., Gauyacq, J.-P., Lagoute, J., and Fölsch, S. (2004) *Phys. Rev. Lett.*, **93**, 206803.

18 Borisov, A.G., Gauyacq, J.P., Kazansky, A.K., Chulkov, E.V., Silkin, V.M., and Echenique, P.M. (2001) *Phys. Rev. Lett.*, **86**, 488.

19 Bauer, M., Pawlik, S., and Aeschlimann, M. (1999) *Phys. Rev. B*, **60**, 5016.

20 Nordlander, P. and Tully, J.C. (1988) *Phys. Rev. Lett.*, **61**, 990.

21 Sánchez-Portal, D., Menzel, D., and Echenique, P.M. (2007) *Phys. Rev. B*, **76**, 235406.
22 Bardeen, J. (1961) *Phys. Rev. Lett.*, **6**, 57.
23 Tersoff, J. and Hamann, D.R. (1983) *Phys. Rev. Lett.*, **50**, 1998.
24 Diez Muiño, R., Sánchez-Portal, D., Silkin, V.M., Chulkov, E.V., and Echenique, P.M. (2011) *PNAS*, **108**, 971.
25 Wu, R. and Freeman, A.J. (1992) *Phys. Rev. Lett.*, **69**, 2867.
26 Lang, N. and Kohn, W. (1970) *Phys. Rev. B*, **1**, 4555; Lang, N. and Kohn, W. (1973) *Phys. Rev. B*, **3**, 1215; Lang, N. and Kohn, W. (1973) *Phys. Rev. B*, **7**, 3541.
27 Serena, P.A., Soler, J.M., and García, N. (1986) *Phys. Rev. B*, **34**, 6767.
28 Deutscher, S.A., Yang, X., and Burgdörfer, J. (1995) *Nucl. Instrum. Methods B*, **100**, 336–341; Deutscher, S.A., Yang, X., Burgdörfer, J., and Gabriel, H. (1996) *Nucl. Instrum. Methods B*, **115**, 152.
29 Nordlander, P. and Tully, J.C. (1990) *Phys. Rev. B*, **42**, 5564.
30 Martin, F. and Politis, M.F. (1996) *Surf. Sci.*, **356**, 247.
31 Teillet-Billy, D. and Gauyacq, J.P. (1990) *Surf. Sci.*, **239**, 343.
32 Kürpick, P., Thumm, U., and Wille, U. (1997) *Nucl. Instrum. Methods B*, **125**, 273.
33 Merino, J., Lorente, N., Pou, P., and Flores, F. (1996) *Phys. Rev. B*, **54**, 10959.
34 Sánchez-Portal, D. (2007) *Prog. Surf. Sci.*, **82**, 313.
35 Butti, G., Caravati, S., Brivio, G.P., Trioni, M.I., and Ishida, H. (2005) *Phys. Rev. B*, **72**, 125402.
36 Achilli, S., Trioni, M.I., Chulkov, E.V., Echenique, P.M., Sametoglu, V., Pontius, N., Winkelmann, A., Kubo, A., Zhao, J., and Petek, H. (2009) *Phys. Rev. B*, **80**, 245419.
37 Inglesfield, J.E. and Benesh, G.A. (1988) *Phys. Rev. B*, **37**, 6682.
38 Taylor, M. and Nordlander, P. (2001) *Phys. Rev. B*, **64**, 115422; Niedfeldt, K., Nordlander, P., and Carter, E.A. (2006) *Phys. Rev. B*, **74**, 115109.
39 Smith, F.T. (1960) *Phys. Rev.*, **118**, 349.
40 Chulkov, E.V., Silkin, V.M., and Echenique, P.M. (1999) *Surf. Sci.*, **437**, 330.
41 Föhlisch, A., Feulner, P., Hennies, F., Fink, A., Menzel, D., Sánchez-Portal, D., Echenique, P.M., and Wurth, W. (2005) *Nature*, **436**, 373.
42 Neaton, J.B., Hybertsen, M.S., and Louie, S.G. (2006) *Phys. Rev. Lett.*, **97**, 216405.
43 Thygesen, K.S. and Rubio, A. (2009) *Phys. Rev. Lett.*, **102**, 046802.
44 Quek, S.-Y., Venkataraman, L., Choi, H.-J., Louie, S.G., Hybertsen, M.S., and Neaton, J.B. (2007) *Nano Lett.*, **7**, 3477.
45 Quijada, M., Borisov, A.G., Nagy, I., Díez Muiño, R., and Echenique, P.M. (2007) *Phys. Rev. A*, **75**, 042902.
46 Usman, E.Yu., Urazgil'din, I.F., Borisov, A.G., and Gauyacq, J.P. (2001) *Phys. Rev. B*, **64**, 205405.
47 Petek, H., Weida, M.J., Nagano, H., and Ogawa, S. (2000) *Science*, **288**, 1402.
48 Borisov, A.G., Kazansky, A.K., and Gauyacq, J.P. (2001) *Phys. Rev. B*, **64**, 201105.
49 Kohanoff, J. (2006) *Electronic Structure Calculations for Solids and Molecules*, Cambridge University Press.
50 Tavernelli, I., Curchod, B.F.E., and Rothlisberger, U. (2010) *Phys. Rev. A*, **81**, 052508 and references therein.
51 Tully, J.C. (1998) *Faraday Discuss.*, **110**, 407.
52 Runge, E. and Gross, E.K.U. (1984) *Phys. Rev. Lett.*, **52**, 997.
53 Lindenblatt, M. and Pehlke, E. (2006) *Phys. Rev. Lett.*, **97**, 216101.
54 Horsfield, A.P., Bowler, D.R., Fisher, A.J., Todorov, T.N., and Montgomery, M.J. (2004) *J. Phys. Condens. Matter*, **16**, 3609.
55 Tully, J.C. (1990) *J. Chem. Phys.*, **93**, 1061.
56 Craig, C.F., Duncan, W.R., and Prezhdo, O.V. (2005) *Phys. Rev. Lett.*, **95**, 163001.
57 Duncan, W.R. and Prezhdo, O.V. (2007) *Annu. Rev. Phys. Chem.*, **58**, 143.
58 Hush, N.S. (1958) *J. Chem. Phys.*, **28**, 962; Cotton, F.A. (ed.) (1967) *Progress in Inorganic Chemistry*, vol. 8, Interscience, New York, p. 391; Hush, N.S. (1968) *Electrochim. Acta*, **13**(5), 1005.

59 Levich, V.G. and Dogonadze, R.R. (1959) *Dokl. Akad. Nauk SSSR*, **124**, 123; Levich, K.J. and Dogonadze, R.R. (1959) *Proc. Acad. Sci. USSR, Phys. Chem. Sect.* (English Transl.), **124**, 9.

60 Rips, I. and Jortner, J. (1987) *J. Chem. Phys.*, **87**, 2090; Rips, I. and Jortner, J. (1987) *J. Chem. Phys.*, **87**, 6513; Rips, I. and Jortner, J. (1988) *J. Chem. Phys.*, **88**, 818.

61 Nitzan, A. (2006) *Chemical Dynamics in Condensed Phases: Relaxation, Transfer, and Reactions in Condensed Molecular Systems*, Oxford University Press, New York.

62 Marcus, R.A. and Sutin, N. (1985) *Biochim. Biophys. Acta*, **811**, 265.

63 Miller, R.J.D., McLendon, G.L., Nozik, A.J., Schmickler, W., and Willig, F. (1995) *Surface Electron Transfer Processes*, VCH, New York.

64 Stähler, J., Meyer, M., Bovensiepen, U., Zhu, X.Y., and Wolf, M. (2007) *New J. Phys.*, **9**, 394.

65 Fonseca, T. (1989) *J. Chem. Phys.*, **91**, 2869.

66 Fonseca, T. (1989) *Chem. Phys Lett.*, **162** (6), 491.

67 van der Zwan, G. and Hynes, J.T. (1982) *J. Chem. Phys.*, **76**, 2993; van der Zwan, G. and Hynes, J.T. (1986) *J. Chem. Phys.*, **90**, 3701.

68 Zusman, L.D. (1980) *Chem. Phys.*, **49**, 295.

69 Kang, T.J., Jarzeba, W., Barbara, P.F., and Fonseca, T. (1990) *Chem. Phys.*, **149**, 81.

70 Wolf, M., Knoesel, E., and Hertel, T. (1996) *Phys. Rev. B*, **54**, 5295.

71 Berthold, W., Rebentrost, F., Feulner, P., and Höfer, U. (2004) *Appl. Phys. A*, **78**, 131.

72 Echenique, P.M., Pitarke, J.M., Chulkov, E.V., and Rubio, A. (2000) *Chem. Phys.*, **251**, 1.

73 Marinica, D.C., Ramseyer, C., Borisov, A.G., Teillet-Billy, D., Gauyacq, J., Berthold, W., Feulner, P., and Höfer, U. (2002) *Phys. Rev. Lett.*, **89**, 046802.

74 Ge, N.-H., Wong, C.M., Lingle, R.L., Jr., McNeill, J.D., Gaffney, K.J., and Harris, C.B. (1998) *Science*, **279**, 202.

75 Li, B., Zhao, J., Onda, K., Jordan, K.D., Yang, J., and Petek, H. (2006) *Science*, **311**, 1436.

76 Zhu, X.-Y., Yang, Q., and Muntwiler, M. (2009) *Acc. Chem. Res.*, **42**, 1779.

77 Johns, J.E., Muller, E.A., Fréchet, J.M.J., and Harris, C.B. (2010) *J. Am. Chem. Soc.*, **132**, 15720.

78 Stähler, J., Meyer, M., Kusmierek, D.O., Bovensiepen, U., and Wolf, M. (2008) *J. Am. Chem. Soc.*, **130** (27), 8797.

79 Seitsonen., A.P. (2000) Theoretical investigations into adsorption and co-adsorption on transition-metal surfaces as models to heterogeneous catalysis. Dissertation, Technische Universität Berlin, http://edocs.tu-berlin.de/diss/2000/seitsonen_ari.htm.

80 Smith, N.V. (1985) *Phys. Rev. B*, **32**, 3549.

81 Stähler, J., Gahl, C., Bovensiepen, U., and Wolf, M. (2006) *J. Phys. Chem. B*, **110**, 9637.

82 Bovensiepen, U., Gahl, C., Stähler, J., Bockstedte, M., Meyer, M., Baletto, F., Scandolo, S., Zhu, X.-Y., Rubio, A., and Wolf, M. (2009) *J. Phys. Chem. C*, **113**, 979.

83 Holstein, T. (1959) *Ann. Phys.*, **8**(3), 325.

5
Electromagnetic Interactions with Solids

Ricardo Díez Muiño, Eugene E. Krasovskii, Wolfgang Schattke, Christoph Lienau, and Hrvoje Petek

In this chapter, we discuss various processes that contribute to absorption of light at solid surfaces. We particularly focus on the optical response of metals including single-particle interband electron–hole (e-h) pair excitations, which are common to semiconductors and insulators, as well as the free-electron response, which is particular to metals and doped semiconductors. Free electrons participate both in the collective screening response and in the single-particle intraband absorption. A characteristic property of metals conferred by free electrons is the dielectric function with large negative real and small positive imaginary parts in the infrared and visible spectral regions. A time-dependent electromagnetic field incident on a metal surface at frequencies below the material-specific plasma frequency is strongly screened by the free-carrier polarization. The screening limits the penetration of the electric field to the skin depth of the metal, typically corresponding to 20–30 nm in the visible spectrum. Moreover, the coherent oscillation of free carriers with the opposite phase as the incoming field (see Section 5.4.2), as imposed by the negative real part of the dielectric function, leads to specular reflection of a π-phase-shifted field with minimal loss to e-h pair excitation. We examine the factors that define the optical properties of metals and how light deposits energy into elementary excitations at a metal surface. Absorption of light generates hot carriers within the metal bulk or at surfaces in intrinsic or adsorbate localized surface states. These excitations undergo relaxation through elastic and inelastic carrier scattering processes or induce chemistry, as discussed elsewhere in this book. The absorption of light is detected through linear or nonlinear optical spectroscopy; alternatively, it can be detected through photoelectron emission. Although both optical and photoemission measurements have the same requirements for momentum conservation in the process of photon absorption, the former confers momentum integrated information and is therefore unsuitable for band structure mapping. We discuss the optical absorption process with the object of specifying the hot electron and hole distributions in the energy and momentum space $[E(\mathbf{k})]$ that are generated by absorption of light through interband and intraband transitions. These distributions are important for the secondary processes induced by hot carriers, such as femtochemistry, as well as to describe photoemission. By contrast

Dynamics at Solid State Surfaces and Interfaces: Volume 2: Fundamentals, First Edition.
Edited by Uwe Bovensiepen, Hrvoje Petek, and Martin Wolf.

to optical spectroscopy, photoemission spectroscopy has provided a wealth of information on the band structures of the occupied states of solids, and with the advent of tunable ultrafast lasers, it holds a further promise for band mapping of their unoccupied states. To this end, we describe recent theoretical proposals for the band mapping by resonant multiphoton photoemission. The coherent response of metals has been used to great advantage in the field of plasmonics. We describe the nature of localized and propagating plasmonic modes of metal surfaces, how they can be manipulated through morphology and dielectric environment, and how they couple with other excitations such as excitons in molecular aggregates.

5.1 Dielectric Function of Metals

For sufficiently low intensities, absorption of the electromagnetic radiation by a solid is described by perturbation theory; transitions between the occupied and unoccupied stationary states represent a negligible deviation of the system from its ground state. In the simplest case, when many-body interactions are described by a mean field, the probability that a photon of energy $\hbar\omega$ be absorbed is reduced to summing up the probabilities of all possible one-particle excitations for all crystal momenta $\mathbf{k}$. The probability of each individual transition obeys the Fermi Golden Rule and, in the long-wavelength limit, the dipole selection rules [1]. In a crystal, the electron states can be chosen to be eigenstates of the translation operator, so that the non-vanishing ones are only the transitions between electron states having the same $\mathbf{k}$. In the dipole approximation, the transition probability amplitude is given by the matrix element $M^{\mathrm{e}}_{mn}(k) = \langle \mathbf{k}n| -i\nabla_{\mathrm{e}} |\mathbf{k}m\rangle$, where ∇_{e} is the projection of the gradient operator in the direction of the electric field and m and n are the band numbers of the occupied and unoccupied Bloch states, respectively. The $\mathbf{k}$-conserving transitions are often referred to as vertical or direct transitions, with the momentum of a photon being negligible compared to the range of momenta spanned by the Brillouin zone vectors. The total probability of being absorbed (per second and per unit volume) is given by an integral over the Brillouin zone, and it can be converted into the optical constants of the material using the macroscopic relationships of the classical electromagnetic theory [2]. A more consistent approach is to calculate the linear longitudinal response of the many-electron system within the random phase approximation (RPA). The simplest version of the RPA [3], which neglects variations in the electric field on the scale of the unit cell, leads to the same result. The imaginary part of the macroscopic dielectric function ε'' corresponding to interband transitions, which is of primary interest in this section, is then simply connected to the band structure:

$$\varepsilon''(\omega) = \frac{4\pi^2 e^2}{m^2\omega^2} \sum_{mn} \int_{BZ} \frac{2dk}{(2\pi)^3} |M^e_{mn}(k)|^2 \delta(E_n(k) - E_m(k) - \hbar\omega). \tag{5.1}$$

The terms $m \neq n$ result in broad absorption spectra reflecting transitions between electronic bands specific to each solid. This interband term is often referred to as the

Lorentz term and can be represented as a sum of several Lorentz oscillators corresponding to specific regions of phase space with a large joint density of initial and final states that are connected by optical transitions.

The one-particle model also describes the fundamental difference between metals and semiconductors: the normal incidence reflectivity of the former approaches unity as $\omega \to 0$. This is caused by the intraband term, $m = n$, which contributes to the optical response of metals due to the presence of the Fermi surface. In the absence of damping, the intraband term gives rise to a singularity of $\varepsilon''(\omega)$ at $\omega \to 0$:

$$\varepsilon''_{\text{intra}}(\omega) = \frac{\pi}{2}\omega_{\text{p}0}^2 \frac{\partial \delta(\omega)}{\partial \omega}. \tag{5.2}$$

The Kramers–Kronig transform of Eq. (5.2) is the Drude contribution to the real part of the dielectric function:

$$\varepsilon'_{\text{intra}}(\omega) = 1 - \frac{\omega_{\text{p}0}^2}{\omega^2}. \tag{5.3}$$

The parameter $\omega_{\text{p}0} = (4\pi n e^2/m)^{1/2}$ in Eqs. (5.2) and (5.3) is the plasma frequency of a free-electron gas, which depends entirely only on the electron density n.

The real part of the dielectric function is also seen to diverge as $\omega \to 0$, and consequently, the normal incidence reflectivity of metals R approaches unity in this limit. As long as the real part remains large and negative, the reflectivity is high; above $\omega_{\text{p}0}$, the free electrons are no longer able to screen the external field and the metal becomes transparent. At the plasma frequency, light can propagate through a metal as a strongly damped collective charge density fluctuation [4]. In real metals, the frequencies of plasma oscillations, which occur when $\varepsilon'(\omega)$ crosses zero and $\varepsilon''(\omega)$ is small, are modified by interband transitions described by the Lorentz term. The Drude parameter ω_{p}^2 is given by the integral over the Fermi surface

$$\omega_{\text{p}}^2 = \frac{1}{\pi^2} \sum_n \int_{\text{FS}} \mathrm{d}S \left(\frac{(\mathbf{e} \cdot \mathbf{v}_n(\mathbf{k}))^2}{|\mathbf{v}_n(\mathbf{k})|} \right). \tag{5.4}$$

Here, $\mathrm{v}_n(\mathbf{k})$ is the group velocity of the Bloch state $|nk\rangle$ and the vector $\mathbf{e}$ points in the direction of polarization of the external field. Fermi surfaces of realistic metals have very complicated shapes, so the actual values of ω_{p}^2 are usually much smaller than the free electron result $\omega_{\text{p}0}^2$.

5.1.1 Calculations of Dielectric Functions

The one-particle approach provides a simple scheme for calculating optical properties with *ab initio* methods of band theory. In the majority of applications, the solutions of the Kohn–Sham equations of the density functional theory are used for $|\mathbf{k}m\rangle$ and $E_m(\mathbf{k})$. For metals, this usually leads to a satisfactory agreement with experiment. The typically high quality of the above approach is illustrated in Figure 5.1a for platinum (Pt): below 30 eV, the measured optical spectra [5] are well described by the theory [6].

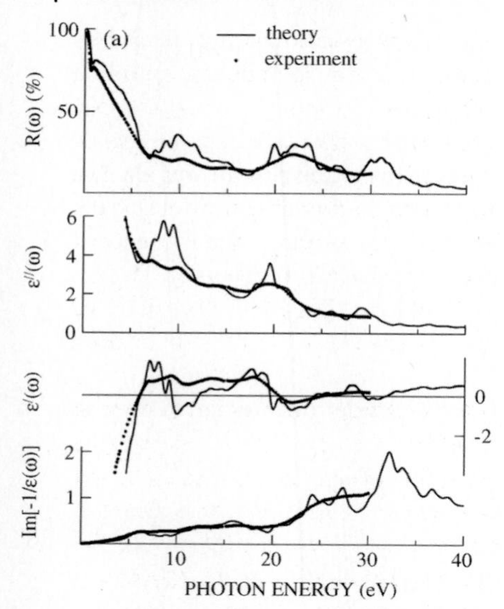

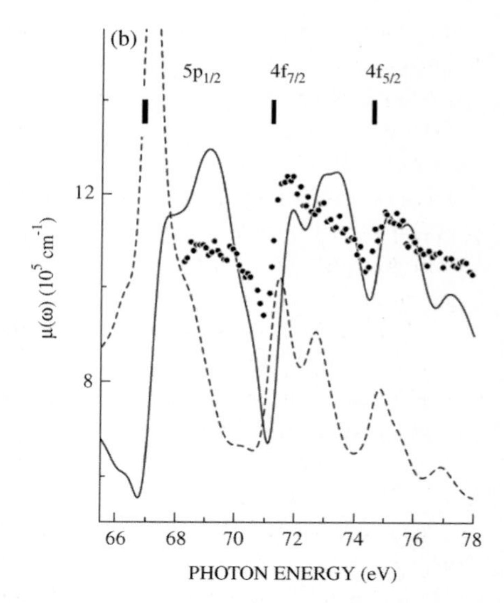

Figure 5.1 (a) The reflectivity $R(\omega)$, the real and imaginary parts of the macroscopic dielectric function $\varepsilon(\omega)$, and the loss function $\mathrm{Im}\left[-\varepsilon(\omega)^{-1}\right]$ of Pt: theory [6] (lines) and experiment [5] (dots). (b) Absorption coefficient $\mu(\omega)$ of Pt: theory with (solid line) and without (dashed line) local fields and experiment [9] (circles). Theoretical energies of 5p and 5f bands are adjusted to XPS measurements (indicated by vertical bars). Reprinted with permission from Ref. [6]. Copyright (2001) by the American Physical Society.

The theory fails at higher energies, however, where absorption is due to highly localized semicore states (see the dashed line in Figure 5.1b). It is shown in Ref. [7] that the problem is caused not only by the errors in quasi-particle energies $E_m(\mathbf{k})$, but also by the simplified form (Eq. (5.1)) of the RPA.

Random phase approximation can be refined by taking into account local field effects [8], that is, by including the short-wavelength response to a long-wavelength perturbation, where $\varepsilon(\omega)$ is $\mathbf{k}$ independent. Thus, one improves the description of electron–electron interaction, while remaining within a one-particle approach. For crystals, the problem is formulated in terms of the microscopic dielectric matrix in the reciprocal space representation $\varepsilon_{\mathbf{GG}'}(\omega)$, whose elements are labeled by the reciprocal lattice vectors $\mathbf{G}$ and $\mathbf{G}'$, each element being a tensor in the real space. The macroscopic dielectric function $\varepsilon(\omega)$ (an element of the tensor) is given by the matrix inverse of the dielectric matrix

$$\varepsilon(\omega) = \frac{1}{\left[\hat{\varepsilon}^{-1}(\omega)\right]_{G=0,G'=0}}, \tag{5.5}$$

$$\varepsilon_{\mathbf{GG}'}(\omega) = \delta_{\mathbf{GG}'} - \lim_{q\to 0} \frac{e^2}{4\pi^2|\mathbf{q}+\mathbf{G}||\mathbf{q}+\mathbf{G}'|} \sum_{mn} \int_{\mathrm{BZ}} \mathrm{d}\mathbf{k} \\ \frac{\langle \mathbf{k}+\mathbf{q}n|\mathrm{e}^{i(\mathbf{q}+\mathbf{G}')r}|\mathbf{k}m\rangle\langle \mathbf{k}m|\mathrm{e}^{-i(\mathbf{q}+\mathbf{G})r}|\mathbf{k}+\mathbf{q}n\rangle}{E_n(\mathbf{k}+\mathbf{q}) - E_m(\mathbf{k}) - \hbar\omega + i\hbar\eta}, \tag{5.6}$$

where $\mathbf{q}$ is the wave vector in the Brillouin zone and η is a broadening parameter. The strong effect of the nondiagonal dielectric response on the far UV optical absorption in Pt is illustrated by Figure 5.1b: the inclusion of microscopic fields is seen to change both the shape and the amplitude of the absorption coefficient $\mu(\omega)$ curve in Figure 5.1b and to shift the absorption bands to higher energies by several eV. The absorption coefficient derived from the true macroscopic dielectric function nearly perfectly agrees with the experiment. Note, however, that in contrast to Figure 5.1a, which was obtained with *ab initio* Kohn–Sham solutions in the local density approximation (LDA) for the valence band of Pt, the one-particle energies of the 5p and 4f bands had to be corrected according to independent X-ray photoemission measurements [6].

Although Kohn–Sham eigenvalues cannot be identified with true quasi-particle energies, experience shows that they are often adequate for delocalized valence bands but are less reliable for core states. Correct quasi-particle equations include many-body effects through a self-energy operator Σ. A state-of-the-art, but still technically challenging approach to Σ in moderately correlated solids is the GW approximation [10]. This approach has been instrumental in calculating the band structure of semiconductors [11], where the LDA is known to strongly underestimate the width of the band gap.

In contrast to metallic systems, in semiconductors the e-h interaction also plays a fundamental role in the formation of optical spectra. It leads to bound excitons and apart from that strongly affects interband transitions. Thus, knowledge of the two-particle Green's function is required. Modern *ab initio* methods treat the problem by solving a Bethe–Salpeter equation (BSE) for the polarization function $X_{\mathbf{GG}'}(\mathbf{q},\omega)$ [12]. The macroscopic dielectric function in the limit of small momentum transfer is then given by

$$\varepsilon(\omega) = \lim_{q\to 0} v(\mathbf{q})\chi_{00}(\mathbf{q},\omega). \tag{5.7}$$

The result can be reduced (after certain reasonable simplification) to a formula similar to Eq. (5.1), but with true excitation energies instead of energy differences $E_n(\mathbf{k})-E_m(\mathbf{k})$ and with a coherent sum over direct e-h pairs instead of the incoherent sum over direct transitions [13, 14]. Thus, apart from obtaining excitation energies within the quasi-particle gap, the two-particle theory also takes into account interference between matrix elements $M^{\mathrm{e}}_{mn}(\mathbf{k})$ coming from different $\mathbf{k}$ points. The latter effect may affect the interband absorption spectrum much more strongly than the former, in particular for semiconductors. It has been shown that for GaAs [14] and GaN [15] the dramatic shift of the absorption band to lower energies resulting from the e-h interaction is entirely due to a constructive interference of the transition amplitudes at lower energies and destructive interference at higher energies. In wide band gap insulators, the Bethe–Salpeter equation approach reliably describes high-energy excitons. For example, GW–BSE methods have successfully provided a unified description of the optical spectra of rutile and anatase polymorphs of TiO_2, clarifying the role of electronic and optical gaps in materials important for solar energy conversion [16]. Excellent agreement of *ab initio* calculations with experiment has been achieved for a wide range of semiconductors and insulators [13–15, 17] (see also the review article [18] and references therein).

5.2
Band Mapping of Solids by Photoemission Spectroscopy

Photoemission has become a preeminent tool to understand real solids in terms of their electronic structure [19, 20]. In a simplified picture and for sufficiently weak incident radiation, photoemission is a linear process in which one electron is emitted for each photon absorbed. Because the mean free path of photoelectrons ranges typically from a few angstroms to a few tens of angstroms, photoemission measurements are surface sensitive. Depending on the photon energy $\hbar\omega$, the photoemitted electron can be extracted either from atomic-like core levels or from more weakly bound valence levels. In the former case, photoelectron diffraction offers considerable information about structural properties. Identification of the chemical environment is also possible as well from chemical shifts in X-ray photoemission spectra. In the latter case, spectroscopy techniques give rise to valuable information on momentum-resolved band structure: angle-resolved photoelectron spectroscopy (ARPES) is the preferred method for complete band mapping of solids. A schematic drawing of the photoemission process is shown in Figure 5.2.

Although photoemission is actually a many-body process, state-of-the-art theoretical descriptions of the photocurrent I rely on Fermi's Golden Rule in a one-electron picture:

$$I\left(E, \mathbf{k}_{\|}^{0}\right) \propto \sum_{\nu k} \left\langle \psi_f\left(E, \mathbf{k}_{\|}^{0}\right) | \mathbf{O} \middle| \psi_\nu(\mathbf{k}) \right\rangle \delta(E - E_\nu(\mathbf{k}) - \hbar\omega), \tag{5.8}$$

where there is a sum over the continuum of initial states $|\psi_\nu(\mathbf{k})\rangle$ with energy $E_\nu(\mathbf{k})$. The indexes ν and $\mathbf{k}$ refer to the band index and the three-dimensional momentum, respectively. In the one-step photoemission theory, the final state $|\psi_{\mathrm{f}}\rangle$ is a time-reversed low-energy electron diffraction (LEED) state defined by the surface parallel projection of the Bloch vector $\mathbf{k}_{\|}^{0}$ and energy E. The LEED wave function is a scattering solution for a plane wave incident from vacuum. The LEED state incorporates the effect of the inelastic scattering of the photoelectron, which is quantitatively treated by adding an imaginary part (optical potential) to the potential in the crystal half-space. Therefore, $|\psi_{\mathrm{f}}\rangle$ is an eigenfunction of a non-Hermitian Hamiltonian with a real eigenvalue E.

The delta function in Eq. (5.8) guarantees total energy conservation. The final state energy E is connected with the measured kinetic energy E_{kin} by $E_{\mathrm{kin}} = E - \Phi$, where Φ is the work function of the solid. Momentum conservation is reduced to the parallel momentum $\mathbf{k}_{\|}^{0}$, while the perpendicular component of the momentum is not conserved due to the symmetry breaking by presence of the surface. The operator $\mathbf{O} = \mathbf{A} \cdot \mathbf{p} + \mathbf{p} \cdot \mathbf{A}$ can be written in terms of the vector potential of the photon field $\mathbf{A}$ and the momentum operator $\mathbf{p}$. The $\mathbf{A} \cdot \mathbf{p}$ term describes photon absorption in a homogeneous medium, that is, within bulk of a solid, whereas the $\mathbf{p} \cdot \mathbf{A}$ term has a

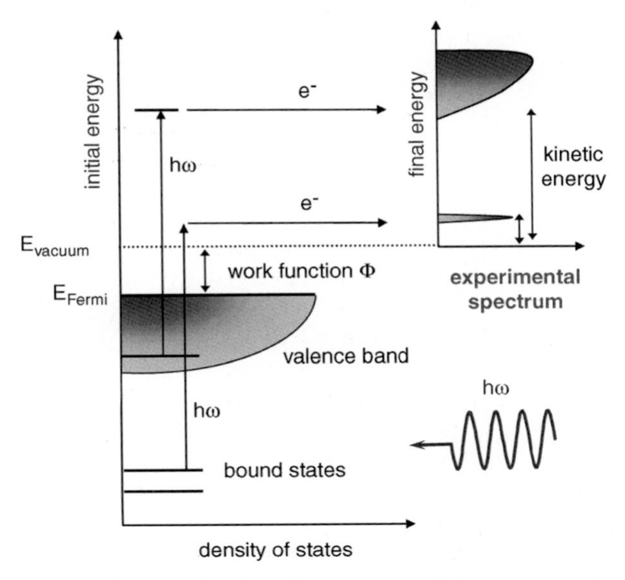

Figure 5.2 Schematic representation of the photoemission process.

significant contribution at a surface, owing to a strong gradient in the vector potential. In the limit of long wavelength and at high photon energy, the vector potential can be approximated as constant (dipole approximation). In metals, below the plasma frequency, this approximation may be unsatisfactory because the exciting field has a strong contribution from the nonlocal dielectric response at the surface, corresponding to the nondiagonal terms in the dielectric matrix of Eq. (5.5). The spatial dependence of the total electric field at the surface, however, can be calculated by modern band structure methods and included into the microscopic theory of photoemission [21].

For ultraviolet (UV) radiation, photoelectrons are extracted from the valence band of the solid. Figure 5.3 shows that an accurate description of the photoemission final state by a time-reversed LEED state $|\psi_f\rangle$ makes it possible to obtain an excellent agreement between theoretical and measured photoelectron spectra for off-normal emission in $TiTe_2$ [22].

The mapping of the solid valence band can be simplified if the final state $|\psi_f\rangle$ in Eq. (5.8) is approximated by a Bloch eigenstate of the bulk system, which conserves the three-dimensional crystal momentum [23, 24]. A further simplification arises if the final state band structure is completely neglected. In this case, and because of the continuum of photoelectron final states, energy conservation is always fulfilled and the matrix elements are constant. Therefore, the sum over initial states in Eq. (5.8) turns into the density of initial states. In this approximation, the spectrum reduces to a one-dimensional density of states (DOS) along the surface normal direction. The example in Figure 5.3, however, shows a crucial role of final-state effects in the

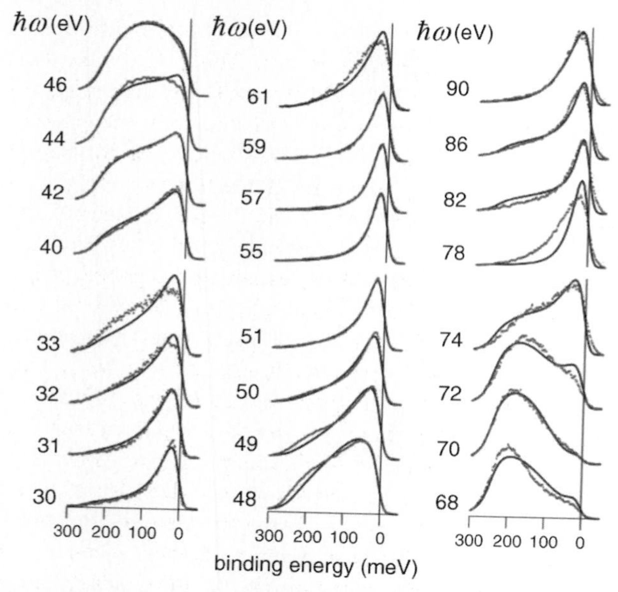

Figure 5.3 Experimental and theoretical energy distribution curves for Ti 3d photoemission in $TiTe_2$. Spectra are marked by photon energies. Reprinted with permission from Ref. [22]. Copyright (2007) by the American Physical Society. (Please find a color version of this figure on the color plates.)

formation of the spectra even for very narrow bands. Their accurate *ab initio* treatment is essential in understanding the photoemission lineshape and its dependence on the photon energy. Modern photoemission theory is capable of taking into account the subtle effects of interference between branches of the complex band structure of the final states and providing a good agreement with the experiment.

Band mapping, such as described in Section 3.4, of occupied states can be considered as a basic application of photoelectron spectroscopy. A more complex task is the derivation of dynamical information from measured lineshapes, linewidths, and peak intensities. Scattering of photoexcited electrons with electrons in the Fermi sea, phonons, and defects, as well as hole screening and decay, can significantly contribute to spectral linewidths [25]. In fact, bulk direct transition linewidths are in most cases dominated by final-state damping. Only in the case of surface states, it is possible to determine the initial state spectral function, without influence of the final state contributions. This feature has been exploited to dissect different contributions from the e-e, e-p, and defect scattering to the linewidths of Shockley surface states of noble metals [26]. Excellent agreement has been obtained between the experimental linewidths and *ab initio* calculations at a level beyond the GW, taking into account the spatial dependence of the screened interaction at a surface [27].

For higher photon energies, electrons can be extracted from core levels for which the theoretical description of the initial electron wave functions is simpler than that for delocalized bands. Spectroscopy in this regime is used for chemical identification, as well as detecting element-specific energy shifts at surfaces. The spatial localization of the initial excitation makes each atom in the solid a point-like source of photoelectrons. Core-level emission is thus a powerful tool for holography and electron diffraction-based techniques [20]. In core-level photoemission, the theoretical description of the photocurrent can be simplified through the three-step model, in which the initial excitation process, the diffraction at the crystal lattice, and the crossing of the surface are considered as three independent processes. Multiple scattering methods have proven to be successful in describing the photoelectron spectra in this regime. The fitting of simulated theoretical spectra to experimental results provides very accurate information on structural parameters of clean and decorated systems, as well as more complex systems [28, 29]. Spin-resolved photoemission provides further insight into the spin structures of surfaces [30–32].

The role of the electron spin is relevant in photoemission processes from magnetic systems, as well as in systems with strong spin-orbit coupling. Spin-dependent phenomena can be investigated through photoelectron spin analysis if the signal levels are sufficient [33]. Even without spin resolution in the photoelectron detection, magnetic dichroism can be used for studying these systems. Magnetic dichroism is based on the change in photoelectron intensity upon reversal of the sample magnetization or the direction of circular polarization of the excitation light [34, 35]. Furthermore, photoemission techniques also provide valuable information on electron correlation. The extreme energy resolution achieved in photoemission measurements in the last years has opened the way to detailed studies of many-body effects in the spectra. This is particularly interesting for the study of strongly correlated systems and superconducting materials, as described in more detail in Chapter 1. Such measurements rely heavily on the ability of photoemission spectroscopy to measure lineshape changes and energy gap shifts with sub-meV resolution [36].

5.2.1
Nonlinear Photoemission as a Band Mapping Tool for Unoccupied States

Nonlinear photoemission processes induced by light sources of high intensity have been shown to be well suited for the study of electronic excitations in solid-state materials. Two-photon photoemission (2PPE) has contributed greatly to the study of ultrafast electron dynamics and adsorbate dynamics at solid surfaces [37, 38]. The lifetime of electronic surface and image states, for instance, has been accurately determined in many systems through pump–probe experiments [39]. In the case of surface states or adsorbate excitations, momentum conservation is relaxed due to the presence of the surface or the localization of the wave functions on the adsorbates [40, 41]. In most cases, 2PPE experiments involve nonresonant excitation or excitations from bulk to surface states that do not involve strong resonance enhancement. This is in part because the excitation lasers have poor tunability, as well as because resonances can complicate interpretation of the spectra.

The spectroscopy of bulk electronic states by 2PPE has not received comparable attention, although it can effectively be used to extend spectroscopy to electronic states below and above the vacuum threshold [42–44]. Other photoemission-based methods, such as inverse photoemission spectroscopy [45, 46], have never attained accuracy comparable to that of ARPES for the occupied states; therefore, establishing a general method with high spectral and temporal resolution for the study of the unoccupied states would be highly desirable. In fact, such methods could have significant advantage for band mapping over conventional ARPES because interband excitations within the bulk of solids rigorously conserve momentum.

In one-photon photoemission, the one-step model (see Eq. (5.8)) is known to provide a firm basis for *ab initio* calculations. A similar approach is expected to be appropriate for the case of multiphoton photoemission. Following perturbation theory, one can show that the second-order contribution to the photoemission current is [47]

$$I \propto \sum_{\nu\mathbf{k}} |h_{\mathrm{f}\mathbf{k}}|^2 \delta(E - E_\nu(\mathbf{k}) - \hbar\omega) + \frac{1}{4} \sum_{\nu\mathbf{k}} \left| \sum_{\mu\mathbf{k}'} \frac{h_{\mathrm{f},\mathbf{k}'\mu}\, h_{\mathbf{k}'\mu,\mathbf{k}\nu}}{E_\mu(\mathbf{k}') - E_\nu(\mathbf{k}) - \hbar\omega - i\eta} \right|^2 \delta(E - E_\nu(\mathbf{k}) - 2\hbar\omega), \tag{5.9}$$

where the first sum corresponds to the first-order process of Eq. (5.8). The second term involves matrix elements between initial and intermediate states $h_{\mathbf{k}'\mu\mathbf{k}\nu}$ and between intermediate and final states $h_{\mathrm{f}\mathbf{k}'\mu}$. The sum over unoccupied intermediate states runs over the band index μ and the three-dimensional momentum $\mathbf{k}'$. The final state $|\psi_\mathrm{f}\rangle$ is again a time-reversed LEED state. In contrast to the relaxed momentum conservation in the transition matrix elements $h_{\mathrm{f}\mathbf{k}'\mu}$ to the inverse LEED state, full three-dimensional momentum is conserved in the intermediate transition $h_{\mathbf{k}'\mu\mathbf{k}\nu}$. This property is crucial to obtain an accurate description of the band dispersion of unoccupied levels. Distinct peaks in the two-photon photoemission yield can therefore be associated with initial to intermediate state resonances that correspond to momentum-conserving transitions. Figure 5.4 illustrates this effect in the case of two-photon transitions in Si(001) [47]. Figure 5.4b shows the photoemission intensity versus the energy (measured with respect to the valence band maximum). In Figure 5.4a, a magnified view of a portion of the band structure that contributes to the photoemission spectrum is shown, and the full band structure is included in the inset as well. Thickened grey lines denote the complex band structure constituents of the LEED states; the thickness gives the contribution of the bulk Bloch wave to the outgoing photoelectron wave function.

Taking full advantage of the band mapping capability of multiphoton photoemission requires broad excitation laser tunability and momentum mapping capability. Owing to the experimental demands, only a few examples of such studies have been reported in the literature [48, 49]. One should also keep in mind that multiphoton excitation processes can occur through multiple resonant and nonresonant pathways, which contribute coherently to the photoemission yield, and therefore one should expect such spectra to be influenced by multiple pathway interference effects. Such multiphoton processes are discussed in Ref. [50].

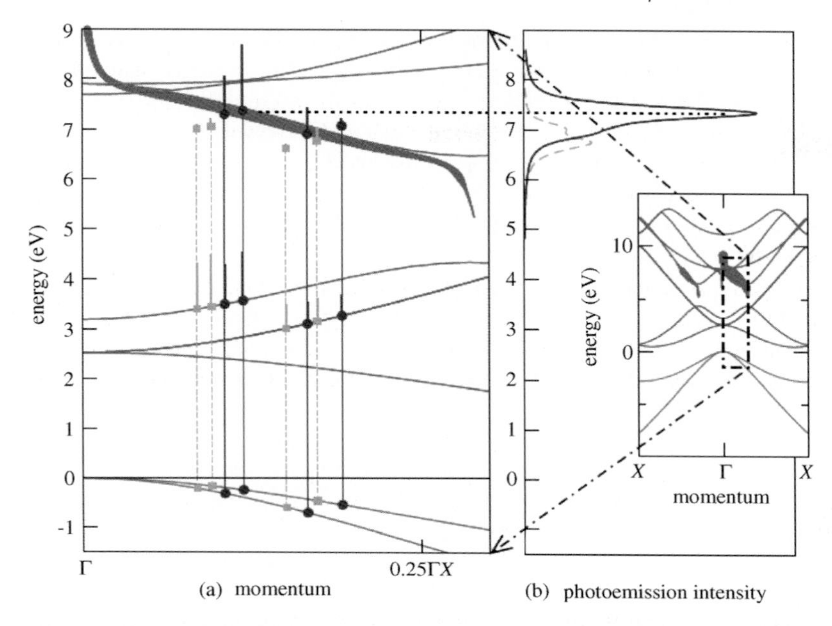

Figure 5.4 Two-photon photoemission yield and some relevant two-photon transitions in Si(001). Right panel: Grey dashed line (dark solid line) corresponds to the spectrum for $\hbar\omega = 3.6$ (3.8) eV. The full band structure is shown in the inset as well. Left panel: A selected part of the band structure is magnified. The transitions relevant for the spectrum are shown by squares (circles) for $\hbar\omega = 3.6$ (3.8) eV. Thicker vertical bars rising from the squares and circles show the magnitude of the coupling matrix elements for the transitions. See text for further explanation. Reprinted with permission from Ref. [47]. Copyright (2008) by the American Physical Society.

5.3
Optical Excitations in Metals

The dielectric function, as described in Eqs. (5.1)–(5.6), is defined by both the electronic structure and carrier phase relaxation phenomena in a solid. Photoemission spectroscopy measures the electronic bands of a solid with energy and momentum resolution. Optical measurements, such as reflectivity and transmission, represent the full electromagnetic response of a solid, which is useful for describing the macroscopic properties, but the nonspecific nature makes it difficult to extract information on the band structure and the specific carrier scattering rates. Nevertheless, optical methods have played an essential role in the development of our understanding of the fundamental properties of metals [51, 52], as well as for ultrafast studies of electron dynamics in solids [53–56]. A valuable theoretical comparison of the spectroscopic and dynamical information content of nonlinear photoemission and optical spectroscopy measurements is given in Ref. [57]. The goal of this section is to analyze the optical response of metals in greater detail in order to describe the processes contributing to different components of the dielectric response and to describe the carrier distributions they generate.

Optical response of a metal involves both single-particle and collective excitations that are described by the complex dielectric function. In the limit of small momentum transfers relevant to the visible/UV spectral response, the complex dielectric function is written as a sum of three terms [8, 51, 58–60]:

$$\varepsilon(\omega) = 1 - \frac{\omega_{p0}^2}{\omega^2 + i\omega/\tau} + \varepsilon^i(\omega), \qquad (5.10)$$

where τ represents the electron–hole pair scattering time, to be described in more detail. The first two terms in Eq. (5.10) correspond to the Drude term already introduced in Eqs. (5.2) and (5.3) in the limit of no damping ($\tau \to \infty$). The Drude term describes the intraband single-particle absorption and the plasma excitation. The remaining term corresponds to the Lorentz term given in Eq. (5.1) and describes single-particle interband absorption. Thus, the dielectric function in Eq. (5.10) combines contributions to photon absorption in a metal from electric dipole interband transitions between electronic bands, which conserve energy and momentum and are described by the Lorentz term, with contributions of higher order processes where e-h pair excitation is accompanied by electron, phonon, or impurity/defect (henceforth, just impurity) scattering to conserve overall energy and momentum. All these higher order processes contribute to the Drude term, and their specific contribution is generally difficult to isolate. Photon absorption under ultrafast laser excitation can also occur through multiple photon absorption between bands or through virtual states. The multiphoton excitation process must conserve energy and momentum only for the transition between initial and final states rather than for the individual excitation steps, as already discussed in Sections 3.4 and 5.2.1 [47, 49, 61]. In multiphoton photoemission spectroscopy, the final state is usually the photoemission continuum, where the inverse LEED states automatically satisfy the energy and perpendicular momentum conservation, whereas the parallel momentum remains rigorously that of the initial state (Section 5.2).

Whether interband dipole transitions dominate light absorption depends primarily on the excitation wavelength and the electronic band structure of absorbing material. In metals, there is an energy threshold for interband transitions that corresponds either to **k**-conserving transitions from an occupied band to a point where another partially occupied band crosses the Fermi surface or from Fermi surface crossings to another unoccupied band. Below the interband threshold, light absorption will be determined by a variety of secondary factors, some of which may be under experimental control, such as sample purity, crystallinity, adsorbate coverage, and temperature. The dominant channel for photoabsorption will determine the initial distribution of electrons and holes excited in the sample. This knowledge of the photoexcitation process is often a missing ingredient for describing the primary photoexcited carrier distributions and how they couple to other degrees of freedom, for instance, the unoccupied resonances of surface adsorbed molecules that may undergo hot electron chemistry [62].

5.3.1
Optical Response of Noble Metals

Because of their useful and instructive physical properties, detailed optical, photoemission, and 2PPE studies have been performed on noble metals [44, 51, 52, 63, 64]. As a specific example of photoexcitation in metals, we consider silver, which has well-known and relatively simple optical and electronic properties and its role has been central in developing theories of interaction of electromagnetic fields with solids. Moreover, because of its favorable optical properties, silver features prominently in the field of plasmonics, which will be discussed in Section 5.4. The electronic structure and optical properties of Ag are known from *ab initio* theoretical calculations, and its dynamical response has been examined at the highest level of theory [58, 65–67]. Figure 5.5 shows the calculated electronic band structure of Ag in the high symmetry directions, which defines its complex dielectric function in Figure 5.6 [66]. Even for a relatively simple metal such as silver, the electronic band structures obtained from calculations at the DFT and GW levels show significant deviations from the optical and photoemission measurements [66, 67]. Significant

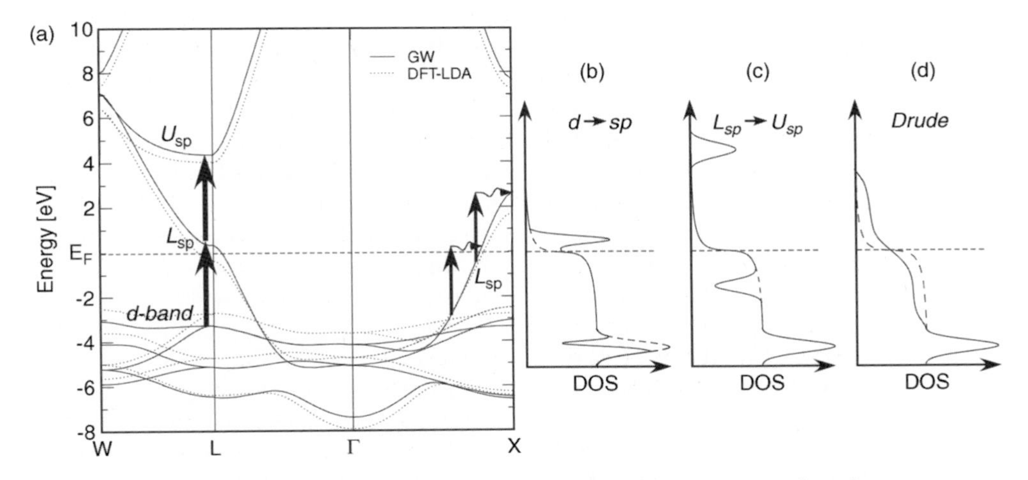

Figure 5.5 (a) The calculated electronic band structure of Ag using the GW and DFT–LDA methods [65, 66] and the possible interband and intraband transitions excited with near-ultraviolet light. The GW calculations place the sp bands at too high energy, whereas the DFT calculations underestimate the binding energy of the d bands. Near the L point, both d → sp and $L_{sp} \rightarrow U_{sp}$ transitions are possible. Intraband transitions within L_{sp} occur through a second-order process involving a scattering process necessary to conserve momentum. (b–d) The corresponding modification of the state occupations (DOS × the distribution function) through the interband and the intraband excitation. The dashed lines indicate the initial population before it is modified by absorption of a photon.

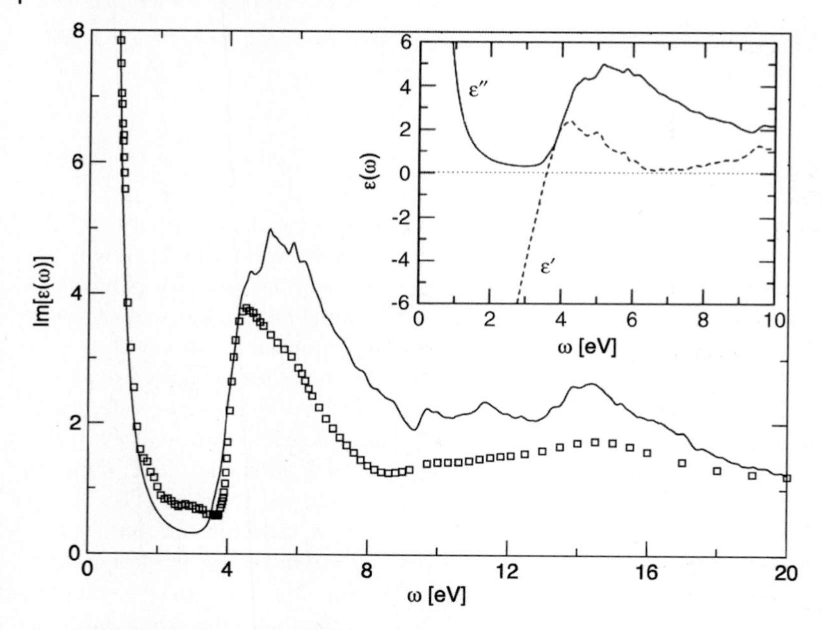

Figure 5.6 The calculated imaginary part of the dielectric constant ε'' of Ag from a GW calculation of Ref. [66] (line). The squares are the experimentally measured values [80]. The inset shows the real and imaginary parts of the dielectric function on the same scale at the onset of the interband excitations, where $\varepsilon' \approx 0$. Inset reprinted with permission from Ref. [66]. Copyright (2002) by the American Physical Society.

improvement between experiment and theory for the optical spectra can be obtained by describing the dynamical excitonic effects through BSE [66]. Even though screening of the Coulomb interaction is efficient in metals, dynamical screening of the interaction between the photoexcited e-h pair is important because of the participation of the d bands with relatively low binding energies in the optical response. Excitonic effects have also been predicted to be increasingly important as the charge density decreases from the bulk value to a few angstroms above a metal surface, where the density-dependent screening response slows down from sub-femtosecond in the bulk to a few femtosecond for image potential (IP) states at surfaces [68].

According to Figure 5.6, the imaginary part of the dielectric constant ε'' of Ag is characterized by the Drude region, which is responsible for decreasing the component that extends from the IR to the near-UV region, and the interband region with a threshold at 3.84 eV. The interband threshold approximately coincides with the crossing of the real part of the dielectric function from negative to positive, that is, $\varepsilon'(\omega) = 0$, which can be seen to occur in Figure 5.6 at 3.92 eV. The $\varepsilon'(\omega) = 0$ condition is responsible for the sharp maximum in the loss function $\mathrm{Im}[-\varepsilon(\omega)^{-1}]$ in Figure 5.7, where the electromagnetic field resonantly drives the collective charge density fluctuations in silver. We describe light absorption in the interband region, and then in the subsequent section in the Drude region.

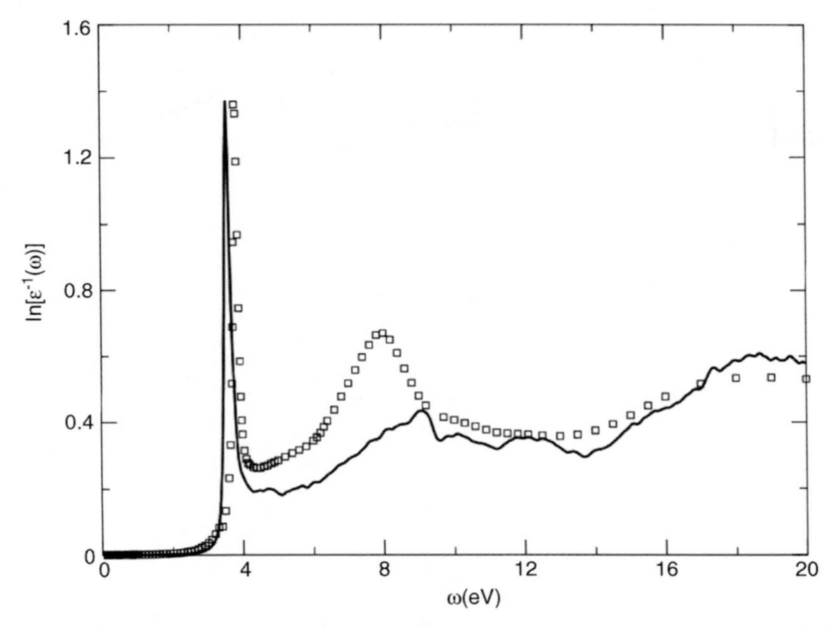

Figure 5.7 The experimental electron energy loss spectrum of silver (boxes) and the calculated (GW) loss spectrum. The major peaks at 3.8 and 7.95 eV are due to the out-of-phase and in-phase screening by sp- and d-band electrons. Reprinted with permission from Ref. [66]. Copyright (2002) by the American Physical Society.

5.3.2
Interband Absorption

At energies below 3.8 eV, direct interband dipole transitions in Ag are not possible according to the band structure in Figure 5.5. Photon absorption by a dipole transition cannot occur, because according to the band structure in Figure 5.5, there are no vertical transitions that couple occupied with unoccupied states. According to the band structure of Ag (Figure 5.5), the onset of interband transitions at 3.84 eV involves two types of excitations [58]. The electronic bands of Ag consist of the partially occupied nearly free electron sp band, which is characterized by a Fermi energy of 5.5 eV (it crosses E_F at several k-space points in high symmetry directions as shown in Figure 5.5). In addition to the sp band, there are fully occupied d bands that extend up to approximately −3.8 eV below E_F. In noble metals, the interband threshold corresponds to momentum-conserving transitions from near the top of the d bands to the sp band at the point where it crosses E_F. Such transitions near the X and L points establish the interband absorption threshold at 3.84 eV [58, 65, 66]. Because the d bands are nearly flat, their contribution to the joint density of states, which are coupled by optical transitions near E_F, is large [63, 69]. In the case of Ag, the threshold region includes an additional resonance at 4.03 eV arising from transition across the L projected band gap from the lower sp band (L_{sp} in Figure 5.5) near its crossing with E_F to the upper sp band (U_{sp} in Figure 5.5) [58, 65, 66]. In the case of copper and gold, the d bands are closer to E_F than for Ag, and therefore their interband thresholds occur at $\sim$2 eV, whereas the $L_{sp} \rightarrow U_{sp}$ excitations across the sp band gap are relatively unchanged from the Ag case.

From the threshold at 3.8 eV, the interband transitions dominate the complex dielectric function of Ag. The band structure in Figure 5.5 portends that in Ag threshold interband photoexcitation generates dominantly energetic holes at the top of d bands near X and L points and electrons near E_F (Figure 5.5b). By contrast, the $L_{sp} \rightarrow U_{sp}$ transition at 4.03 eV generates holes near E_F and electrons at the bottom of U_{sp} (Figure 5.5c). The d band processes dominate the hot carrier generation in noble metals because of their larger joint DOS. Owing to the crystalline symmetry, these interband excitations occur according to Eq. (5.1) along several equivalent lines in BZ that satisfy simultaneously the energy and momentum conservation [63, 69]. Therefore, the initial electron and hole distributions generated by interband excitation are usually strongly localized in energy and momentum according to the specific band structure of the absorbing material and the photon energy, as indicated schematically in Figure 5.5b and c. Such nascent electron distributions can be characterized by angle-resolved two-photon photoemission [43, 61].

Above the interband threshold, the dielectric function exhibits additional structure that reflects the joint DOS in dipole-mediated optical transitions. In particular, features appear at higher energies associated with other critical points that contribute high joint DOS [70]. We also note that for photon energies above the vacuum level (E_{vac}), the conservation of perpendicular momentum is relaxed by the presence of the surface, as noted in Section 5.1.3. The potential discontinuity at the surface allows Bloch states within the bulk to couple to free-electron states in the vacuum without the requirement for conservation of momentum perpendicular to the surface. Under such circumstances, photoemission can occur by separate pathways involving bulk or surface excitations, but starting and terminating at the same initial and final states through interband transitions within the bulk or through surface-mediated processes. Such parallel pathways lead to quantum interference effects that can be diagnosed by their characteristic asymmetric Fano lineshapes in photoemission spectra [44, 71].

The circumstances for interband absorption in transition metals are somewhat different and more complicated than for noble metals [72, 73]. Because both d and sp bands are partially filled for transition metals, in principle, interband transitions could occur with very low photon energies, depending on the separation between the coupled bands at the Fermi surface crossings. Excitation of complex carrier distributions in the phase space is likely, because multiple d bands can participate as the initial and final states. Further complications arise from the spin-orbit splitting of the d bands that are responsible for separate excitation pathways for electrons with the majority and minority spins [74–76]. Excitations to low-lying unoccupied states have been investigated by 2PPE methods for Gd and Ru, among metals, and from d bands to image potential states for Fe and Co [77–79].

5.3.3
Intraband Absorption

In the year 1900, Drude derived a relation $\sigma_0 = (ne^2\tau)/m$ for the DC electrical conductivity of a metal [81]. This is the starting point for the description of the collective electron response responsible for both intraband absorption to be described in this section and the plasmonic phenomena that will follow in Section 5.4. In the Drude model, the electrons in the metal are treated as a freely moving gas with a density n, mass m, and charge e, which can be accelerated during a certain phenomenological relaxation time τ in the direction of an externally applied DC field $\mathbf{E}$ until random scattering processes randomize their momentum. This sequence of acceleration and scattering generates a drift motion of the electrons with velocity $\mathbf{v} = -e\mathbf{E}\tau/m$, which is manifested in the net current density $\mathbf{j} = -ne\mathbf{v} = \sigma_0\mathbf{E}$. Later, Sommerfeld provided a quantum mechanical extension of this model by accounting for the Pauli exclusion principle and the Fermi–Dirac statistics of the electron gas [82]. Corrections to the mass of the conduction electrons due to the finite Coulomb interaction among the electrons and between the electrons and the ionic lattice are often incorporated by replacing m with the effective mass m^*. Within this model, the room temperature DC conductivity of the noble metals gold and silver, which have some of the highest conductivities among metals, is well described by parameters $n_{\mathrm{Au}} = 5.90 \times 10^{-22}$ cm^{-3}, $n_{\mathrm{Ag}} = 5.86 \times 10^{-22}$ cm^{-3}, $\tau_{\mathrm{Au}} = 30$ fs, and $\tau_{\mathrm{Ag}} = 40$ fs [83]. The effective masses in the conduction bands of Au and Ag are close to unity, $m^*/m_{\mathrm{Au}} = 0.99$, $m^*/m_{\mathrm{Ag}} = 0.96$ [59].

Although the Drude model fails to account for the Fermi–Dirac statistics and Pauli exclusion, it has been remarkably successful in describing the optical response of free electrons in a metal. We will consider in detail the intraband component in the dielectric function within the framework of the extended Drude model (EDM) [84] and its relation to the Lorentz component. Charge carriers in a metal subject to an applied oscillating field are continuously accelerated by the field, but over a cycle of oscillation cannot absorb energy in absence of momentum changing collisions [85]. Such collisions can involve e-e, e-p, or impurity scattering. As Drude absorption requires momentum scattering, the electron distributions excited are not localized in k-space and mainly depend on the photon energy, as well as total density of the occupied and unoccupied states that conserve energy and momentum through a secondary momentum scattering process. The degree to which electron scattering is isotropic depends on the material and the scattering process [85]. On the one hand, if a quasielastic scattering process mediates the optical transition, the energy distribution excited within a solid will be determined primarily by the photon energy. Drude absorption will excite hole states between $E_{\mathrm{F}} - \hbar\omega$ to E_{F} and electron states from E_{F} to $E_{\mathrm{F}} + \hbar\omega$. If the occupied and unoccupied DOS vary slowly in the optically coupled region, as can be expected when sp bands cross E_{F}, the energy distribution should resemble that shown in Figure 5.5d. Because sp bands deviate from the free-electron dispersion most strongly near the Brillouin zone boundaries, the Drude distributions could be quite anisotropic in k-space [86]. Isotropic distributions in

k-space corresponding to a DOS that is constant in energy are often assumed by default in photoexcitation of metals, but such distributions can be justified only for free-electron metals, such as alkali metals, when the Drude absorption is the dominant excitation channel. To observe the nascent electron distributions from Drude absorption, however, would require laser pulses with durations that are shorter than the electron scattering times (typically ~20 fs for electrons at the Fermi level for noble metals). More typically, pulses of 50–100 fs duration are employed, so the observed photoelectron distributions in 2PPE experiments, for instance, are found to follow a Fermi–Dirac distribution described by a rapidly evolving temperature that decreases with pump–probe delay on account of electron–electron (e-e) and electron–phonon (e-p) scattering [77, 78, 87]. If, on the other hand, momentum conservation in an optical transition is achieved through e-e scattering, the photon energy is distributed between at least two electrons and two holes that conserve overall energy and momentum. Therefore, the hot electron distribution will already partially relax in the process of photon absorption through the first cycle in an e-e scattering cascade. Such Auger-like processes could be particularly efficient for intraband absorption above the interband threshold of metals and have been described in case of copper in Figure 2.32 [88, 89].

To describe intraband absorption, Eq. (5.10) for the complex dielectric function includes a phenomenological damping term. The dielectric function can equivalently be expressed in terms of complex optical conductivity

$$\varepsilon(\omega) = \varepsilon'(\omega) + i\varepsilon''(\omega) = \varepsilon_\infty + \frac{4\pi i \sigma(\omega)}{\omega}. \tag{5.11}$$

In Eq. (5.11), $\varepsilon_\infty = 1 + 4\pi N\alpha$ is the effective dielectric constant with the second term describing the core polarization in terms of the ionic polarizability α and the number density N. The core polarization includes the renormalization of the low-frequency response by the high-frequency interband excitations. In the case of noble metals, this term is dominated by the d- to sp-band excitations, that is, the interband contribution of the Lorentz term. As for Eqs. (5.2) and (5.3), the real and imaginary parts of the dielectric constant are related by causality; that is, they are constrained by the Kramers–Kronig relationship [65, 90].

For a free-electron gas (FEG), the intraband optical conductivity of a metal is given in terms of the Drude parameters, that is, the free-electron plasma frequency, ω_{p0}, and the phenomenological damping constant, τ:

$$\sigma(\omega) = \frac{\omega_{p0}^2}{4\pi(\tau^{-1} - i\omega)}. \tag{5.12}$$

Treating Ag as an FEG with n given by the density of 5s electrons of 5.85×10^{22} cm^{-3} predicts a plasma frequency of 8.98 eV [65]. The actual plasma frequency should be different because of electron–electron interactions and screening by the interband excitations.

Alternatively, the plasma frequency can be obtained by using the sum rule for the real part of the conductivity [90]:

$$\int_0^\infty \sigma'(\omega)\,\mathrm{d}\omega = \frac{\omega_{\mathrm{p}0}^2}{8}. \tag{5.13}$$

In practice, σ' is known only in a finite frequency range and has to be extrapolated to frequency regions where measurements are not available. Another estimation of $\omega_{\mathrm{p}0}$ is available from the high-frequency limit of the real part of the dielectric constant,

$$\varepsilon'(\omega) \approx \varepsilon_\infty - \frac{\omega_{\mathrm{p}0}^2}{\omega^2}, \tag{5.14}$$

under the assumption that $\omega\tau \gg 1$. The same approximation gives the relaxation time from the imaginary part of the dielectric constant as

$$\varepsilon''(\omega) \approx \frac{\omega_{\mathrm{p}0}^2}{\omega^3\tau}. \tag{5.15}$$

For noble metals in the IR/visible spectral range, the assumption that $\omega\tau \gg 1$ holds true [52, 84].

Whereas the plasma frequency has a clear interpretation, the scattering time has contributions from many possible momentum scattering processes. These processes can be extrinsic, involving impurity scattering, or intrinsic, involving e-e, e-p, or other quasi-particle scattering [91–94]. Within the Fermi liquid theory, it is well established that e-e scattering rate has a quadratic dependence on the electron energy [95–97]. Therefore, one expects the scattering time to be frequency dependent. The characteristic frequency dependence of $\tau(\omega)$ is often used as a diagnostic of the material-dependent dominant scattering processes contributing to the Drude absorption, as discussed further in Section 5.3.5 [52, 93, 94].

5.3.4 Extended Drude Model

Causality demands that if $\tau(\omega)$ is frequency dependent, so is the plasma frequency $\omega_{\mathrm{p}}(\omega)$ [65, 84]. The consequences of the frequency dependence of $\tau(\omega)$ are described within the extended Drude model. First, we will develop the EDM and then discuss some origins of the frequency-dependent scattering rate.

In EDM, we derive $\tau(\omega)$ and the frequency-dependent plasma frequency, $\omega_{\mathrm{p}}(\omega)$, from the complex optical conductivity, constraints imposed by the sum rule of Eq. (5.13), and causality. The components of optical conductivity can be obtained from Eq. (5.12) in terms of the frequency-dependent $\tau(\omega)$ and $\omega_{\mathrm{p}}(\omega)$:

$$\sigma'(\omega) = \frac{\omega_{\mathrm{p}}^2\tau^{-1}}{4\pi(\tau^{-2}+\omega^2)}, \qquad \sigma''(\omega) = \frac{\omega_{\mathrm{p}}^2\omega}{4\pi(\tau^{-2}+\omega^2)}. \tag{5.16}$$

The expressions in Eq. (5.16) satisfy causality, which requires that $\sigma'(-\omega) = \sigma'(\omega)$ and $\sigma''(-\omega) = -\sigma''(\omega)$ are even and odd functions of ω. They can be solved for $\tau(\omega)$

and $\omega_p(\omega)$. The relaxation time is obtained from the ratio of the real and the imaginary conductivity

$$\frac{1}{\tau(\omega)} = \frac{\omega\sigma'(\omega)}{\sigma''(\omega)}. \tag{5.17}$$

Taking the real and imaginary parts of Eq. (5.12) gives another pair of equations,

$$\frac{1}{\omega_p(\omega)^2} = \frac{1}{4\pi\omega}\mathrm{Im}\left(\frac{-1}{\sigma(\omega)}\right), \qquad \frac{1}{\tau(\omega)} = \frac{\omega_p(\omega)^2}{4\pi}\mathrm{Re}\left(\frac{1}{\sigma(\omega)}\right), \tag{5.18}$$

which together with Eq. (5.17) yield a set of equations that give $\tau(\omega)$ and $\omega_p(\omega)$ even if the frequency dependence of $\omega_p(\omega)$ is unknown [84].

When considering a Fermi liquid, as described in Chapter 2, rather than a free-electron gas, the quasi-particle mass and therefore $\tau(\omega)$ and $\omega_p(\omega)$ are renormalized through e-e interaction. The renormalization can be expressed through a mass enhancement factor λ, such that $\lambda = m^*/m^{-1}$. Then the renormalized quantities become

$$\omega_p(\omega) = \frac{\omega_{p0}^2}{1+\lambda(\omega)}, \qquad \tau(\omega) = (1+\lambda(\omega))\tau_0(\omega), \tag{5.19}$$

where $\tau_0(\omega)$ is the free-electron scattering time. With ω_{p0} calculated from the carrier density, Eq. (5.19) can be substituted into Eq. (5.18) to obtain $\lambda(\omega)$ and $\tau_0(\omega)$

$$1+\lambda(\omega) = \frac{\omega_{p0}^2}{4\pi\omega}\mathrm{Im}\left(\frac{-1}{\sigma(\omega)}\right), \qquad \frac{1}{\tau_0(\omega)} = \frac{\omega_{p0}^2}{4\pi}\mathrm{Re}\left(\frac{1}{\sigma(\omega)}\right). \tag{5.20}$$

From the symmetry properties of the real and the imaginary optical conductivity, it is evident that $\lambda(\omega)$, $\tau_0(\omega)$, $\omega_p(\omega)$, and $\tau(\omega)$ are even functions of ω.

We can now compare the Drude parameters from EDM with the experimental loss function of Ag in Figure 5.7. The loss function shows a broad peak with high spectral weight at 8.0 eV, which is to be compared with the renormalized plasmon frequency of 9.2 eV (here the renormalization factor from EDM is $1+\lambda(\omega) = 0.95$) calculated from the conduction band electron density [65]. In addition to the experimentally observed plasmon at 8.0 eV, we already noted the prominent sharp peak that occurs when $\varepsilon'(\omega) = 0$ with a smaller spectral weight at 3.9 eV [66]. This peak is too far removed from the renormalized plasmon frequency to be attributed exclusively to the response of the conduction electrons. Its origin can be rationalized by noting that plasmonic excitations occur when $\varepsilon(\Omega) = 0$, where Ω is complex-valued frequency with a small imaginary part, and for $q = 0$. Neglecting the damping effects, the dielectric function can be written as

$$\varepsilon(\omega) = \varepsilon_\infty - \frac{\omega_p^2}{\omega^2}. \tag{5.21}$$

The solution for $\varepsilon(\Omega) = 0$ occurs for $\omega_p^* = \omega_{p0}/\sqrt{\varepsilon_\infty}$ eV, which we understand as the plasma mode of sp band electrons dressed by a virtual cloud of interband excitations of d and sp bands. Thus, in Ag, as in other noble metals, the free-electron plasma peak occurs within a continuum of interband excitations, leading to its significant broadening and renormalization (see Figure 5.7). Because in Ag the $\varepsilon(\Omega) = 0$ condition occurs right below the interband threshold, where the imaginary part of Ω is small, the loss spectrum also has a sharp component associated with the plasma response renormalized by the interband excitations [65]. Thus, two peaks in the energy loss spectrum of Figure 5.7 at 7.95 and 3.9 eV can be considered as the collective oscillations of the sp- and d-band electrons, where the two components oscillate in phase for the former and out of phase for the latter [58, 65, 98]. In Cu and Au, the $\varepsilon(\Omega) = 0$ condition occurs within the interband excitation continuum, and therefore a sharp out-of-phase peak is not observed. Nevertheless, the plasmonic responses of Ag and Au, to be described in Section 5.4, arise from the out-of-phase responses of the sp- and d-band electrons of these metals.

5.3.5
Frequency-Dependent Scattering Rate

By applying EDM to the experimentally measured complex, frequency-dependent dielectric function of a metal, we can derive several useful dynamical properties, such as the frequency-dependent plasma frequency and scattering time, and the renormalization of the free-carrier response by the interband excitations. Perhaps the most interesting and difficult quantity to calculate is $\tau(\omega)$, which we discuss in more detail.

A plot of $\tau^{-1}(\omega)$ versus ω^2 obtained by plotting Eqs. (5.17) or (5.18) can often be used as a diagnostic for the quasi-particle interactions in simple and correlated metals [52, 84, 85, 91–94, 99–102]. In Figure 5.8, we show such a plot for Cu, Ag, and Au. The frequency dependence of $\tau^{-1}(\omega)$ can be assumed to have the general form given by [52, 92]

$$\tau^{-1}(\omega) = \tau_0^{-1}(T) + a(T)\omega + b(T)\omega^2. \tag{5.22}$$

The form of Eq. (5.22) is expected to be valid in a frequency band above the low-frequency region, which is dominated by electron–boson (e.g., phonon) coupling, and below the interband threshold. Outside these limits, $\tau^{-1}(\omega)$ is strongly material dependent, whereas in the intermediate region, it is expected to follow the simple frequency and temperature dependence of Eq. (5.22). The exact behavior depends on the dominant mechanism for momentum scattering of e-h pairs coupled by the optical transition. The frequency dependence of $\tau^{-1}(\omega)$ has therefore been used to analyze the scattering processes that define the optical conductivity in metals, as well as the optical and transport properties in strongly correlated materials [52, 91–93, 101, 102].

In addition to Eq. (5.22), we also assume the validity of Matthiessen's rule, which allows us to write the total scattering rate as a sum of independent contributions from e-e, e-p, surface, and impurity scattering,

$$\frac{1}{\tau(\omega, T)} = \frac{1}{\tau_{\text{e-e}}(\omega, T)} + \frac{1}{\tau_{\text{e-p}}(\omega, T)} + \frac{1}{\tau_s} + \frac{1}{t_i}. \tag{5.23}$$

The last two terms in Eq. (5.23) arise from the breakdown of the translational invariance of a crystal in the presence of a surface (τ_s^{-1}) or impurities (τ_i^{-1}). We

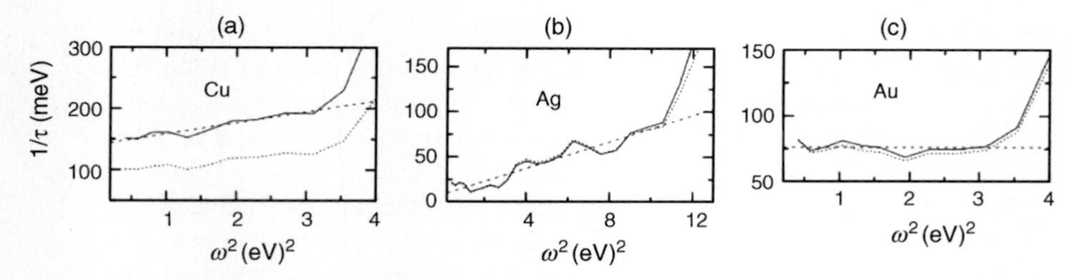

Figure 5.8 Plot of the dependence of $\tau^{-1}(\omega)$ on ω^2 for (a) Cu, (b) Ag, and (c) Au below the interband excitation threshold. The ω^2 dependence is found for Cu and Ag but not Au. Reprinted with permission from Ref. [85]. Copyright (2007) John Wiley and Sons.

assume that both these processes are independent of ω and T [52, 100], and therefore, contribute only to the intercept of a plot of τ^{-1} versus ω or T.

The surface contribution depends on the optical skin depth. As required by momentum conservation, surface absorption will have significant contribution only if $qv_{\mathrm{F}} > \omega$, where q is a scattering wave vector. If the skin depth δ is larger than the distance v_{F}/ω traveled by an electron during the period of the radiation field oscillation, the surface absorption will be negligible [99]. The surface absorption should be significant only at low frequencies and low temperatures, but the frequency range falls outside the range of validity in Eq. (5.22).

The contribution of e-p scattering can be estimated through a FEG approach [99]. The FEG scattering rate depends on the probability of the second-order process involving simultaneous absorption of a photon and absorption or emission of a phonon to satisfy energy and momentum conservation of the overall process. Therefore, the Debye frequency and temperature will determine ω and T range where τ_{ep}^{-1} has a significant variation. The phonon contribution to Drude absorption according to FEG is

$$\frac{1}{\tau_{\mathrm{ep}}(\omega, T)} = \frac{2\pi}{\omega} \int_0^{\omega} \mathrm{d}\Omega(\omega - \Omega)\alpha_{\mathrm{tr}}^2 F(\Omega). \tag{5.24}$$

Here, Ω is the phonon frequency and $\alpha_{\mathrm{tr}}^2 F(\Omega)$ is the phonon DOS weighted by the amplitude for large-angle scattering on the Fermi surface (see Section 2.2.3). Umklapp processes can be included explicitly in the calculation of $\alpha_{\mathrm{tr}}^2 F(\Omega)$. The scattering time is frequency dependent in the region spanned by the phonon spectrum. The phonon contribution to the scattering time goes to infinity as ω^{-5} for frequencies below the Debye frequency, similar to the phonon scattering contribution to electrical conductivity. The low-frequency behavior of the Drude scattering time, $\tau^{-1}(\omega)$, is useful in studies of superconductivity because it provides information on e-p interaction strength through $\alpha_{\mathrm{tr}}^2 F(\Omega)$ and on the superconducting gap 2Δ. For a superconducting system, the upper limit of integration in Eq. (5.24) is changed to $\omega - 2\Delta$ and the frequency dependence of τ_{ep}^{-1} reflects the corresponding changes in the integral of Eq. (5.24) [99].

The conventional e-p contribution to $\tau^{-1}(\omega)$ appears only in the temperature-dependent intercept $\tau_0^{-1}(T)$ because Eq. (5.22) is valid only above the Debye frequency. The surface and defect scattering contributions can be distinguished as a temperature-independent offset of $\tau_0^{-1}(T)$ from the extrapolated $T=0$ value of the $\tau_{\mathrm{ep}}^{-1}(T)$.

Numerous studies of the optical properties of the relatively simple noble metals have been used to validate the Drude theory. In the frequency range from 0.6 eV up to the interband absorption onset, the frequency dependence of Eq. (5.22) is dominated by the $b\omega^2$ term for Cu and Ag, as can be seen in Figure 5.8. Whether this ω^2 dependence arises from e-p or e-e scattering, however, is controversial and uncertain [52, 84, 91, 92, 103, 104]. By extending the Fermi liquid theory to describe the e-e scattering contribution to optical conductivity [105, 106], Christy and coworkers proposed that the e-e scattering should lead to a scattering rate of the form

$$\tau_{\mathrm{ee}}^{-1}(\omega, T) = b\omega^2 + c(k_{\mathrm{B}}T^2). \tag{5.25}$$

As in the case of electrical conductivity, the temperature dependence of Eq. (5.25) should be observed only at sufficiently low temperatures where the e-p contribution is frozen out [106]. This term makes a contribution to the intercept of Eq. (5.22), which is too small to identify considering the uncertainties in performing and evaluating the optical measurements. The $b\omega^2$ term is a consequence of the energy dependence of the phase space for e-e scattering in a three-dimensional Fermi liquid [96]. The e-e scattering time according to the Fermi liquid theory is given by

$$\tau_{\mathrm{ee}}(E) = 263 r_{\mathrm{s}}^{-5/2}(E - E_{\mathrm{F}})^{-2}\mathrm{fs\,eV}^2, \tag{5.26}$$

where r_{s} is the electron density parameter (see also Section 2.4.3.2) [97, 107]. Equation (5.26) gives the e-e scattering time for an electron at a specific energy above E_{F}, whereas the frequency dependence of Eq. (5.25) arises from the scattering of an e-h pair coupled by photons of energy $\hbar\omega$. Therefore, the optical scattering rate with the $b\omega^2$ dependence is an average of the scattering rates of holes in the energy range from $E_{\mathrm{F}} - \hbar\omega$ to E_{F} and electrons from E_{F} to $E_{\mathrm{F}} + \hbar\omega$ [52].

Although it is quite reasonable that above the Debye frequency the Drude scattering rate of noble metals should be dominated by e-e scattering, the measured parameter b is several times too large in comparison to the e-e scattering rates deduced from transport measurements. In order to explain this discrepancy, Smith, Cisneros, and coworkers pointed out that the Umklapp contribution of e-p scattering should also follow the quadratic dependence in frequency and therefore contribute to the $b\omega^2$ term [52, 104]. The e-p mechanism, however, predicts b to be temperature dependent, but this is not found in the temperature-dependent measurements of $\tau^{-1}(\omega, T)$ for Cu, Ag, and Au [92].

The interpretation of the frequency and temperature dependence of $\tau^{-1}(\omega, T)$ for noble metals is thus still not fully understood, yet Drude theory is actively being applied as a diagnostic for electron scattering processes in, for instance, strongly correlated materials. For example, in high-temperature superconductors, where the $a\omega$ term in Eq. (5.22) often dominates, the linear dependence has been examined in terms of the bosonic interactions that give rise to exotic superconductivity and the marginal Fermi liquid behavior [93, 94, 101]. The applications of EDTand particularly the frequency dependence of the scattering rate on highly correlated materials are discussed at length in the review by Basov and Timusk [93].

5.3.6
Surface Absorption

So far we have considered only the role of bulk dielectric properties of metals in the absorption of light. The effect of the surface appeared only as a source of momentum due to the finite optical skin depth. A surface can also influence absorption on account of the intrinsic or adsorbate-induced electronic structure with implicit relaxation momentum conservation on account of reduced dimensionality and disorder.

Surface absorption is well documented under UHV conditions where surface science methods are used to prepare atomically ordered samples. Perhaps the most detailed studies have been performed on the anisotropic absorption at (110) surfaces of noble metals [108–112]. The anisotropy of the surface makes it possible to separate the surface contribution from the isotropic bulk response. For example, the reflectivity of a Cu(110) surface has a sharp drop at 2.0 eV, attributed to a transition between an occupied Shockley surface state at −0.4 eV to an unoccupied surface state at 1.6 eV (Figure 5.9). This surface is peculiar in having two low-energy surface states within the $\bar{Y}$ projected band gap. These surface states are relatively sharp because their wave function penetration into the bulk is small, and, therefore, the inelastic scattering of the surface state electrons and holes with the bulk carriers is constrained. Moreover, the surface states have similar dispersions, leading to a large joint DOS for optical transitions. Both these factors contribute to a sharp absorption feature that is observed in the anisotropic reflectivity spectrum. This resonance is sensitive to the adsorption of impurities at low coverages because the extremely efficient scattering of surface state electrons leads to the resonance broadening [110, 113]. When adsorbates form highly ordered layers, however, such as the oxygen-induced (2×1) reconstruction of a Cu(110) surface, adsorbate-localized excitations can also contribute to sharp spectroscopic features in surface reflection spectra [109, 111].

In addition to surface localized interband transitions, a complete description of light absorption at surfaces should also include the surface-to-bulk and bulk-to-surface optical transitions. Such transitions are known to happen from two-photon photoemission spectroscopy. For instance, studies of coherent control of photoinduced current involving the $n = 1$ image potential state of a Cu(001) surface rely on the coherent excitation from the occupied bulk sp band near E_F [114]. Although such processes can happen with high efficiency, as evidenced by the intense IP state signal from Cu(001), the coupling of 3D bulk with 2D surface bands is unlikely to contribute distinct features in the optical spectra because such transitions have relatively low joint DOS and they occur above the interband threshold for bulk excitations. In the case of Cu(001), the threshold for IP state excitation at ~4.0 eV overlaps with the more intense d- to sp-band interband spectrum. Transitions between bulk and surface

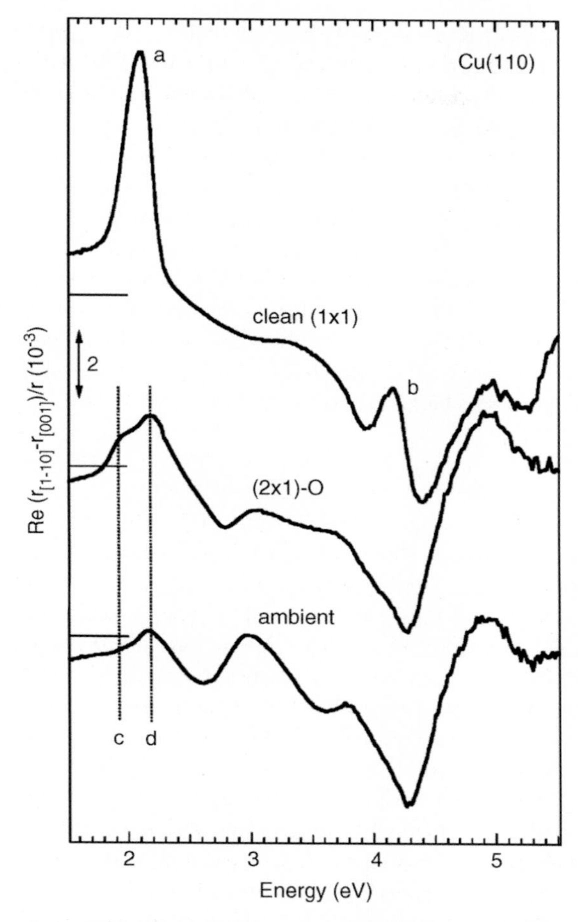

Figure 5.9 Plot of the anisotropic reflectivity from Cu(110) and Cu(110)-(2 × 1)-O surfaces in ultrahigh vacuum and air. Peaks a–d involve transitions between surface states. Reprinted with permission from Ref. [110]. Copyright (2000) by the American Physical Society.

bands might have more significant contributions to the dielectric function below the interband threshold. For instance, the onset of excitations from the sp band to an unoccupied part of the Shockley surface state should have an onset above ~0.3 eV for Ag(111) and Cu(111) surfaces [115, 116]. Such transitions have recently been found in 3PPE spectra of Cu(111) [117].

Finally, we turn our attention to isolated adsorbates on metal surfaces. UHV surface studies have shown that adsorbates can have significant effect on the dielectric functions of metals through resonant or nonresonant effects [110, 118, 119]. Surface adsorbates are impurity scattering centers, similar to bulk defects, and therefore can enhance intraband absorption by acting as other nonresonant sources of momentum. Electronic states of impurity atoms and molecules can also act as resonant absorbers of electromagnetic energy. This is most clearly seen in two-photon photoemission spectra of alkali atoms on noble metal surfaces, where resonant excitation from substrate to adsorbate-localized states is evident in optical dephasing of coherent two-photon absorption [120–123]. Although there are many examples in the literature claiming hot electron excitation of adsorbates on metals, where presumably electron–hole pairs are absorbed by metal and hot electrons scattered into adsorbate resonances [124, 125], most experimental studies of adsorbate resonance-mediated two-photon photoemission are consistent with the direct excitation from bulk bands of the substrate to adsorbate-localized resonances [126]. Because excitation occurs to localized states, such charge transfer processes need to conserve only energy. The transition probability depends on the optical transition moment and the overlap between the Bloch waves of the substrate and localized resonance wave functions of adsorbate. Unfortunately, such interfacial charge transfer excitation processes have yet to be described by theory.

5.3.7
Summary

In this section, our goal was to describe photon absorption processes at metal surfaces. Traditionally, the optical response of metals has been understood on the basis of the complex dielectric function. The dielectric function incorporates the electronic band structure and the dynamical many-body response of a system to an external electromagnetic perturbation. Far more detailed information on the material response available to linear optical methods such as ellipsometric measurements can now be obtained by means of nonlinear optical spectroscopy and, in particular, by angle, energy, spin, and time-resolved photoemission [43, 49, 76, 112, 114, 127]. Such measurements can be typically understood from the static band structure of metals, but experiments and theory are advancing to the attosecond timescales, where effects of the coherent many-body response of metals should have measurable effects [68, 128, 129]. We believe that further progress in the understanding of the optical response of metals can be made by quantitative time-resolved two-photon photoemission measurements in partnership with electronic structure theory.

5.4
Plasmonic Excitations at Surfaces and Nanostructures

In the previous section, we described the optical excitation processes in metals. The focus has been primarily on single-particle excitations, though we found that the collective response associated with the plasma excitations plays an important role in defining the dielectric properties of metals. This section describes the collective plasma response, particularly in how it affects the optical properties of metallic particles with different geometries and in different dielectric environments. Specifically, effects of the shape of metallic nanostructures on their linear optical properties will be addressed. Starting with very simple geometries, the fundamental optical excitations of extended interfaces (surface plasmon polaritons [130]) and of nanoparticles (localized surface plasmons [131, 132]) will be introduced. Structures with novel and interesting optical functionality can be designed by fabricating nanostructures supporting both propagating and localized surface plasmon polariton fields and will be discussed for specific examples of subwavelength gratings and adiabatic metallic tapers. Finally, some recent ideas for coupling surface plasmon fields to excitonic excitations in hybrid metal/dielectric nanostructures will be presented. Much of the discussion will focus on optical properties in the visible and near-infrared range, treating the local dielectric function of the metal in terms of the Drude model introduced earlier in this chapter.

5.4.1
Drude Model for Optical Conductivity

When applying a harmonic time-varying electric field $\vec{E}(t) = \vec{E}_0\, \mathrm{e}^{-i\omega t}$ of frequency ω, collective oscillations of the free-electron gas are induced. Their amplitude $\vec{x}(t)$ is given by the equation of motion $m(\ddot{\vec{x}} + (\dot{\vec{x}}/\tau)) = -e \cdot \vec{E}$, inducing a dipole moment per electron of $\vec{p}(\omega) = -e\vec{x}(\omega) = (e^2/m)(1/(\omega^2 + (i\omega/\tau))$ and hence a macroscopic polarization $\vec{P} = ne\vec{x} = (\varepsilon{-}1)\varepsilon_0\vec{E}$. In this approach, the frequency-dependent dielectric function $\varepsilon(\omega)$ is thus given by the intraband contribution to Eq. (5.10) [133, 134]. Using Ohm's law, $\vec{j}(\omega) = ne\dot{\vec{x}}(\omega) = \sigma(\omega)\vec{E}(\omega)$, gives the relationship between the external field and the induced current density in terms of optical conductivity, as given in Eq. (5.12). Therefore, the collective response of the electron plasma in the metal results in a dielectric function with a large negative real part and a much smaller positive imaginary part in the infrared and visible spectral ranges, as described in Section 5.3. In the high-frequency limit, ε approaches unity.

Experimental values for the dielectric function of gold and silver up to the interband thresholds, taken from Ref. [60], are shown in Figure 5.10, together with a fit to Eq. (5.10), ignoring $\varepsilon_i(\omega)$ and assuming plasma frequencies of $\omega_{\mathrm{p,Au}} = 9.1$ eV and $\omega_{\mathrm{p,Ag}} = 9.2$ eV, respectively. The existence of the plasma response at these frequencies is evident from the dielectric and loss functions of Ag in Figures 5.6 and 5.7, which extend into the higher energy range than Figure 5.10. The phenomenological damping times, taken to be frequency independent, are $\tau_{\mathrm{Au}} = 9.3$ fs and $\tau_{\mathrm{Ag}} = 31$ fs, respectively. It is evident that the dielectric functions of silver and gold essentially follow the predictions of the Drude model for energies up to their interband thresholds at about 3.8 and 2.0 eV if values $\varepsilon_{\infty,\mathrm{Ag}} = 4.0$ and $\varepsilon_{\infty,\mathrm{Au}} = 7$ are assumed. In Figure 5.10 (solid lines), the interband contribution to the dielectric function $\varepsilon_i(\omega)$ is modeled phenomenologically as a sum over transitions at critical points in the joint density of states that are coupled by optical excitation as described in Eq. (5.1) [135]. For gold, a reasonable agreement with the experiment is reached by including two transitions, $i = 1, 2$, whereas for silver more resonances are needed [135, 136]. Effects connected with a nonlocal (**k**-dependent) response of the metal [137, 138], relevant for particle sizes of less than 10 nm, will not be considered here. It is currently believed that such a local Drude-like dielectric function can

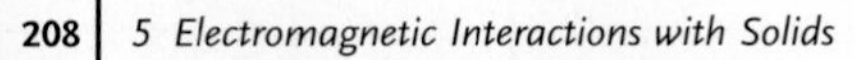

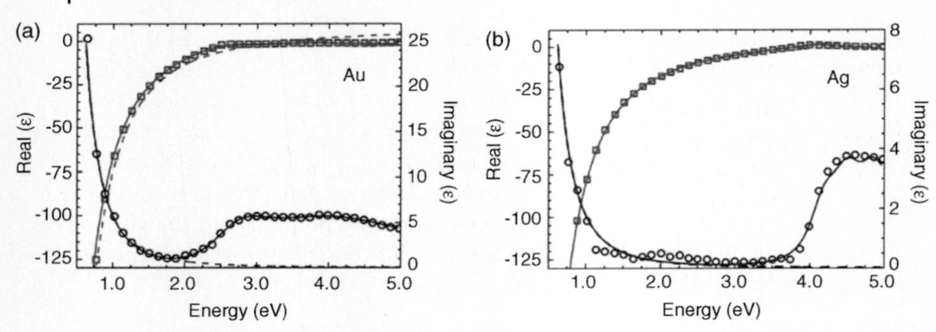

Figure 5.10 The real and imaginary parts of the dielectric function of (a) gold and (b) silver. The experimental data (open circles) are taken from Ref. [60]. The dashed lines indicate fits to the Drude model (Eq. (5.10)) without the interband term $\varepsilon_i(\omega)$. The solid lines represent fits to the Drude model, including a phenomenological description of $\varepsilon_i(\omega)$.

account for the majority of collective optical phenomena at metal surfaces [139–142]. More refined description of the optical response of noble metals can be obtained by EDM, as described in Section 5.3.1.

5.4.2
Interaction of Light with a Planar Metallic Surface

In combination with Maxwell's equations, the dielectric function of metals as represented by the Drude model directly accounts for the mirror-like reflectivity at frequencies below the plasma frequency. We consider a planar interface, located at $z = 0$, between a semi-infinite and nonmagnetic metal with a local dielectric function $\varepsilon_1(\omega)$ and a dielectric with $\varepsilon_2(\omega)$. The metal lies in the negative half-space, $z < 0$. The corresponding complex refractive indices $n_i = n'_i + in''_i = \sqrt{\varepsilon_i} = \sqrt{\varepsilon'_i + i\varepsilon''_i}$ of the two layers are then obtained from

$$n'_i = \left(\frac{1}{2}\left(\varepsilon'_i + \sqrt{\varepsilon'^2_i + \varepsilon''^2_i}\right)\right)^{1/2}$$

and $n''_i = \varepsilon''_i/2n'_i$, where the prime and double-prime functions indicate the amplitudes of the real and imaginary components. Consider a monochromatic plane wave $\vec{E}(\vec{r},t) = \vec{E}_0 e^{i(\vec{k}_1\vec{r}-\omega t)}$ propagating through the dielectric toward the metal. The plane of incidence is defined by $y = 0$, that is, $\vec{k}_1 = (k_x, 0, k_{1z}) = (\omega n_1/c)(\sin(\theta_1), 0, \cos(\theta_1))$, with θ_1 denoting the angle of incidence and c being the speed of light in vacuum. Snell's law gives $n_1 \sin(\theta_1) = n_2 \sin(\theta_2)$ and the continuity equations require that the wave vectors of the reflected and transmitted beams be given by $\vec{k}_{1r} = (k_x, 0, -k_{1z})$ and $\vec{k}_2 = (k_x, 0, k_{2z}) = (\omega/c)(n_1 \sin(\theta_1), 0, n_2 \cos(\theta_2))$, respectively.

For s-polarized incident light, the electric field vector lies perpendicular to the plane of incidence and $\vec{E}_0 = (0, E_0, 0)$. The reflection and transmission coefficients ϱ_s and τ_s for the incident beam are given by Fresnel's equations [143, 144]

$$\varrho_s = \frac{1-a}{1+a}, \qquad \tau_s = \frac{2}{1+a}, \tag{5.27}$$

respectively, with $a = k_{2z}/k_{1z} = \tan(\theta_1)/\tan(\theta_2)$. In this case, the amplitudes of the reflected and transmitted fields are $\vec{E}_{0r} = \varrho_s \vec{E}_0$ and $\vec{E}_{0t} = \tau_s \vec{E}_0$, respectively.

For p-polarized incident light, $\vec{E}_0 = E_0(-\cos(\theta_1), 0, \sin(\theta_1))$ lies in the plane of incidence and Fresnel's equations give

$$\varrho_p = \frac{1-b}{1+b}, \qquad \tau_p = \frac{2(n_1/n_2)}{1+b} \tag{5.28}$$

with $b = (n_1/n_2)^2(k_{2z}/k_{1z}) = (n_1/n_2)^2 a$. The reflected and transmitted field amplitudes then are $\vec{E}_{0r} = \varrho_p E_0(\cos(\theta_1), 0, \sin(\theta_1))$ and $\vec{E}_{0t} = \tau_p \vec{E}_0(-\cos(\theta_2), 0, \sin(\theta_2))$, respectively. The resulting angle-dependent reflectance curves $R_{s,p}(\theta) = |\varrho_{s,p}|^2$ of an Ag/air interface irradiated with s- and p-polarized light at $\lambda = 1000$ nm ($\omega = 1.24$ eV) are shown in Figure 5.11.

In this case, $\omega \ll \omega_p$ so that the dielectric function of the metal is governed by its large and negative real part ($\varepsilon_2 = -51.1 + 0.943i$). Hence, the induced coherent polarization $\vec{P}$ of the electron gas is phase shifted by 180° with respect to the driving field. The electromagnetic field inside the metal is thus effectively screened and decays within the skin depth $\delta = \lambda/(2\pi\,\mathrm{Im}\,(n_2))$, which is about 25 nm for gold and silver in a fairly broad range of wavelengths from the visible up to the mid-infrared. In the present example, $\delta = 22$ nm. The second important consequence of the coherent free-carrier oscillation, that is, the same coherent polarization, is the almost loss-free specular reflection of the incident light wave. It is important to realize that s- and p-polarized light waves experience different phase shifts $\varphi = \arg(\varrho)$, and hence reflection from a metallic mirror at an off-normal angle of incidence modifies the polarization state of the reflected field.

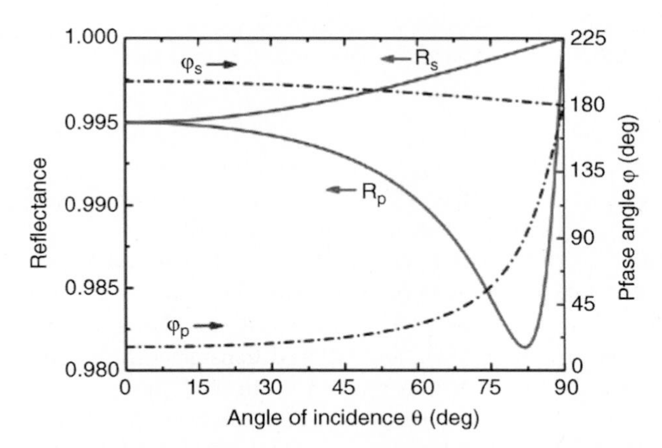

Figure 5.11 The reflectance and phase shift for the reflection of a light wave at $\lambda = 1000$ nm at a silver mirror. The dielectric function of silver is taken as $\varepsilon_2 = -51.1 + 0.943i$. The data are plotted for s- and p-polarized light as a function of the angle of incidence θ.

5.4.3
Surface Plasmon Polariton Fields

5.4.3.1 Planar Interfaces

It has been known since the pioneering work of Ritchie that metal/dielectric interfaces support collective charge density oscillations and surface plasmons [32, 145]. These surface charge oscillations couple strongly to light and the resulting coupled modes between charges and fields are termed surface plasmon polaritons (SPPs). Surface plasmon polaritons propagate freely along planar interfaces and are evanescently confined in the direction perpendicular to the interface. They penetrate into the dielectric on a scale given by their wavelength, whereas the skin depth gives their penetration depth into the metal. Throughout most of this depth, the dipolar charge oscillations preserve charge neutrality. A net charge density exists only within the Thomas–Fermi screening length of about 1 Å at the surface. For a perfectly planar surface, such SPP fields must have a finite component of the electric field normal to the surface; s-polarized SPP modes with an electric field vector oriented parallel to the surface do not exist [139, 142]. SPP modes on planar interfaces are necessarily p-polarized electromagnetic waves with a magnetic field vector $\vec{H}$ pointing parallel to the interface. Some of the fundamental SPP properties are readily derived directly from Maxwell's equations in the absence of external sources [130, 142].

We again consider a planar interface, located at $z = 0$, between a semi-infinite and a nonmagnetic metal in half-space $z < 0$, with a local dielectric function $\varepsilon_1(\omega)$ and a dielectric with $\varepsilon_2(\omega)$. The metallic layer is irradiated with a monochromatic wave of frequency ω propagating along the x-axis with vectors

$$\vec{E}_i = (E_{ix}, 0, E_{iz})e^{-\kappa_i|z|}\, e^{(ik_{ix}x-\omega t)}, \tag{5.29}$$

$$\vec{H}_i = (0, H_{iy}, 0)e^{-\kappa_i|z|}\, e^{(ik_{ix}x-\omega t)}, \tag{5.30}$$

where k_{ix} gives the component of the wave vector parallel to the interface in medium $i = 1, 2$. Ampère's law connects the electric and magnetic field amplitudes (+ sign for $i = 1$)

$$i\kappa_i H_{iy} = \pm\varepsilon_i\varepsilon_0\omega E_{ix}, \tag{5.31}$$

$$k_{ix}H_{iy} = -\varepsilon_i\varepsilon_0\omega E_{iz} \tag{5.32}$$

and hence gives $E_{ix}/E_{iz} = \pm\kappa_i/ik_{ix}$. It requires, together with Faraday's law, the magnitude of the wave vector components along the z-axis to be

$$\kappa_i = \sqrt{k_{ix}^2 - \varepsilon_i k_0^2} \tag{5.33}$$

with $k_0 = \omega/c$ being the magnitude of the light wave vector. Since the tangential components of E and H are continuous at the interface, $H_{1y} = H_{2y}$ and $E_{1x} = E_{2x}$. Equation (5.10) then gives the surface plasmon condition

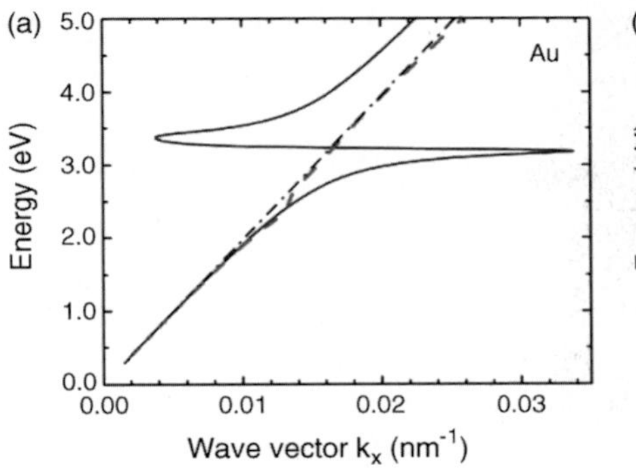

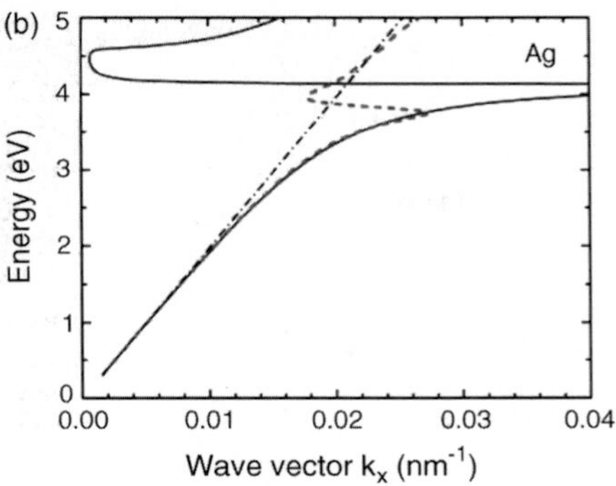

Figure 5.12 SPP dispersion relations $\omega(\text{Re}\,(k_x))$ based on Eq. (5.35) for planar (a) gold/air and (b) silver/air interfaces. The curves are calculated based on the Drude model neglecting interband absorption ($\varepsilon_i = 0$, solid lines) and including interband absorption (dashed lines). The light line $\omega = c \cdot k_x$ is shown as a dash-dotted line.

$$\frac{\varepsilon_1}{\kappa_1} + \frac{\varepsilon_2}{\kappa_2} = 0. \tag{5.34}$$

The boundary conditions at the interface also demand the continuity of the tangential component of the wave vector $k_{1x} = k_{2x} = k_x$. Together with Eqs. (5.33) and (5.34), the continuity equation gives the surface plasmon dispersion relation for a planar interface [32]

$$k_x = \frac{\omega}{c}\left(\frac{\varepsilon_1 \varepsilon_2}{\varepsilon_1 + \varepsilon_2}\right)^{1/2}. \tag{5.35}$$

The dispersion relations $\omega(\text{Re}\,(k_x))$ are shown in Figure 5.12 for air/gold and air/silver interfaces. The curves have been calculated by using the Drude dielectric function (Eq. (5.10)) in the presence (dashed lines) and absence (solid lines) of interband absorption terms. For frequencies above the plasma frequency $\omega > \omega_p$, $\text{Re}\,(\varepsilon_1) > 0$ and the metal becomes transparent. In this region, the diagram displays the dispersion relation of light inside the metal. For sufficiently high frequencies, $\varepsilon_1 \rightarrow 1$ and the curve approaches the light line $\omega = c \cdot k_x$ (c is the speed of light in vacuum) indicated by a dash-dotted line. The high k_x asymptotic limit given by $\varepsilon_1 + \varepsilon_2 = 0$ defines the classical surface plasmon frequency, which in the case of a lossless Drude model for the dielectric function at a metal/air interface is $\omega_{SP} = \omega_p/\sqrt{2}$.

Of interest is the SPP dispersion relation at frequencies below ω_p. At sufficiently low frequencies, in the so-called retarded region, the SPP dispersion lies only slightly outside the light line. In the absence of interband screening, this region roughly covers the small wave vector part ($k_x < \omega_p/(\sqrt{2}c)$) of the dispersion curve at energies $\omega < \omega_{SP} = \omega_p/\sqrt{2}$. In this situation, SPPs are light-like quasi-particles, propagating at essentially the speed of light and having evanescent SPP field amplitudes (Eqs. (5.29) and (5.30)), which decay exponentially on either side of the interface. In this region, the spatial extent of the electromagnetic field is very different on the

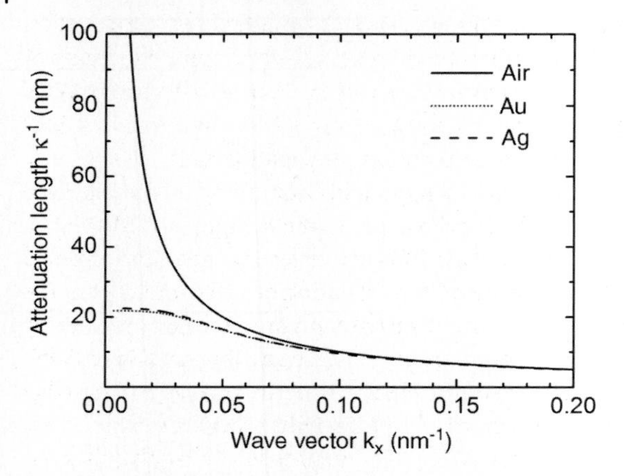

Figure 5.13 The attenuation lengths $l_i = |1/\kappa_i|$ of the electromagnetic field on either side of Au/air and Ag/air interfaces deduced from Eq. (5.34). An idealized loss-free Drude model ($\varepsilon_\infty = 1$, $\tau \to \infty$, $\varepsilon_i = 0$) is taken for the dielectric function of the metal. The solid black line shows the attenuation length l_1 in air, whereas the dotted and dashed lines denote l_2 in gold and silver, respectively.

two sides. Equations (5.34) and (5.35) give the following expression for the SPP decay constants k_i perpendicular to the interface:

$$\kappa_i = \frac{\omega}{c}\sqrt{\frac{-\varepsilon_i^2}{\varepsilon_1 + \varepsilon_2}}. \tag{5.36}$$

The resulting attenuation lengths $l_i = |1/\kappa_i|$ at which the electromagnetic field falls off to $1/e$ are shown in Figure 5.13. For clarity, values resulting from a only lossless Drude model are shown. In the long-wavelength limit ($q \to 0$), the attenuation length in the metal is given by the skin depth. The field on the air side, however, is much less confined and extends over more than $c/\omega = \lambda/(2\pi)$, with λ being the wavelength of light in vacuum. In this regime, the interface supports SPP modes, but the fields are only weakly confined to the interface and field localization effects are basically absent. Therefore, retardation effects arising from the finite SPP phase velocity are important for optical properties of the interface.

This changes in the nonretarded regime of large in-plane wave vectors $k_x > \omega_{SP}/c$ in the frequency range $\omega < \omega_{SP}$. Here, for a lossless Drude metal, the dispersion relation $\omega(k_x)$ is a monotonically increasing function of k_x that approaches asymptotically ω_{SP}. SPP waves in the nonretarded regime have interesting properties. Their in-plane wave vector k_x is much larger than that of a light wave of the same frequency. In the fictitious lossless case, the corresponding in-plane wavelength can be reduced to essentially arbitrarily small values. Therefore, SPP waves can be localized in volumes much smaller than λ^3, breaking the diffraction limit in conventional far-field optics. SPP waves in the nonretarded regime are strongly confined to the interface. It is seen in Figure 5.13 that the attenuation length in air becomes similar to that in the metal and approaches $1/k_x$ as ω approaches ω_{SP}. As $k_z \gg k_0$, the finite propagation speed of the SPP modes is of minor importance and quasi-static approximations of Maxwell's equations can describe the optical properties of nonretarded SPP modes reasonably well. Consequently, the magnetic field associated with nonretarded SPP waves is much weaker than that of a corresponding propagating field.

More important, since $k_x > k_0$, propagating light fields impinging on the interface cannot directly excite SPP modes. Coupling of far-field light to SPP at a metal/vacuum interface can occur in a total internal reflection geometry from a high-index substrate material for a thin metal film, creating evanescent fields at the interface, as demonstrated by Otto [146] and Kretschmann and Raether [147], or by scattering light off surface roughness or gratings, as pioneered by Teng and Stern [148]. In the latter approach, the scattering at a subwavelength asperity such as a sharp edge or a slit with smaller features, the wavelength of exciting light provides the necessary momentum to couple to the SPP modes [149]. Also, near-field excitation schemes, directly providing evanescent fields, can be employed [150]. Most recently, nonlinear four-wave mixing has been demonstrated as a versatile method for exciting SPP modes [151].

For real, that is, lossy metals, the situation is obviously less ideal. As illustrated in Figure 5.12, the finite imaginary part of ε_1 removes the singularity in Eq. (5.36) and hence limits the maximum value of Re (k_x), as well as the field confinement to the interface. For Im $(\varepsilon_1) \neq 0$, also the quasi-bound, leaky part of the dispersion relation with $\omega_{SP} < \omega < \omega_p$ is allowed. It is evident that for both gold and silver structures, the interband contribution to the dielectric function greatly affects the SPP modes in the visible range.

In case of a planar interface, the SPP dispersion is fully governed by the frequency-dependent dielectric function of the metal and the dielectric. It will be illustrated in Section 5.4.4.3 that structuring of the metal provides an additional degree of freedom for tailoring SPP dispersion and fields. This led to artificially designed surface plasmon resonances or, more precisely, the optical properties of structures supporting SPP fields by patterning metallic films with closely packed geometric arrangements of structures with dimensions and separations much smaller than the vacuum wavelength [152, 153]. This emerging field is now sometimes called designer or spoof plasmonics and evidently bears many similarities with other fields such as band gap engineering of semiconductors [154], or the design of photonic crystals [155] or metamaterials [156, 157].

An additional important consequence of Eq. (5.35) is the finite propagation length of SPP wave packets along planar interfaces. The finite extent of the SPP field into the metal results in damping of SPP waves due to unavoidable Ohmic losses. For a complex value of ε_1, the propagation constant k_x is also a complex number and the SPPs are damped with a propagation length $L_d = 1/(2\,\mathrm{Im}\,(k_x))$. Typical propagation lengths are <10–100 μm in the visible spectrum for Ag films (Figure 5.14).

The propagation lengths increase greatly with longer wavelength due to the concomitant increase in k_2. For planar interfaces, the lengths can be increased further by sandwiching a thin metal film between the two dielectric layers. In these multilayers, SPPs at both interfaces are coupled giving rise to symmetric and

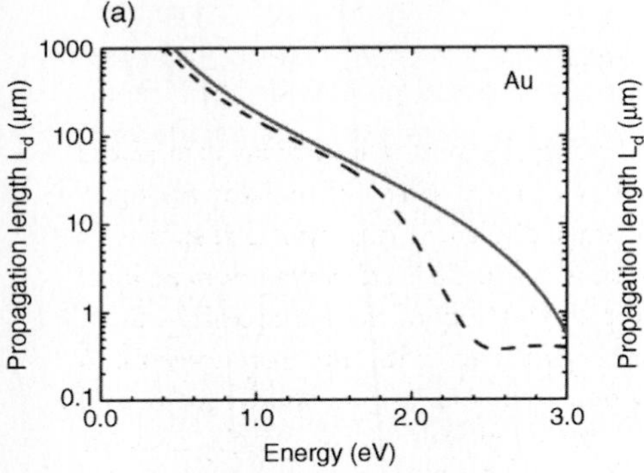

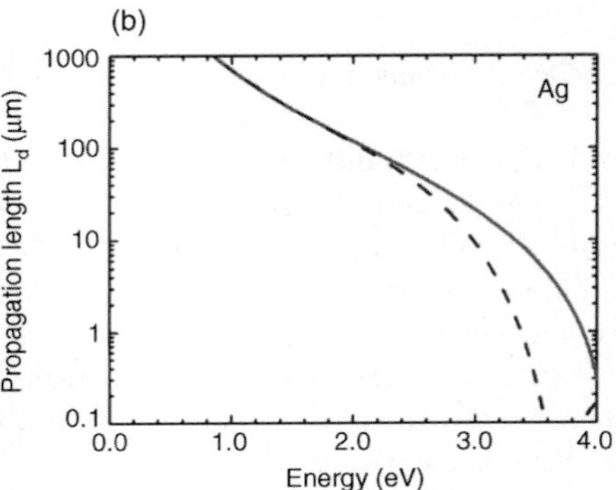

Figure 5.14 SPP propagation lengths $L_d = 1/(2\,\mathrm{Im}\,(k_x))$ as a function of SPP energy calculated for (a) gold/air and (b) silver/air interfaces based on the dispersion relation (Eq. (5.33)) for Drude models neglecting ($\varepsilon_i = 0$, solid lines) and including interband absorption (dashed lines). The propagation lengths are plotted on a logarithmic scale.

antisymmetric modes. The antisymmetric modes have reduced Ohmic losses because they are only weakly confined to the metal. These long-range SPP solutions [158] show drastically increased propagation lengths, extending to the mm range. Attempts to geometrically pattern the metal surface will almost inevitably result in a reduction of the propagation length [159–162] because the collective charge oscillations in these geometrically confined metallic structures can now also emit radiation into the far field. Radiation damping therefore appears as an additional loss mechanism [163]. Hence, the reduction of SPP losses in passive or active hybrid metallic–dielectric nanostructures is a very active field of current research [164, 165].

The preceding description of SPPs is based on a local dielectric function $\varepsilon(\omega)$, independent of the SPP wave vector. Possible nonlocality of the electronic response and the microscopic spatial distribution of the electron density have been neglected. Such microscopic effects are thought to be of minor importance for sufficiently long SPP wavelengths $2\pi/k_x$, but will become important if the SPP wavelength approaches the Fermi wave vector, $k_x \approx k_\mathrm{F}$. For a detailed discussion, we recommend Ref. [142].

The SPP coupling at discontinuities and propagation in flat metal films can be imaged by near-field microscopy or photoemission electron microscopy (PEEM) [149, 166–168]. Such measurements rely on the interference between the external excitation field and the SPP wave packet generated at a vacuum/metal interface. Near-field microscopy has the advantage of being a linear technique, but it does not directly measure the total field present in the metal. By contrast, PEEM measurements detect the spatial distribution of two-photon photoemission, without introducing a near-field perturbation. Therefore, PEEM measurements provide a nonlinear map of the polarization gratings excited by different field components. Moreover, because PEEM in an imaging method, it is relatively simple to perform femtosecond time-resolved measurements on surface plasmon dynamics [169]. Proposals have been put forward to use PEEM in the attosecond domain in order to resolve plasmonic fields with a time resolution of less than one optical cycle [170].

5.4.4
Surface Plasmons in Nanostructured Metal Films

A fundamentally different approach for localizing electromagnetic fields at surfaces relies on the optical excitation of individual or coupled arrays of metallic nanoparticles. Metallic nanostructures of arbitrary shape show a strong optical response at certain resonance frequencies. This response is connected with a pronounced local enhancement of the electromagnetic field in the vicinity of the nanoparticle. Such enhancements are important in, for instance, light harvesting for solar conversion applications, or surface-enhanced spectroscopies, such as Raman spectroscopy, and surface photochemistry [171–173]. The optical properties of single metallic nanoparticles have already been discussed in textbooks and reviews [131, 132, 139, 174, 175], and only very basic properties will be summarized here as prototypical examples for how the interplay between metallic dielectric function and geometric shape gives rise to new optical properties and greatly enhanced local electromagnetic fields.

5.4.4.1 Spherical Nanoparticles

In the quasi-static approximation, the isotropic polarizability α of a spherical nanoparticle of radius a is found by solving the Laplace equation for the scalar potential Φ, $\Delta\Phi = 0$ [143], and is given as

$$\alpha = 4\pi a^3 \frac{\varepsilon_1 - \varepsilon_2}{\varepsilon_1 + 2\varepsilon_2} \tag{5.37}$$

for a metal particle described by ε_1 in an isotropic and nonabsorbing medium with dielectric constant ε_2.

An incident electromagnetic field $\vec{E}_0(\omega)$ will polarize the nanoparticle and create a spatially homogeneous field $\vec{E}_{\text{in}}(\omega) = (3\varepsilon_2/(\varepsilon_1 + 2\varepsilon_2))\vec{E}(\omega)$ inside the sphere. This approximation can be valid only for particles smaller in size than the skin depth because for larger particles $\vec{E}_{\text{in}}$ will necessarily decay in the interior of the sphere. The electromagnetic field $\vec{E}_{\text{out}} = \vec{E}_0 + \vec{E}_1$ outside the sphere is given as the sum of the incident field $\vec{E}_0(\omega)$ and the field $\vec{E}_1(\omega)$ that is reradiated by a fictitious point-like dipole located at the center of the sphere ($\vec{r} = 0$) and having a dipole moment

$$\vec{p}(\omega) = \varepsilon_2 \varepsilon_0 \alpha(\omega) \vec{E}_0(\omega). \tag{5.38}$$

For a monochromatic incident field at frequency ω, the field $\vec{E}_1(\vec{r}, t)$ at position $\vec{r}$ outside the sphere is thus given by

$$\vec{E}_1(\vec{r}, t) = \frac{1}{4\pi\varepsilon_2\varepsilon_0} \left[k^2(\vec{n} \times \vec{p}) \times \vec{n} \frac{e^{ikr}}{r} + [3(\vec{n} \cdot \vec{p}) - \vec{p}] \left(\frac{1}{r^3} - \frac{ik}{r^2} \right) e^{ikr} \right] e^{-i\omega t} \tag{5.39}$$

with $\vec{n} = \vec{r}/|\vec{r}|$ and $k = \sqrt{\varepsilon_2}\omega/c$. The different symmetries of the optical near field given by the last two terms and the optical far field given by the first term in Eq. (5.39) are illustrated in Figure 5.15 for a particle with a 10 nm radius [176].

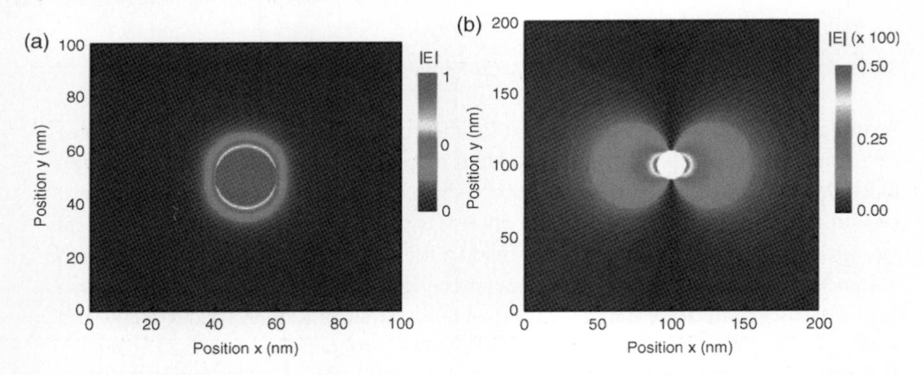

Figure 5.15 Magnitude of the electric field $|\vec{E}(\vec{r})|$ (arbitrary units) near a small nanoparticle with a radius of 10 nm. The field profile is calculated in the x–y plane using Eq. (5.39), assuming that it is given by that of a point-like oscillating dipole $\vec{p} = p_0 \cdot \vec{e}_y \cdot e^{i\omega t}$ at the center of the sphere. (b) Electric field magnitude $|\vec{E}(\vec{r})|$ given by the far-field term in Eq. (5.39). Close to the particle, the far-field amplitude is up to two orders of magnitude smaller than that of the corresponding near field. (Please find a color version of this figure on the color plates.)

The near field is preferentially oriented along the direction of the incident field, whereas the far field vanishes along the dipole axis. For such a small particle, the field intensity in the direct vicinity of the particle is up to 10 000 times larger than that given by the far-field term in Eq. (5.39). This is a manifestation of the large enhancement of the local field by the nanoparticle.

Equation (5.37) shows that the polarizability experiences a resonant enhancement if the magnitude of the denominator $|\varepsilon_1 + 2\varepsilon_2|$ tends toward zero. For a sufficiently small Im (ε_1), the resonance (Fröhlich) condition is simply $\text{Re}(\varepsilon_1) = -2\varepsilon_2$ and the associated quasi-particle is known as the dipole surface plasmon. For a spherical particle, which has a lossless Drude dielectric function and is embedded in air, the condition is met for $\omega_0 = \omega_p/\sqrt{3}$. With increasing ε_2, the resonance shifts to the red. Higher order resonances occur when [174]

$$l(\varepsilon_1) + (l+1)\varepsilon_2 = 0, \quad l = 1, 2, \ldots \tag{5.40}$$

These are the quadrupole ($l = 2$) and higher resonances.

Within this quasi-static model, the scattering cross section of the nanoparticle $C_{\text{sca}}(\omega)$ is obtained by calculating the total radiated power emitted by the dipole and dividing it by the intensity of the incident plane wave:

$$C_{\text{sca}}(\omega) = \frac{k^4}{6\pi}|\alpha(\omega)|. \tag{5.41}$$

Similarly, the absorption cross section $C_{\text{abs}}(\omega)$ is obtained from the power $P_{\text{abs}} = (\omega/2)\,\text{Im}\left[\vec{p} \cdot \vec{E}_0^*\right]$ absorbed by a point dipole:

$$C_{\text{abs}} = k\,\text{Im}\,[\alpha(\omega)]. \tag{5.42}$$

The extinction cross section is given by the sum $C_{\text{ext}} = C_{\text{sca}} + C_{\text{abs}}$.

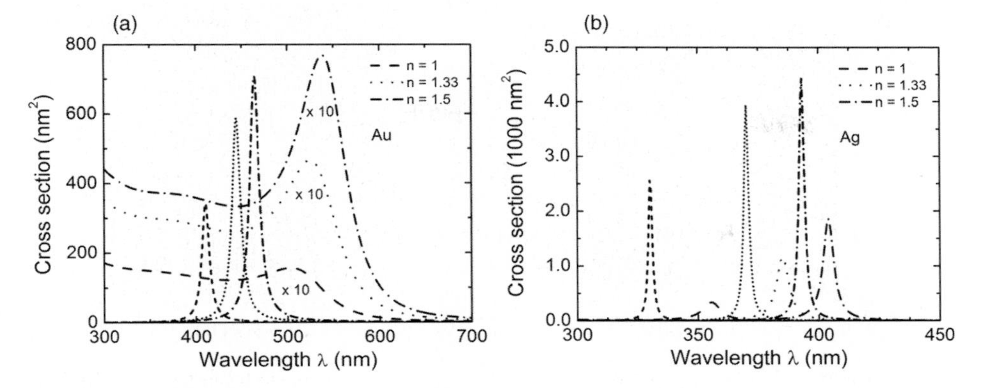

Figure 5.16 Absorption spectra $C_{\text{abs}}(\lambda)$ calculated using Eq. (5.42) for (a) spherical gold nanoparticles and (b) spherical silver particles with a radius of 5 nm embedded in different dielectric media with refractive indices $n = 1$, 1.33, and 1.5. The dielectric function of the metal is based on the dispersion relation (Eq. (5.35)) for Drude models neglecting ($\varepsilon_i = 0$, short dashed and dotted lines) and including (long dashed and dotted lines) interband absorption. The cross sections for gold particles in presence of interband damping have been enlarged by a factor of 10.

Absorption cross sections of gold and silver particles calculated for a Drude model in the absence of interband resonances and the full dielectric function in Eq. (5.10) are presented in Figure 5.16. The resonant enhancement depends critically on the details of the nanoparticle dielectric function (and of course its geometric shape) in the region around $\varepsilon_1 \approx -2\varepsilon_2$, making nanoparticle scattering spectra sensitive to the dielectric function of the environment [177]. This feature has been exploited in sensing applications [172].

The scaling of $C_{\text{abs}}(\omega)$ and $C_{\text{sca}}(\omega)$ with particle size is very different. The scattering cross section is proportional to $|\alpha|^2$ and hence to the sixth power of the particle radius, $C_{\text{sca}} \propto a^6$. By contrast, the absorption cross section scales only with a^3. This is immediately relevant for spectroscopic studies of individual nanoparticles, which are of particular importance for elucidating the complex interplay between nanoparticle size, shape, and environment on their optical properties. The scaling suggests that absorption studies are more sensitive to small particles than light scattering. Indeed photothermal imaging techniques have successfully detected individual sub-10 nm particles [178]. Light scattering from single small particles is challenging to resolve. After initial attempts to use near-field spectroscopic techniques [179], a variety of different far-field techniques, often relying on dark-field excitation schemes and combined with interferometric detection or nonlinear optical techniques, have been developed for this purpose [180–183]. PEEM-based measurements provide an alternative method for imaging emission from regions of high field enhancement that are associated with the excitation of plasmonic modes of metallic nanoparticles [169, 184, 185].

The quasi-static approach outlined above completely neglects the radiative damping of the particle dipole as well as retardation effects due to phase changes in the driving and scattered fields within the volume of the particle. Both effects are included in the rigorous electrodynamic model developed by Mie [132, 174].

Mie theory gives an approximate value for the polarizability of a sphere of volume V as [186, 187]

$$\alpha(\omega) = \frac{1-0.1\,(\varepsilon_1+\varepsilon_2)x^2}{((1/3)+\varepsilon_2/(\varepsilon_1-\varepsilon_2))-(1/30)(\varepsilon_1+10\varepsilon_2)x^2-i(4\pi^2\varepsilon_2^{3/2}/3)(V/\lambda^3)} \tag{5.43}$$

with $x = \pi a/\lambda$ being the size parameter and λ the vacuum wavelength. The new term in the numerator characterizes the retardation of the excitation field across the particle. The x^2 term in the denominator is due to the retardation of the depolarization field inside the particle [186]. Both terms lead to a spectral redshift of the resonance. The imaginary term in the denominator accounts for the radiation damping. Higher terms in x may be included in Eq. (5.43) and will lead to additional multipolar resonances in the spectra.

The radiation damping contributes to the finite lifetime T_2 of the photoinduced dipole moment $\vec{p}$ [188] and hence to a finite homogeneous linewidth $\Gamma = 2\hbar/T_2$ of the linear optical spectrum of a single nanoparticle. The radiative damping rate $1/T_{2,\mathrm{rad}} = \Gamma_{\mathrm{rad}}/2\hbar = \kappa V$ scales with the volume of the particle (κ is the proportionality coefficient) [132, 189, 190]. Therefore, the homogeneous spectral linewidth $\Gamma = \Gamma_{\mathrm{rad}} + \Gamma_{\mathrm{b}}$ of large spherical nanoparticles is expected to be larger than the linewidth Γ_{b} deduced within the quasi-static approximation and should increase with increasing particle diameter. The effect of particle size, geometry, and environment on Γ has been the subject of numerous primarily frequency domain studies of single metal nanoparticles during the last decade. For gold nanoparticles [189, 190], the linewidth Γ has indeed been found to increase from 200 meV ($T_2 = 6.5$ fs) for 20 nm particles to more than 800 meV ($T_2 = 1.7$ fs) for larger particles of more than 100 nm diameter. Values for the coefficient κ of 4×10^{-7} fs^{-1} nm^{-3} to 6×10^{-7} fs^{-1} nm^{-3} have been deduced [132, 189, 191]. For silver nanoparticles, a similar increase in linewidth has been reported [189, 191]. Solid agreement between the observed size dependence of Γ and Mie theory predictions based on ε values from Ref. [60] has been found. When reducing the size of the nanoparticle much below 50 nm, the contribution from radiative damping to the linewidth vanishes and the T_2 time approaches that deduced from quasi-static models. In this regime, for particle diameters between 50 and 20 nm, nonradiative contributions to the plasmon damping arising from intra- and interband electron–electron scattering, electron–phonon scattering, or impurity scattering dominate. Their effects on the dielectric function of a metal are discussed in more detail in Section 5.3.

When decreasing the particle diameter below 20 nm, the linewidth tends to increase again [132, 192, 193]. The linewidth increase is found to be inversely proportional to the particle radius, $\Gamma = \Gamma_{\mathrm{b}} + A \cdot a^{-1}$. Different microscopic mechanisms can contribute to this increase, in particular a reduction in the surface plasmon mean free path due to scattering at the nanoparticle surface [132, 194]. This modifies the decay of the surface plasmon into electron–hole pairs (Landau damping). In quantum mechanical terms, this is understood as an enhanced electron scattering induced by the quantum confinement of electronic states inside the nanoparticle [195, 196]. Also, the inelastic scattering of plasmon excitations at adsorbate or interface states (chemical interface damping) contributes [132, 187, 192, 197]. For particles with a well-controlled environment, for example, a defined dielectric shell, chemical interface damping can be suppressed and the effect of quantum confinement on surface plasmon damping can be quantitatively measured [193].

5.4.4.2 **Elliptical Nanoparticles**

To illustrate the effect of the shape of the particle on its optical properties, we briefly discuss nanoparticles whose shape in Cartesian coordinates is given by an ellipsoid of the form

$$\frac{x^2}{a_1^2}+\frac{y^2}{a_2^2}+\frac{z^2}{a_3^2}=1 \tag{5.44}$$

The principal axes are chosen such that $a_1 \leq a_2 \leq a_3$. In quasi-static approximation, the analytical solution for the polarizability tensor $\overleftrightarrow{\alpha}=\alpha_1\vec{e}_x+\alpha_2\vec{e}_y+\alpha_3\vec{e}_z$ is [174]

$$\alpha_i = 4\pi a_1 a_2 a_3 \frac{\varepsilon_1(\omega)-\varepsilon_2}{3\varepsilon_2+3L_i(\varepsilon_1(\omega)-\varepsilon_2)} \tag{5.45}$$

with $i = 1, 2, 3$. The geometric factors L_i are

$$L_i = \frac{a_1 a_2 a_3}{2}\int_0^\infty \frac{\mathrm{d}q}{(a_i^2+q)f(q)} \tag{5.46}$$

with $f(q)=\sqrt{(q+a_1^2)(q+a_2^2)(q+a_3^2)}$. For spherical particles, $L_i = 1/3$ and the sum rule $\sum_i L_i = 1$ is fulfilled for all shapes. These polarizabilities can then be taken to calculate the fields $\vec{E}_1(\vec{r})$ near the outer surface of the ellipsoid by using Eq. (5.39) and $\vec{p}(\omega)=\varepsilon_2\varepsilon_0\overleftrightarrow{\alpha}(\omega)\vec{E}_0(\omega)$. This immediately gives the scattering and absorption cross sections of the nanoparticle following the same approach as outlined above.

The optical spectra of such particles hence depend critically on the orientation of the particle with respect to the incident field. This is best illustrated for prolate $(a_1 = a_2 < a_3)$ or oblate $(a_1 < a_2 = a_3)$ spheroidal particles having only two different nonzero elements of $\overleftrightarrow{\alpha}$ and hence showing two distinct spectral resonances in their optical spectra. For a prolate spheroid, the short-wavelength resonance is excited with incident light polarized along one of the short axes, whereas light polarized along the long axes couples to α_3. As illustrated in Figure 5.17, the resonance wavelength of α_3 is sensitive to the aspect ratio $r = a_3/a_1$ and gradually shifts to longer wavelengths as a_3 is increased. Also, an increase in aspect ratio greatly enhances the polarizability along the long axis and transforms the particle from an isotropic into a strongly anisotropic light scatterer, with its preferential polarization direction oriented along the long axis.

Variation in the shape of nanoparticles therefore is an important means to tailor their optical spectra. It affects not only the resonance energies but also the SP lifetimes [189]. SP dephasing times T_2 as long as 20 fs have been reported [189–191].

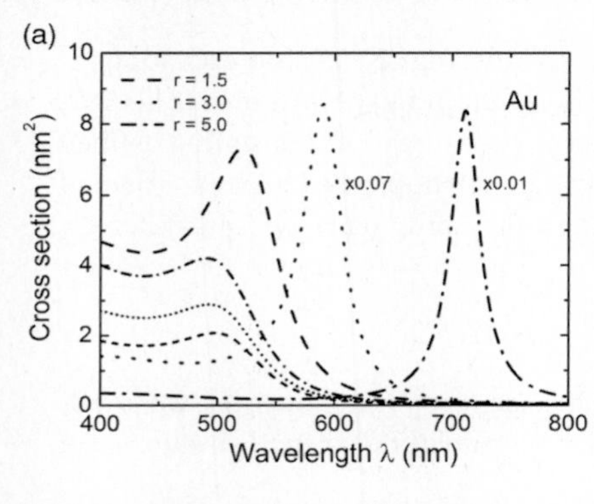

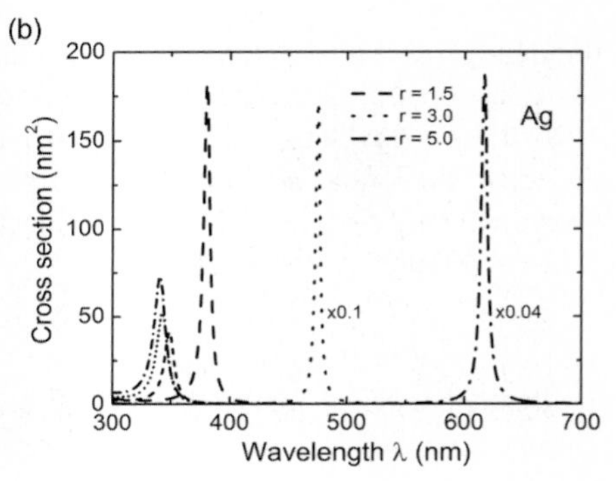

Figure 5.17 Simulations of absorption spectra $C_{abs}(\lambda)$ of (a) elliptical gold and (b) elliptical silver nanoparticles with axes $a_1 = a_2 = 5$ nm and $a_3 = r \cdot 5$ nm. The short and long dashed and dotted lines lines represent spectra for light linearly polarized along the long and short axes, respectively. Interband absorption is included in the model for the dielectric function of the metal. With increasing ellipticity, the cross section for linearly polarized light along the long axes increases. The spectra have been scaled for clarity.

The radiative damping is weak and thought not to affect the T_2 time because the volume of the investigated nanoparticles is small. The comparatively long dephasing times are attributed to the suppression of interband absorption resulting from the redshift of the plasmon resonances. To first approximation, the electric field distribution in the vicinity of the particle is obtained from Eq. (5.39). Spatially resolved imaging of those fields may be achieved by different experimental techniques, including near-field scanning optical microscopy (NSOM) [175, 198–200], two-photon photoemission microscopy [169], tip-enhanced electron emission microscopy [201], cathodoluminescence imaging [202], or electron energy loss spectroscopy (EELS) [203]. Empirical extensions of the polarizability model given in Eq. (5.43) to larger sized elliptical nanoparticles have been discussed in the literature [192].

5.4.4.3 **Diffraction Gratings**

Diffraction gratings present another prime example illustrating the influence of the geometric shape of a metallic nanostructure on its optical properties. These have fascinated researchers since Wood published his observations on anomalous spectrally narrow dark bands in their reflectivity spectra [204, 205]. It took 40 years until Fano assigned these anomalies to the resonant excitation of surface waves [206], later termed surface plasmon polaritons. Renewed interest in diffraction gratings emerged in 1998 when Ebbesen *et al.* discovered the extraordinary enhancement of transmission of light through periodic two-dimensional arrays of subwavelength holes at certain resonance frequencies (Figure 5.18) [207]. To the surprise of many, they reported transmission coefficients much larger than $T = (64/27\pi^2)(k\,r)^4$, the so-called Bethe limit, for a normally incident plane with wavevector k through a single aperture with radius r in a perfectly conducting metal film [208, 209], and even exceeding the hole filling fraction.

A dominant effect of the grating in such experiments is to exchange momentum with the incident light beams. When scattering a monochromatic plane wave with

(a)
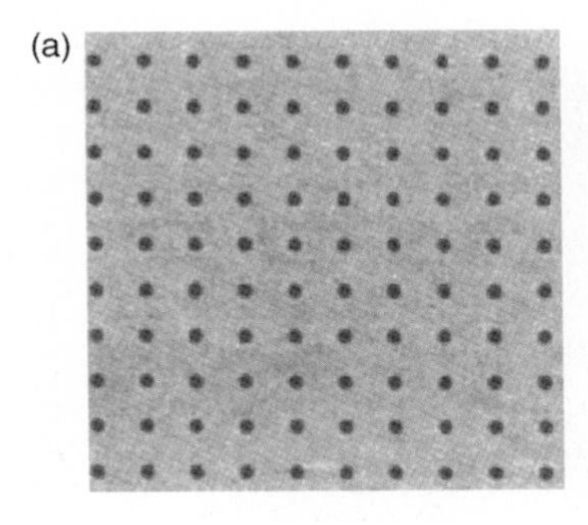
(b)
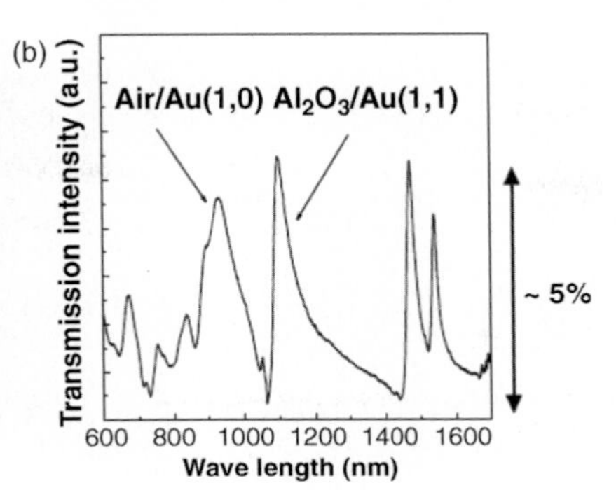

Figure 5.18 (a) Scanning electron microscope image of a periodic nanohole array in an optically thick gold film. The array period is 850 nm and the hole diameter 150 nm. (b) Representative far-field transmission spectrum through such an array in a gold film deposited on a sapphire substrate. The spectrum is recorded at near-normal incidence and transmission resonances at the air/gold or the sapphire/gold interface are indicated. Reprinted with permission from Ref. [210]. Copyright (2002), American Institute of Physics.

in-plane momentum $\vec{k}_{||} = (k_x, k_y, 0)$ off the grating, momentum conservation requires that the diffraction orders have in-plane wave vectors $\vec{k}_{p,q} = \vec{k}_{||} + p \cdot \vec{G}_x + q \cdot \vec{G}_y$. The integers p and q denote the diffraction orders and $\vec{G}_x = (2\pi/a_x)\vec{e}_x$ and $\vec{G}_y = (2\pi/a_y)\vec{e}_y$ are the reciprocal lattice vectors of a two-dimensional grating with periods $a_{x,y}$ along the x and y directions, respectively. Regular propagating diffraction orders are given if $|\vec{k}_{p,q}| \leq \omega/c$, that is, if the wave vector of the diffracted beam lies inside the light cone. Evanescent surface plasmon polariton fields can be excited at the grating interface if $\vec{k}_{p,q}$ lies outside the light cone, that is, if $|\vec{k}_{p,q}| > \omega/c$. Efficient grating coupling to SPP requires that energy and momentum conservation be fulfilled according to Eq. (5.35)

$$\omega(\vec{k}_{p,q}) = c \cdot |\vec{k}_{p,q}| \left(\frac{\varepsilon_1 + \varepsilon_2}{\varepsilon_1 \varepsilon_2} \right)^{1/2}. \tag{5.47}$$

Subwavelength diffraction gratings ($a_x, a_y < \lambda$) are of particular interest for coupling to SPP modes. In this case, close to normal incidence, only the zero-order mode with wave vector $k_{0,0}$ is a propagating mode, whereas all higher diffraction orders are evanescent SPP modes.

The resonance condition expressed in Eq. (5.47) gives approximate values for the transmission resonances in Figure 5.18 when considering that for thin transmission gratings deposited on a dielectric substrate, SPP modes can be excited either at the air side ($\varepsilon_2 = 1$) or at the dielectric side ($\varepsilon_2 = \varepsilon_d$) of the grating. Consequently, different transmission resonances appear in Figure 5.18 for excitation of SPP modes at the air or the dielectric side. Excitation of SPP modes in diffraction gratings can be verified by mapping resonances in angle-resolved or spectrally resolved reflectivity measurements or, more directly, by microscopically imaging the resulting SPP fields. A representative NSOM image of the intensity of light transmitted through a two-dimensional array of nanoholes in a thin gold film is shown in Figure 5.19 [210].

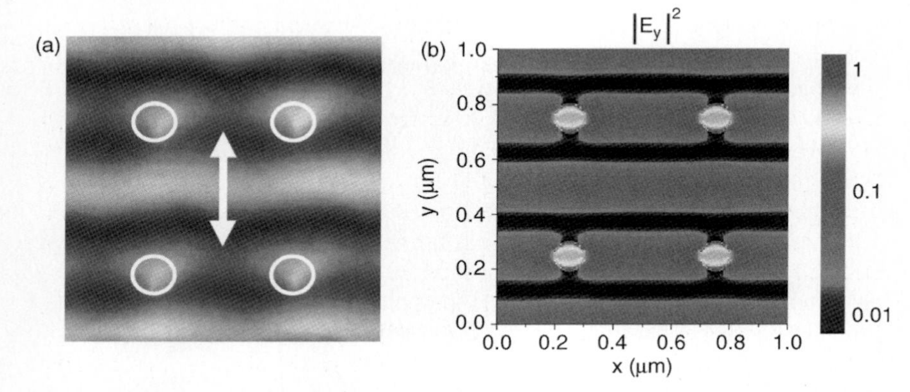

Figure 5.19 (a) NSOM image of the light transmitted through a subwavelength diffraction grating. The grating is fabricated by milling an 850 nm period array of 150 nm diameter nanoholes into a thin gold film deposited on a sapphire substrate. The grating is illuminated from the sapphire side with y-polarized light (see arrow) at 877 nm. The light at the air side of the grating is collected with a metal-coated NSOM fiber probe. Light regions correspond to high intensity, whereas the intensity drops to zero in the dark regions. Standing SPP waves at the grating interface are mapped. (b) Near-field intensity $|E_y|^2$ obtained from a three-dimensional finite difference time domain simulation. Reprinted with permission from Ref. [210]. Copyright (2002), American Institute of Physics. (Please find a color version of this figure on the color plates.)

The film thickness (about 100 nm) is much larger than the skin depth and hence is sufficient to fully suppress photon tunneling through the unstructured metallic film. The image reveals pronounced field intensity maxima in the region outside the holes (marked as white circles). The stripe-like pattern is oriented perpendicular to the polarization direction of the incident field, giving evidence for the excitation of longitudinal SPP modes. The clear standing wave pattern at the metal surface mainly results from the interference of SPP modes with wave vectors $\vec{k}_{||} = \pm\vec{G}_y$. A rotation of the incident polarization also changes the orientation of the standing wave pattern [210], giving additional support that these measurements probe SPP fields. The presence of the holes has different effects on the SPP fields propagating along the metal interface. They partly scatter the SPP field back into the far field and hence give rise to a radiative damping of the SPP modes [163]. Also, they scatter the SPP field into the holes. Photon tunneling couples the electromagnetic fields at the front and back sides of the film because electromagnetic waves do not propagate through sub-wavelength-sized cylindrical apertures [211]. This evanescent coupling between SPP fields at both interfaces is important for enhancing the transmission through the film and can be optimized by choosing identical dielectrics on both sides [211]. If air and dielectric modes are tuned into resonance, photon tunneling through the holes results in a coherent coupling between both modes and gives rise to the formation of new coupled SPP modes extending across both interfaces. This coupling leads to anticrossings between the interacting SPP resonances in angle-resolved linear optical spectra [212]. One contribution to the enhanced transmission, therefore, stems from the coupling of the incident light to SPP modes at the front side of the film, their coupling to evanescent modes inside the hole channels, and to SPP modes at the back side of the film and the outcoupling of the confined fields to the far field. This contribution is enhanced by tuning the excitation laser into resonance with the SPP modes. In addition, a certain fraction of the incident light can directly tunnel through the hole channels without coupling to SPP modes. Experimentally, both contributions can be readily distinguished in time-resolved pulse transmission experiments using ultrafast lasers with a pulse duration that is shorter than the SPP lifetime [213]. The linear optical transmission spectra are then given by the interference between the electric fields transmitted through the resonant SPP channel and the continuous direct transmission channel [214]. This interference gives rise to the asymmetric, Fano-like lineshapes shown in Figure 5.18b [215].

The scattering of SPPs at the holes also couples SPP modes on the same side of the metal film. The effect of this coupling on the linear optical spectra is readily seen in angle-resolved transmission spectra of an 150 nm thick gold film deposited on a sapphire substrate and perforated with an array of 50 nm wide slits with a period of 650 nm (Figure 5.20).

A clear anticrossing with a splitting of 70 meV is observed at angles of about 36° due to the coupling between SM[+ 1] and SM[−2] [For one-dimensional arrays, the allowed in-plane SPP wave vectors are $2\pi/a$, with a denoting the grating period. The SPP modes at the air and sapphire interface are then denoted as AM[p] and SM[p], respectively]. The coherent coupling between both modes thus leads to the opening of a band gap in the SPP dispersion relation. In addition, it results in a pronounced modification of the linewidths of the SPP resonances on the two sides of the crossing. The SPP coupling evidently affects the radiative lifetimes of the coupled modes because these linewidths are governed by the radiative SPP damping. This is qualitatively understood by considering the symmetries of the coupled modes. In the so-called strong coupling limit where the coupling strength is larger than the

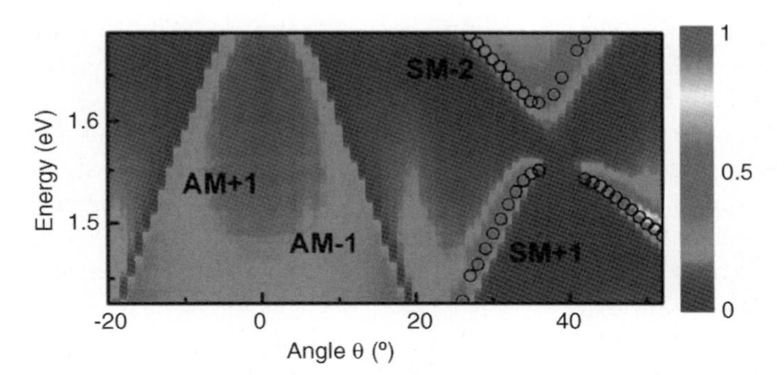

Figure 5.20 Experimentally measured angle-resolved transmission spectrum for a gold nanoslit array with a period of $a_0 = 650$ nm and a slit width of 50 nm. Open circles: Calculated SPP band structure near the crossing of SM[+ 1] and SM[−2] resonances. A band gap splitting of 72 meV is revealed. Note the spectral narrowing of the transmission spectrum and the decrease in transmission intensity in the lower energy region of the SM[+ 1]/SM[−2] crossing. (Please find a color version of this figure on the color plates.)

damping times of the individual resonances [216, 217], the coupling results in coupled SPP modes whose spatial mode profile is symmetric and antisymmetric with respect to the slit center [213]. As verified by near-field imaging, the spatial overlap of both modes with the slit scattering centers is very different and results in a very strong suppression of radiative damping for the antisymmetric mode [218]. Long SPP lifetimes exceeding 200 fs have been observed that exceed the SPP lifetimes of a perfectly flat metal/dielectric interface [213]. Hence, the coherent coupling between SPP modes not only provides a means to tailor the dispersion relations, but also manifestly alters the SPP lifetimes and the quality factors, as well as local field enhancements of nanoplasmonic resonators.

There have been numerous attempts to theoretically describe the phenomenon of extraordinarily enhanced transmission. For a very recent overview of the different approaches, the reader is referred to Ref. [219]. Initial attempts relied on semianalytical transfer matrix models, describing the metallic nanostructure as a special type of Fabry–Pérot resonator [211, 220]. Rather good agreement has been found between the experimental results of Ref. [210] and a fully vectorial diffraction model initially developed by Lochbihler [221] and later refined by Park and Lee [222]. Simulations based on this model as well as numerically demanding finite difference time domain simulations [223, 224] support the qualitative picture outlined above.

Numerous possible applications of such periodically structured metallic nanostructures have been discussed [141]. Of particular interest seem to be the directional transmission (beaming) of light through a single aperture flanked by periodic corrugations [225] and the ability to localize light in very small volumes, which is of interest for the sensing of single molecules [226].

5.4.4.4 Adiabatic Metallic Tapers

One of the most intriguing physical properties of metallic surface plasmon polariton waveguides is their ability to localize light in extremely small volumes, much below the diffraction limit. Here, we briefly discuss the optical properties of a specific SPP waveguide, a sharp, tapered gold tip, to highlight the interplay between geometrical shape and optical properties and to illustrate the unique light localization capabilities of such nanostructures that is advantageous for spectroscopic, sensing, and microscopic functions in both continuous wave and ultrafast measurements [200, 227, 228].

Consider a surface plasmon polariton wave packet propagating toward the apex of a conical metallic taper (Figure 5.21) with perfectly smooth interfaces. Owing to the evanescent nature of the SPP mode, this wave packet cannot emit radiation into the far field until it reaches the very apex of the taper. The taper apex acts as a scatterer for SPP waves and thus as a point-like light source. The optical properties of such conical tapers have been studied analytically by Babadjanyan *et al.* [230] and within the Wentzel–Kramers–Brillouin approximation [231] by Stockman *et al.* [232]. Several intriguing features of these results merit attention. First, the magnitude of the SPP in-plane wave vector $k \propto 1/r$ is inversely proportional to the distance r between a certain position on the taper surface and the tip apex. Correspondingly, the in-plane SPP wavelength tends to zero near the tip apex. Both phase and group velocities

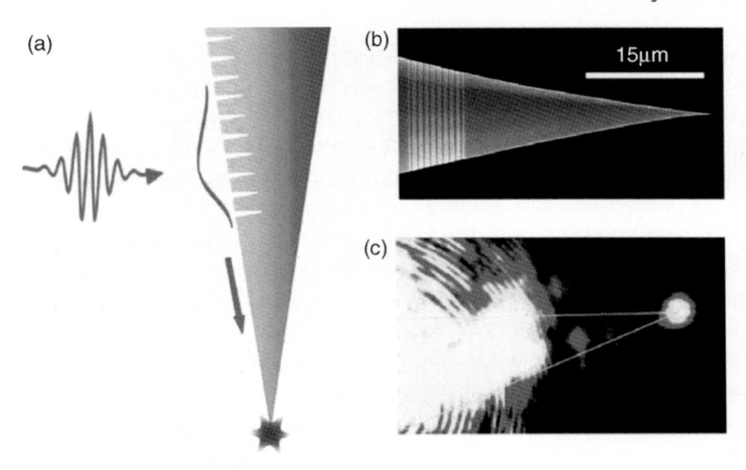

Figure 5.21 (a) Experimental geometry for adiabatic nanofocusing on conical gold tapers. Light from a tunable Ti:sapphire laser is focused by a high-NA objective onto a nanoslit grating milled onto the shaft of a gold taper. Grating coupling launches a SPP wave packet on the taper that propagates toward the tip apex, where it is scattered into the far field. (b) Scanning electron microscopy image of a chemically etched gold taper with a nanoslit grating patterned by focused ion beam milling. (c) Optical microscopy image of the light that is scattered from the tip apex after grating illumination. Reprinted with permission from Ref. [229]. Copyright (2007) American Chemical Society.

decrease to zero, and hence the SPP wave packet is predicted to be slowed down and come to a complete halt as it approaches the tip apex. The adiabatic decrease in group velocity is related to a divergence of the electric field strength near the tip apex and to an extreme concentration of the electromagnetic energy stored in the SPP wave packet into a vanishingly small spot size. Effectively, the spatial extent of the SPP wave packet reduces as it approaches the apex and is transformed adiabatically from a propagating into a localized mode. Even if the field singularity at the tip apex is removed by limiting the minimum tip radius to 2 nm [232], a strong enhancement of the field intensity by more than three orders of magnitude remains. This field localization is expected for adiabatic tapers whose change in taper diameter is small on a scale of the wavelength.

From this early work, the phenomenon of adiabatic nanocompression has received considerable theoretical and experimental attention. Different nanofocusing geometries including wedges [233, 234], cones [235–237], and nanogrooves [238] have been proposed and theoretically analyzed. On the experimental side, first the effect has been studied in two-dimensional tapered waveguides [239, 240]. A possible geometry for demonstrating adiabatic nanofocusing on conical tapers, introduced in Ref. [229], is shown in Figure 5.21 and consists of a chemically etched gold taper with an opening angle from 20° to 30°. The taper is patterned with a slit grating with a period of 800 nm and slit width of 150–300 nm. When illuminated with laser light and a wavelength approximately matching the grating period, a SPP wave packet is launched onto the taper shaft, propagating toward the tip apex. The 15–30 μm distance between the grating edge and the taper apex is chosen such that a significant fraction of the launched wave packet reaches the taper apex with minimal Ohmic loss or scattering into the far field. Only at the tip apex are the SPPs efficiently transformed into radiation. This is readily seen in microscope images taken from such tapers when illuminating the grating [229]. Apart from light scattering from the grating edges, the taper shaft remains dark, while a second and intense light spot is seen at the very apex of the tip. This was taken as indirect evidence for adiabatic nanofocusing on conical tapers. More direct evidence can be given by using such a localized nanoscale light spot at the apex as the light source in a scattering-type near-field scanning optical microscope [200, 227]. When scanning such a tip across a glass substrate covered with small gold metallic nanoparticles while illuminating the grating coupler, strong enhancement of the light scattered into the far field is seen when the tip is positioned directly on top of a nanoparticle (Figure 5.22).

The spatial resolution in these images is only of the order of 40 nm. The enhancement in scattering signal by a single 30 nm diameter particle is strong and amounts to more than 30% of the signal given by the light scattering from the tip apex in the absence of a nanoparticle [200]. This is considered direct evidence for the very efficient adiabatic focusing of far-field light toward a nanometer-sized spot at the tip apex. Such an essentially background-free high spatial resolution near-field scanning

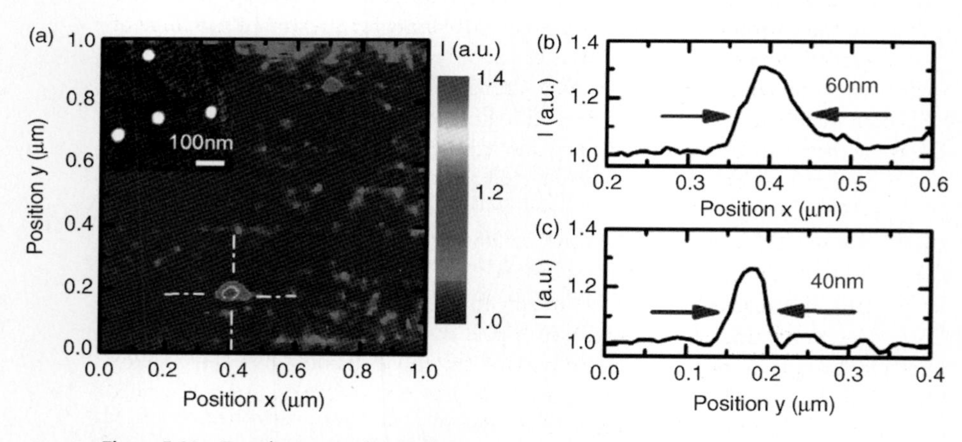

Figure 5.22 Two-dimensional optical images of individual gold nanoparticles on a glass substrate recorded by adiabatic nanofocusing scattering-type near-field scanning optical microscopy (s-NSOM). (a) Optical s-NSOM image of a single gold nanoparticle with <30 nm radius. In these experiments, SPP waves are launched onto a gold taper by grating coupling (Figure 5.21b) and the light scattering from the tip apex is recorded in the far field while scanning the tip across the surface of a dielectric substrate covered with a low concentration of gold nanoparticles. *Inset*: Scanning electron microscope image of the gold nanoparticles. (b) Cross sections of the optical intensity along the x and y directions (marked by dash-dotted lines in (a)). The optical resolution of about 40 nm and the large signal-to-background ratio confirm the efficient adiabatic nanofocusing of SPP waves at the tip apex. Reprinted with permission from Ref. [200]. Copyright (2011) American Chemical Society. (Please find a color version of this figure on the color plates.)

optical microscope is certainly of considerable interest for linear light scattering, Raman and fluorescence imaging of densely packed nanoparticle samples.

5.4.5 Exciton–Plasmon Coupling

So far, we have considered electromagnetic fields at the interface between a metal and a transparent dielectric and have focused on propagating or localized surface plasmon excitations. As all other electromagnetic waves, surface plasmons interact with materials such as semiconductors or molecules that may be present in the near field. When replacing the dielectric with a thin semiconducting layer or a layer of molecules, this layer may itself show optical absorption resonances in the spectral range of the surface plasmon resonances and therefore introduce resonant coupling between the modes. In case of low-dimensional, quantum-confined semiconductors, these resonances arise mostly from dipole-allowed excitonic excitations, optically induced electron–hole pairs bound by the Coulomb interaction [241–243]. In molecular layers, these are molecular excitons, arising from the light-induced transition of an electron from the highest occupied to the lowest unoccupied molecular orbital [244, 245].

In this case, an external light field not only can interact with the surface plasmon polaritons at the interface between the metal and its surrounding, but can also create excitons in the surrounding layer. Hence, the possibility for a resonant exchange of electromagnetic energy between excitons and surface plasmons exists. If this exciton–plasmon coupling is sufficiently strong, it will influence the optical properties of the coupled system. When the excitons are located close to the metal surface, that is, at distances of much less than one wavelength, the energy exchange will mainly be mediated by optical near fields (the last two terms in Eq. (5.39)). For larger distances, propagating fields (the first term in Eq. (5.39)) contribute and for very short distances of only a few angstroms, tunnel coupling of electrons and holes is expected to set in. The resulting charge transfer processes will not be considered here and we will restrict the discussion to electromagnetic couplings between the optically induced excitonic and surface plasmon dipole moments. In molecular systems, such dipole–dipole couplings are important for the excitation transfer in light harvesting systems [246]. Recently, they have also been investigated in coupled semiconductor quantum dot systems [247, 248].

An intuitive insight into the effects of such couplings on the optical properties of plasmonic systems can be gained from the hybridization model introduced by Nordlander and coworkers [249]. On the basis of earlier work [250, 251], the authors have investigated spherical metallic nanoshells covering small dielectric spheres. The theoretical work shows that the optical response of this composite system results from the electromagnetic interaction between the plasmon resonances of the individual constituents, in this case a metallic sphere and a spherical void inside a metallic film. The lowest order dipolar resonance of the sphere is discussed in Section 5.4.4.1. In quasi-static approximation, the polarizability of a spherical inclusion of radius a in a metal film with $\varepsilon_2(\omega)$ and filled with dielectric with ε_1

is given by exchanging ε_1 and ε_2 in Eq. (5.37) [132, 174]:

$$\alpha = 4\pi a^3 \frac{\varepsilon_2 - \varepsilon_1}{\varepsilon_2 + 2\varepsilon_1}. \tag{5.48}$$

Then, the Fröhlich condition for the SP resonance is $2\text{Re}(\varepsilon_1) = \varepsilon_2$. This gives a SP resonance frequency $\omega_\text{v} = \sqrt{(2/3)}\omega_\text{p}$ in a lossless Drude model compared to $\omega_\text{s} = \omega_\text{p}/\sqrt{3}$ for the sphere. Using a hydrodynamic model for the SP response [252], it was shown that ω_v can be understood as the resonance frequency of a harmonic oscillator model for the collective charge density oscillation. The problem of finding the resonance of the hybrid system can then be transformed into solving coupled oscillator equations [252]. In this dipolar approximation, the resonance frequencies of the composite system are then derived as

$$\omega_\pm^2 = \frac{\omega_\text{p}^2}{2}\left[1 \pm \frac{1}{3}\sqrt{1 + 8x^3}\right], \tag{5.49}$$

where $x = a/b$ defines the ratio of the inner and outer radii of the shell. This expression gives the same resonance frequencies as obtained from classical Mie scattering theory [132, 174]. For vanishingly small void volume, $x \to 0$, one retrieves the uncoupled resonances. With increasing void volume, the splitting between the resonances increases, indicating the formation of antisymmetrically coupled (antibonding) ω_+ plasmon mode and a symmetrically coupled (bonding) ω_- plasmon mode. The two modes correspond to out-of-phase and in-phase charge density oscillations at the inner and outer surfaces of the shell, respectively [249]. For sufficiently small x, the splitting between the modes can be approximated as $\omega_+ - \omega_- = \sqrt{(\omega_\text{s} - \omega_\text{v})^2 + 4|V|^2}$, giving a coupling energy (or Rabi splitting) [217, 231] of $\hbar V \approx \hbar\omega_\text{p}(1/3)x^{3/2}$. If the coupling is sufficiently strong, the spectra of the composite particles show new resonances, which can be attributed to the coupled plasmon modes [249].

The same theoretical approach can be used to describe surface plasmon couplings in metallic dimers [253], multishell concentric spheres [249], or exciton–plasmon interactions [254]. When coating a metallic nanoshell with, for example, a thin layer of a J-aggregate molecular dye, the plasmonic excitations of the nanoshell can couple to the excitonic excitation of the molecular layer. This then leads to an exchange of electromagnetic energy between the excitonic and the plasmonic system and – for sufficiently strong coupling – as in the previous line to the formation of new coupled resonances with resonance frequencies that are spectrally shifted with respect to those of the uncoupled ones. J-aggregate dyes [245], formed by dissolving dye molecules, for example, 2,2′dimethyl-8-phenyl-5,6,5′,6′-dibenzothiacarbocyanine chloride (Figure 5.23a), at high concentration in polymer matrices are well suited for studies of exciton–plasmon interactions because of their large oscillator strengths and spectrally narrow absorption resonances (Figure 5.23b) [255].

The optical absorption spectra of ensembles of spherical silver and gold nanoparticles [257] and of gold nanoshells [254] covered with J-aggregate dye molecules have been studied experimentally. Spectral splittings of more than 100 meV have

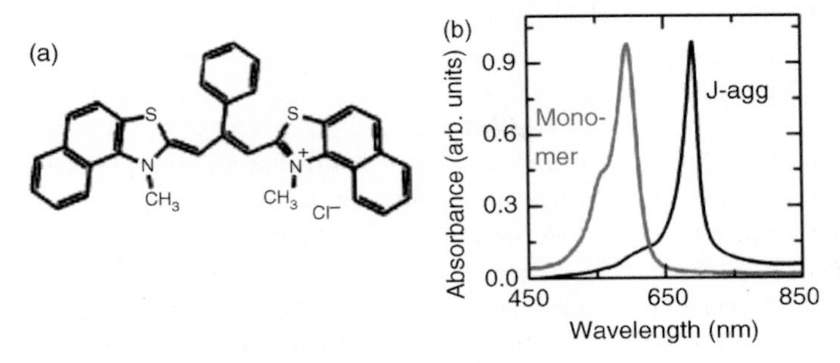

Figure 5.23 (a) Chemical structure of the cyanine dye 2,2′dimethyl-8-phenyl-5,6,5′,6′-dibenzothiacarbocyanine. (b) Room temperature optical absorption spectra of this dye in monomeric form (red line) and in J-aggregate form (black line). Reprinted with permission from Ref. [256]. Copyright (2010) American Chemical Society.

been observed [254] and taken as a signature of the coherent coupling between the plasmon excitations of the nanoparticle and the excitons of the J-aggregate complex. Since these splittings are less than the ensemble-averaged linewidth of the nanoshell absorption spectra, it is not yet clear whether the strong coupling regime, in which the coupling strength V is larger than the spectral width of the uncoupled resonances, can be reached in these systems [216, 217].

Clear evidence for strong exciton–plasmon coupling has been given for the coupling of J-aggregate excitons to surface plasmon polariton excitations on planar metallic films [258] and in one-dimensional nanoslit arrays [256] or two-dimensional hole arrays [259]. As an example, linear optical reflectivity spectra of a nanoslit grating in a gold film covered with a 50 nm thick J-aggregate dye layer are presented in Figure 5.24.

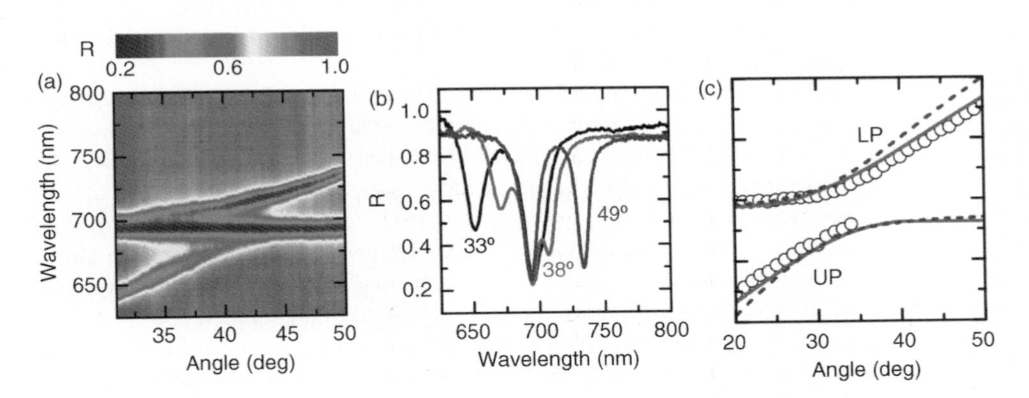

Figure 5.24 (a) Angle-resolved p-polarized linear reflectivity spectra ($T = 77$ K) of the J-aggregate dye deposited on a gold film perforated with a nanoslit grating of 430 nm period and 45 nm width. (b) Reflectivity spectra at different angles of 33°, 38°, and 49°. (c) Polariton dispersion relation obtained from the experimental spectra (open circles), a full vectorial solution of Maxwell's equations (solid line), and a coupled oscillator model (dashed line). Reprinted with permission from Ref. [256]. Copyright (2010) American Chemical Society. (Please find a color version of this figure on the color plates.)

The surface plasmon resonances of such gratings are discussed in Section 5.4.4.3. Here, the grating period of 430 nm was chosen such that the SPP resonance $\omega_{\mathrm{SPP}}(\theta)$ of the grating in the absence of exciton–plasmon coupling (Eq. (5.47)) is shifted from 650 to 750 nm when varying the angle of incidence θ between 30° and 50°. This allows one to angle tune the SPP resonance across the (angle-independent) exciton resonance ω_X of the dye. The width and the depth of the slits at 45 and 30 nm, respectively, were chosen to obtain spectrally narrow SPP resonances with a linewidth of less 10 nm. Near the crossing angle, $\omega_{\mathrm{SPP}}(\theta) = \omega_X$, a clear splitting of the optical spectra into upper and lower polariton resonances were observed. The resonance energies of the newly formed coupled polariton modes were found to be

$$\tilde{\omega}_{\mathrm{UP},LP} = \frac{1}{2}(\tilde{\omega}_X + \tilde{\omega}_{\mathrm{SPP}}) \pm \frac{1}{2}\sqrt{(\tilde{\omega}_X - \tilde{\omega}_{\mathrm{SPP}})^2 + 4V^2}, \tag{5.50}$$

where $\tilde{\omega}_{X,\mathrm{SPP}} = \omega_{X,\mathrm{SPP}} - i\gamma_{X,\mathrm{SPP}}$ denote the complex eigenfrequencies of the uncoupled exciton and SPP resonances. Equation (5.50) matches the observed dispersion relation when choosing a coupling energy of $\hbar V \approx 55$ meV, which was larger than the widths $\hbar\gamma_X$ and $\hbar\gamma_{\mathrm{SPP}}$ of both uncoupled resonances, confirming that the regime of strong exciton–SPP coupling was indeed reached in these experiments. The Rabi splitting $V = \int \vec{\mu}_X(\bar{r}) \cdot \vec{E}_{\mathrm{SPP}}(\bar{r})\, \mathrm{d}\bar{r}$ reflects the overlap integral between the excitonic transition dipole moment density $\vec{\mu}_X(\vec{r})$ and the electric field vector $\vec{E}_{\mathrm{SPP}}(\vec{r})$ of the local SPP mode and therefore measures the rate of exchange of electromagnetic energy between excitons and SPPs. The observation of such large coupling energies is of interest because the coupling and the optical properties of such hybrid structures can be altered by externally manipulating $\vec{\mu}_X(\vec{r})$ or $\vec{E}_{\mathrm{SPP}}(\vec{r})$. Owing to the short anticipated lifetimes of the coupled polariton modes, this provides a new degree of freedom for control of optical properties on fast timescales and potentially also small length scales. This functionality might be of interest for future applications in all-optical switching or nanolasing.

5.4.6 **Summary**

Electromagnetic fields at the interface between a metal and a dielectric are strongly confined to the interface. The coupling between an incident light field and collective charge density oscillations inside the metal results in propagating surface plasmon polariton and localized surface plasmon modes. These modes are fundamentally interesting because they allow the efficient localization of light on the nanoscale, in dimensions that are substantially smaller than the wavelength of light. For simple geometries such as planar interfaces, gratings, or spherical and elliptical nanoparticles, the local electromagnetic fields and the resulting linear optical properties of these structures are reasonably well understood using phenomenological Drude-like models for the local dielectric function of the metal. More complex geometries such as two- or three-dimensional arrays or nanometer-sized taper waveguides carry the potential to act as antennas or waveguides, providing exquisite and unprecedented control of the propagation and localization of light on the nanoscale. Such structures for localization of light on the nanometer scale are interesting for spectroscopic and dynamical imaging of single molecules, sensing, and inducing linear and nonlinear photochemistry of adsorbed molecules. Hybrid nanostructures consist of metals and gain materials add nonlinear ultrafast switching functionality and open the door to a new class of functional nanophotonic devices. A microscopic understanding of their optical properties requires knowledge about the spatiotemporal dynamics of electromagnetic fields near interfaces on short, nanometer-sized length and extremely fast atto- to femtosecond timescales, posing considerable experimental and theoretical challenges. We believe that emerging advances in ultrafast optical and electron microscopy in combination with advanced quantum theoretical modeling will provide more insights in the near future.

References

1 Davydov, A.S. (1976) *Quantum Mechanics*, Pergamon Press.

2 Bassani, F. and Parravicini, G.P. (1975) in *Electronic States and Optical Transitions in Solids* (ed. R.A. Ballinger), Pergamon Press, Oxford.

3 Ehrenreich, H. and Cohen, M.H. (1959) *Phys. Rev.*, **115**, 786.

4 Feibelman, P.J. (1982) *Prog. Surf. Sci.*, **12**, 287.

5 Weaver, J.H., Krafka, C., Lynch, D.W., and Koch, E.E. (1981) *Physics Data: Optical Properties of Metals*, Fachinformationszentrum Energie, Physik, Mathematik GmbH, Karlsruhe.

6 Krasovskii, E.E. and Schattke, W. (2001) *Phys. Rev. B*, **63**, 235112.

7 Krasovskii, E.E. and Schattke, W. (1999) *Phys. Rev. B*, **60**, R16251.

8 Adler, S.L. (1962) *Phys. Rev.*, **126**, 413.

9 Dietz, R.E., McRae, E.G., and Weaver, J.H. (1980) *Phys. Rev. B*, **21**, 2229.

10 Hedin, L. and Lundqvist, S. (1970) in *Solid State Physics*, vol. 23 (eds. D.T. Frederick Seiz and E. Henry), Academic Press, p. 1.

11 Hybertsen, M.S. and Louie, S.G. (1986) *Phys. Rev. B*, **34**, 5390.

12 Hanke, W. and Sham, L.J. (1980) *Phys. Rev. B*, **21**, 4656.

13 Albrecht, S., Reining, L., Del Sole, R., and Onida, G. (1998) *Phys. Rev. Lett.*, **80**, 4510.

14 Rohlfing, M. and Louie, S.G. (2000) *Phys. Rev. B*, **62**, 4927.

15 Benedict, L.X. and Shirley, E.L. (1999) *Phys. Rev. B*, **59**, 5441.

16 Chiodo, L., García-Lastra, J.M., Iacomino, A., Ossicini, S., Zhao, J., Petek, H., and Rubio, A. (2010) *Phys. Rev. B*, **82**, 045207.

17 Arnaud, B., Lebègue, S., Rabiller, P., and Alouani, M. (2006) *Phys. Rev. Lett.*, **96**, 026402.

18 Onida, G., Reining, L., and Rubio, A. (2002) *Rev. Mod. Phys.*, **74**, 601.

19 Hufner, S. (1995) *Photoelectron Spectroscopy: Principles and Applications*, Springer, Berlin.

20 Schattke, W. and Van Hove, M.A. (2003) *Solid-State Photoemission and Related Methods*, Wiley-VCH Verlag GmbH, Berlin.

21 Krasovskii, E.E., Silkin, V.M., Nazarov, V.U., Echenique, P.M., and Chulkov, E.V. (2010) *Phys. Rev. B*, **82**, 125102.

22 Krasovskii, E.E., Rossnagel, K., Fedorov, A., Schattke, W., and Kipp, L. (2007) *Phys. Rev. Lett.*, **98**, 217604.

23 Kane, E.O. (1964) Implications of crystal momentum conservation in photoelectric emission for band-structure measurements. *Phys. Rev. Lett.*, **12**, 97.

24 Feibelman, P.J. and Eastman, D.E. (1974) *Phys. Rev. B*, **10**, 4932.

25 Smith, N.V., Thiry, P., and Petroff, Y. (1993) *Phys. Rev. B*, **47**, 15476.

26 Reinert, F., Nicolay, G., Schmidt, S., Ehm, D., and Hüfner, S. (2001) *Phys. Rev. B*, **63**, 115415.

27 Vergniory, M.G., Pitarke, J.M., and Echenique, P.M. (2007) *Phys. Rev. B*, **76**, 245416.

28 Chen, Y., García de Abajo, F.J., Chassé, A., Ynzunza, R.X., Kaduwela, A.P., Van Hove, M.A., and Fadley, C.S. (1998) *Phys. Rev. B*, **58**, 13121.

29 García de Abajo, F.J., Van Hove, M.A., and Fadley, C.S. (2001) *Phys. Rev. B*, **63**, 075404.

30 Dil, J.H. (2009) *J. Phys. Condens. Matter*, **21**, 403001.

31 Kimura, A. *et al.* (2010) *Phys. Rev. Lett.*, **105**, 076804.

32 Ritchie, R.H. and Eldridge, H.B. (1962) *Phys. Rev.*, **126**, 1935.

33 Winkelmann, A., Hartung, D., Engelhard, H., Chiang, C.T., and Kirschner, J. (2008) *Rev. Sci. Instrum.*, **79**, 083303.

34 Baumgarten, L., Schneider, C.M., Petersen, H., Schäfers, F., and Kirschner, J. (1990) *Phys. Rev. Lett.*, **65**, 492.

35 Chiang, C.-T., Winkelmann, A., Yu, P., and Kirschner, J. (2009) *Phys. Rev. Lett.*, **103**, 077601.

36 Kiss, T. *et al.* (2005) *Phys. Rev. Lett.*, **94**, 057001.

37 Petek, H. and Ogawa, S. (1997) *Prog. Surf. Sci.*, **56**, 239.

38 Weinelt, M. (2002) *J. Phys. Condens. Matter*, **14**, R1099.

39 Chulkov, E.V., Borisov, A.G., Gauyacq, J.P., Sánchez-Portal, D., Silkin, V.M., Zhukov, V.P., and Echenique, P.M. (2006) *Chem. Rev.*, **106**, 4160.

40 Ge, N.-H., Wong, C.M., Lingle, R.L., Jr., McNeill, J.D., Gaffney, K.J., and Harris, C.B. (1998) *Science*, **279**, 202.

41 Andrianov, I., Klamroth, T., Saalfrank, P., Bovensiepen, U., Gahl, C., and Wolf, M. (2005) *J. Chem. Phys.*, **122**, 234710.

42 Ferrini, G., Giannetti, C., Galimberti, G., Pagliara, S., Fausti, D., Banfi, F., and Parmigiani, F. (2004) *Phys. Rev. Lett.*, **92**, 256802.

43 Bisio, F., Nývlt, M., Franta, J., Petek, H., and Kirschner, J. (2006) *Phys. Rev. Lett.*, **96**, 087601.

44 Pontius, N., Sametoglu, V., and Petek, H. (2005) *Phys. Rev. B*, **72**, 115105.

45 Himpsel, F.J. and Fauster, T. (1984) *J. Vac. Sci. Technol. A*, **2**, 815.

46 Claessen, R., Burandt, B., Carstensen, H., and Skibowski, M. (1990) *Phys. Rev. B*, **41**, 8270.

47 Schattke, W., Krasovskii, E.E., Díez Muiño, R., and Echenique, P.M. (2008) *Phys. Rev. B*, **78**, 155314.

48 Shudo, K. and Munakata, T. (2001) *Phys. Rev. B*, **63**, 125324.

49 Winkelmann, A., Lin, W.-C., Chiang, C.-T., Bisio, F., Petek, H., and Kirschner, J. (2009) *Phys. Rev. B*, **80**, 155128.

50 Winkelmann, A., Chiang, C.-T., Tusche, C., Ünal, A.A.,Kubo, A., Wang, L., and PetekA., H. (2012) Ultrafast multiphoton photoemission microscopy of solid surfaces in the real and reciprocal space, chapter in Dynamics of interfacial electron and excitation transfer in solar energy conversion: theory and experiment, (ed. P. Piotrowiak), Royal Society of Chemistry, London.

51 Ehrenreich, H. and Philipp, H.R. (1962) *Phys. Rev.*, **128**, 1622.

52 Smith, J.B. and Ehrenreich, H. (1982) *Phys. Rev. B*, **25**, 923.

53 Bonn, M., Denzler, D.N., Funk, S., Wolf, M., Wellershoff, S.S., and Hohlfeld, J. (2000) *Phys. Rev. B*, **61**, 1101.

54 Del Fatti, N., Bouffanais, R., Vallée, F., and Flytzanis, C. (1998) *Phys. Rev. Lett.*, **81**, 922.

55 Sjodin, T., Petek, H., and Dai, H.-L. (1998) *Phys. Rev. Lett.*, **81**, 5664.

56 Del Fatti, N., Voisin, C., Achermann, M., Tzortzakis, S., Christofilos, D., and Vallée, F. (2000) *Phys. Rev. B*, **61**, 16956.

57 Timm, C. and Bennemann, K.H. (2004) *J. Phys. Condens. Matter*, **16**, 661.

58 Stahrenberg, K., Herrmann, T., Wilmers, K., Esser, N., Richter, W., and Lee, M.J.G. (2001) *Phys. Rev. B*, **64**, 115111.

59 Liebsch, A. (1997) *Electronic Excitations at Metal Surfaces*, Plenum, New York.

60 Johnson, P.B. and Christy, R.W. (1972) *Phys. Rev. B*, **6**, 4370.

61 Hao, Z., Dadap, J.I., Knox, K.R., Yilmaz, M.B., Zaki, N., Johnson, P.D., and Osgood, R.M. (2010) *Phys. Rev. Lett.*, **105**, 017602.

62 Cavanagh, R.R., King, D.S., Stephenson, J.C., and Heinz, T.F. (1993) *J. Phys. Chem.*, **97**, 786.

63 Cooper, B.R., Ehrenreich, H., and Philipp, H.R. (1965) *Phys. Rev.*, **138**, A494.

64 Miller, T., Hansen, E.D., McMahon, W.E., and Chiang, T.C. (1997) *Surf. Sci.*, **376**, 32.

65 Cazalilla, M.A., Dolado, J.S., Rubio, A., and Echenique, P.M. (2000) *Phys. Rev. B*, **61**, 8033.

66 Marini, A., Del Sole, R., and Onida, G. (2002) *Phys. Rev. B*, **66**, 115101.

67 Marini, A. and Del Sole, R. (2003) *Phys. Rev. Lett.*, **91**, 176402.

68 Gumhalter, B., Lazić, P., and Došlić, N. (2010) *Phys. Status Solidi B*, **247**, 1907.

69 Rosei, R. (1974) *Phys. Rev. B*, **10**, 474.

70 Yu, P.Y. and Cardona, M. (2003) *Fundamentals of Semiconductors*, Springer, Berlin.

71 Miller, T., McMahon, W.E., and Chiang, T.C. (1996) *Phys. Rev. Lett.*, **77**, 1167.

72 Allen, J.W. and Mikkelsen, J.C. (1977) *Phys. Rev. B*, **15**, 2952.

73 Rakić, A.D., Djurišić, A.B., Elazar, J.M., and Majewski, M.L. (1998) *Appl. Opt.*, **37**, 5271.

74 Fischer, N., Schuppler, S., Fischer, R., Fauster, T., and Steinmann, W. (1993) *Phys. Rev. B*, **47**, 4705.

75 Schmidt, A.B., Pickel, M., Wiemhofer, M., Donath, M., and Weinelt, M. (2005) *Phys. Rev. Lett.*, **95**, 107402.

76 Pickel, M., Schmidt, A.B., Giesen, F., Braun, J., Minár, J., Ebert, H., Donath, M., and Weinelt, M. (2008) *Phys. Rev. Lett.*, **101**, 066402.

77 Bovensiepen, U. (2007) *J. Phys. Condens. Matter*, **19**, 083201.

78 Lisowski, M., Loukakos, P.A., Bovensiepen, U., Stähler, J., Gahl, C., and Wolf, M. (2004) *App. Phys. A*, **78**, 165.

79 Weinelt, M., Schmidt, A.B., Pickel, M., and Donath, M. (2007) *Prog. Surf. Sci.*, **82**, 388.

80 Palik (ed.) (1998) *Handbook of Optical Constants of Solids*, Academic Press, New York.

81 Drude, P. (1900) *Ann. Phys.*, **306**, 566.

82 Bethe, H.A. and Sommerfeld, A. (1967) *Elektronentheorie der Metalle*, Springer, Berlin.

83 Ashcroft, N.W. and Mermin, N.D. (1976) *Solid State Physics*, Holt, New York.

84 Youn, S.J., Rho, T.H., Min, B.I., and Kim, K.S. (2007) *Phys. Status Solidi B*, **244**, 1354.

85 Smith, N.V. (2001) *Phys. Rev. B*, **64**, 155106.

86 Ogawa, S., Nagano, H., and Petek, H. (1997) *Phys. Rev. B*, **55**, 10869.

87 Fann, W.S., Storz, R., Tom, H.W.K., and Bokor, J. (1992) *Phys. Rev. B*, **46**, 13592.

88 Sakaue, M., Kasai, H., and Okiji, A. (2011) *J. Phys. Soc. Jpn.*, **69**, 160.

89 Petek, H., Nagano, H., Weida, M.J., and Ogawa, S. (2000) *Chem. Phys.*, **251**, 71.

90 Altarelli, M., Dexter, D.L., Nussenzveig, H.M., and Smith, D.Y. (1972) *Phys. Rev. B*, **6**, 4502.

91 Beach, R.T. and Christy, R.W. (1977) *Phys. Rev. B*, **16**, 5277.

92 Parkins, G.R., Lawrence, W.E., and Christy, R.W. (1981) *Phys. Rev. B*, **23**, 6408.

93 Basov, D.N. and Timusk, T. (2005) *Rev. Mod. Phys.*, **77**, 721.

94 Littlewood, P.B. and Varma, C.M. (1991) *J. Appl. Phys.*, **69**, 4979.

95 Quinn, J.J. (1962) *Phys. Rev.*, **126**, 1453.

96 Gasparov, V.A. and Huguenin, R. (1993) *Adv. Phys.*, **42**, 393.

97 Echenique, P.M., Pitarke, J.M., Chulkov, E.V., and Rubio, A. (2000) *Chem. Phys.*, **251**, 1.

98 Marini, A., Onida, G., and Del Sole, R. (2001) *Phys. Rev. Lett.*, **88**, 016403.

99 Allen, P.B. (1971) *Phys. Rev. B*, **3**, 305.

100 Götze, W. and Wölfle, P. (1972) *Phys. Rev. B*, **6**, 1226.

101 Varma, C.M. (1997) *Phys. Rev. B*, **55**, 14554.

102 Mori, T., Nicol, E.J., Shiizuka, S., Kuniyasu, K., Nojima, T., Toyota, N., and Carbotte, J.P. (2008) *Phys. Rev. B*, **77**, 174515.

103 Cisneros, G. and Helman, J.S. (1982) *Phys. Rev. B*, **25**, 6504.

104 Li, H.Y., Zhou, S.M., Li, J., Chen, Y.L., Wang, S.Y., Shen, Z.C., Chen, L.Y., Liu, H., and Zhang, X.X. (2001) *Appl. Opt.*, **40**, 6307.

105 Gurzhi, R.N. (1957) *Zh. Eksp. Teor. Fiz.*, **33**, 660.

106 Lawrence, W.E. and Wilkins, J.W. (1973) *Phys. Rev. B*, **7**, 2317.

107 Quinn, J.J. and Ferrell, R.A. (1958) *Phys. Rev.*, **112**, 812.

108 Hofmann, P., Rose, K.C., Fernandez, V., Bradshaw, A.M., and Richter, W. (1995) *Phys. Rev. Lett.*, **75**, 2039.

109 Stahrenberg, K., Herrmann, T., Esser, N., and Richter, W. (2000) *Phys. Rev. B*, **61**, 3043.

110 Sun, L.D., Hohage, M., Zeppenfeld, P., Balderas-Navarro, R.E., and Hingerl, K. (2003) *Phys. Rev. Lett.*, **90**, 106104.

111 Harl, J., Kresse, G., Sun, L.D., Hohage, M., and Zeppenfeld, P. (2007) *Phys. Rev. B*, **76**, 035436.

112 Urbach, L.E., Percival, K.L., Hicks, J.M., Plummer, E.W., and Dai, H.L. (1992) *Phys. Rev. B*, **45**, 3769.

113 Fauster, T., Weinelt, M., and Höfer, U. (2007) *Prog. Surf. Sci.*, **82**, 224.

114 Güdde, J., Rohleder, M., Meier, T., Koch, S.W., and Höfer, U. (2007) *Science*, **318**, 1287.

115 Becker, M., Crampin, S., and Berndt, R. (2006) *Phys. Rev. B*, **73**, 081402.

116 Vergniory, M.G., Pitarke, J.M., and Crampin, S. (2005) *Phys. Rev. B*, **72**, 193401.

117 Ünal, A.A., Tusche, C., Ouazi, S., Wedekind, S., Chiang, C.-T., Winkelmann, A., Sander, D., Henk, J., and Kirschner, J. (2011) *Phys. Rev. B*, **84**, 073107.

118 Volokitin, A.I. and Persson, B.N.J. (1995) *Phys. Rev. B*, **52**, 2899.

119 Dvorak, J. and Dai, H.-L. (2000) *J. Chem. Phys.*, **112**, 923.

120 Ogawa, S., Nagano, H., Petek, H., and Heberle, A.P. (1997) *Phys. Rev. Lett.*, **78**, 1339.

121 Petek, H., Weida, M.J., Nagano, H., and Ogawa, S. (2000) *Science*, **288**, 1402.

122 Zhao, J. *et al.* (2008) *Phys. Rev. B*, **78**, 085419.

123 Wang, L.-M., Sametoglu, V., Winkelmann, A., Zhao, J., and Petek, H. (2011) *J. Phys. Chem. A*, **115**, 9479.

124 Weik, F., de Meijere, A., and Hasselbrink, E. (1993) *J. Chem. Phys.*, **99**, 682.

125 Frischkorn, C. and Wolf, M. (2006) *Chem. Rev.*, **106**, 4207.

126 Wolf, M., Hotzel, A., Knoesel, E., and Velic, D. (1999) *Phys. Rev. B*, **59**, 5926.

127 Bisio, F., Winkelmann, A., Lin, W.C., Chiang, C.T., Nývlt, M., Petek, H., and Kirschner, J. (2009) *Phys. Rev. B*, **80**, 125432.

128 Cavalieri, A.L. *et al.* (2007) *Nature*, **449**, 1029.

129 Ueba, H. and Gumhalter, B. (2007) *Prog. Surf. Sci.*, **82**, 193.

130 Raether, H. (1988) *Surface Plasmons on Smooth and Rough Surfaces and on Gratings*, Springer, Berlin.

131 van de Hulst, H.C. (1981) *Light Scattering by Small Particles*, Dover Publications, New York.

132 Kreibig, U. and Vollmer, M. (1995) *Optical Properties of Metal Clusters*, Springer, Berlin.

133 Bohm, D. and Pines, D. (1953) *Phys. Rev.*, **92**, 609.

134 Pines, D. (1956) *Rev. Mod. Phys.*, **28**, 184.

135 Etchegoin, P.G., Le Ru, E.C., and Meyer, M. (2006) *J. Chem. Phys.*, **125**, 164705.

136 Etchegoin, P.G. and Le Ru, E.C. (2006) *J. Phys. Condens. Matter*, **18**, 1175.

137 McMahon, J.M., Gray, S.K., and Schatz, G.C. (2009) *Phys. Rev. Lett.*, **103**, 097403.

138 McMahon, J.M., Gray, S.K., and Schatz, G.C. (2009) *Phys. Rev. Lett.*, **103**, 097403.

139 Maier, S.A. (2007) *Plasmonics: Fundamentals and Applications*, Springer, New York.

140 Zayats, A.V., Smolyaninov, I.I., and Maradudin, A.A. (2005) *Phys. Rep.*, **408**, 131.

141 Barnes, W.L., Dereux, A., and Ebbesen, T.W. (2003) *Nature*, **424**, 824.

142 Pitarke, J.M., Silkin, V.M., Chulkov, E.V., and Echenique, P.M. (2007) *Rep. Prog. Phys.*, **70**, 1.

143 Jackson, J.D. (1999) *Classical Electrodynamics*, John Wiley & Sons, Inc., New York.

144 Klein, M.V. and Furtak, T.E. (1986) *Optics*, John Wiley & Sons, Inc., New York.

145 Ritchie, R.H. (1957) *Phys. Rev.*, **106**, 874.

146 Otto, A. (1968) *Z. Phys.*, **216**, 398.

147 Kretschmann, E. and Raether, H. (1968) *Z. Naturforsch. A*, **23A**, 2135.

148 Teng, Y.Y. and Stern, E.A. (1967) *Phys. Rev. Lett.*, **19**, 511.
149 Kubo, A., Pontius, N., and Petek, H. (2007) *Nano Lett.*, **7**, 470.
150 Hecht, B., Bielefeldt, H., Novotny, L., Inouye, Y., and Pohl, D.W. (1996) *Phys. Rev. Lett.*, **77**, 1889.
151 Renger, J., Quidant, R., van Hulst, N., and Novotny, L. (2010) *Phys. Rev. Lett.*, **104**, 046803.
152 Pendry, J.B., Martin-Moreno, L., and Garcia-Vidal, F.J. (2004) *Science*, **305**, 847.
153 Hibbins, A.P., Evans, B.R., and Sambles, J.R. (2005) *Science*, **308**, 670.
154 Bastard, G. (1988) *Wave Mechanics Applied to Semiconductor Heterostructures*, Halsted Press, New York.
155 Joannopoulos, J.D., Meade, R.D., and Winn, J.N. (1995) *Photonic Crystals: Molding the Flow of Light*, Princeton University Press, Princeton, NJ.
156 Pendry, J.B. (2000) *Phys. Rev. Lett.*, **85**, 3966.
157 Pendry, J.B., Schurig, D., and Smith, D.R. (2006) *Science*, **312**, 1780.
158 Sarid, D. (1981) *Phys. Rev. Lett.*, **47**, 1927.
159 Lamprecht, B., Krenn, J.R., Schider, G., Ditlbacher, H., Salerno, M., Felidj, N., Leitner, A., Aussenegg, F.R., and Weeber, J.C. (2001) *Appl. Phys. Lett.*, **79**, 51.
160 Berini, P. (1999) *Opt. Lett.*, **24**, 1011.
161 Berini, P. (2000) *Phys. Rev. B*, **61**, 10484.
162 Berini, P. (2001) *Phys. Rev. B*, **63**, 125417.
163 Kim, D.S. *et al.* (2003) *Phys. Rev. Lett.*, **91**, 143901.
164 Bergman, D.J. and Stockman, M.I. (2003) *Phys. Rev. Lett.*, **90**, 027402.
165 Oulton, R.F., Sorger, V.J., Zentgraf, T., Ma, R.M., Gladden, C., Dai, L., Bartal, G., and Zhang, X. (2009) *Nature*, **461**, 629.
166 Aigouy, L., Lalanne, P., Hugonin, J.P., Julié, G., Mathet, V., and Mortier, M. (2007) *Phys. Rev. Lett.*, **98**, 153902.
167 Meyer zu Heringdorf, F.J., Chelaru, L.I., Möllenbeck, S., Thien, D., and Horn-von Hoegen, M. (2007) *Surf. Sci.*, **601**, 4700.
168 Vesseur, E.J.R., de Waele, R., Lezec, H.J., Atwater, H.A., de Abajo, F.J.G., and Polman, A. (2008) *Appl. Phys. Lett.*, **92**, 083110.
169 Kubo, A., Onda, K., Petek, H., Sun, Z.J., Jung, Y.S., and Kim, H.K. (2005) *Nano Lett.*, **5**, 1123.
170 Stockman, M.I., Kling, M.F., Kleineberg, U., and Krausz, F. (2007) *Nat. Photonics*, **1**, 539.
171 Atwater, H.A. and Polman, A. (2010) *Nat. Mater.*, **9**, 205.
172 Willets, K.A. and Van Duyne, R.P. (2007) *Annu. Rev. Phys. Chem.*, **58**, 267.
173 Kim, K.H., Watanabe, K., Mulugeta, D., Freund, H.-J., and Menzel, D. (2011) *Phys. Rev. Lett.*, **107**, 047401.
174 Bohren, C.F. and Huffman, D.R. (1983) *Absorption and Scattering of Light by Small Particles*, John Wiley & Sons, Inc., New York.
175 Novotny, L. and Hecht, B. (2006) *Principles of Nano-Optics*, Cambridge University Press, Cambridge.
176 Jackson, J.D. (1999) *Classical Electrodynamics*, John Wiley & Sons, Inc., New York.
177 Muller, J., Sonnichsen, C., von Poschinger, H., von Plessen, G., Klar, T.A., and Feldmann, J. (2002) *Appl. Phys. Lett.*, **81**, 171.
178 Boyer, D., Tamarat, P., Maali, A., Lounis, B., and Orrit, M. (2002) *Science*, **297**, 1160.
179 Klar, T., Perner, M., Grosse, S., von Plessen, G., Spirkl, W., and Feldmann, J. (1998) *Phys. Rev. Lett.*, **80**, 4249.
180 Lindfors, K., Kalkbrenner, T., Stoller, P., and Sandoghdar, V. (2004) *Phys. Rev. Lett.*, **93**, 037401.
181 Arbouet, A., Christofilos, D., Del Fatti, N., Vallee, F., Huntzinger, J.R., Arnaud, L., Billaud, P., and Broyer, M. (2004) *Phys. Rev. Lett.*, **93**, 127401.
182 van Dijk, M.A., Lippitz, M., and Orrit, M. (2005) *Phys. Rev. Lett.*, **95**, 267406.
183 Muskens, O.L., Billaud, P., Broyer, M., Fatti, N., and Vallee, F. (2008) *Phys. Rev. B*, **78**, 205410.
184 Kubo, A., Jung, Y.S., Kim, H.K., and Petek, H. (2007) *J. Phys. B*, **40**, S259.
185 Munzinger, M., Wiemann, C., Rohmer, M., Guo, L., Aeschlimann, M., and Bauer, M. (2005) *New J. Phys.*, **7**, 1.
186 Meier, M. and Wokaun, A. (1983) *Opt. Lett.*, **8**, 581.

187 Kuwata, H., Tamaru, H., Esumi, K., and Miyano, K. (2003) *Appl. Phys. Lett.*, **83**, 4625.

188 Heilweil, E.J. and Hochstrasser, R.M. (1985) *J. Chem. Phys.*, **82**, 4762.

189 Sonnichsen, C., Franzl, T., Wilk, T., von Plessen, G., Feldmann, J., Wilson, O., and Mulvaney, P. (2002) *Phys. Rev. Lett.*, **88**, 077402.

190 Sonnichsen, C., Franzl, T., Wilk, T., von Plessen, G., and Feldmann, J. (2002) *New J. Phys.*, **4**, 93.1–93.8.

191 Hu, M., Novo, C., Funston, A., Wang, H.N., Staleva, H., Zou, S.L., Mulvaney, P., Xia, Y.N., and Hartland, G.V. (2008) *J. Mater. Chem.*, **18**, 1949.

192 Stietz, F., Bosbach, J., Wenzel, T., Vartanyan, T., Goldmann, A., and Trager, F. (2000) *Phys. Rev. Lett.*, **84**, 5644.

193 Baida, H. *et al.* (2009) *Nano Lett.*, **9**, 3463.

194 Pustovit, V.N. and Shahbazyan, T.V. (2006) *Chem. Phys. Lett.*, **420**, 469.

195 Kawabata, A. and Kubo, R. (1966) *J. Phys. Soc. Jpn.*, **21**, 1765.

196 Yannouleas, C. and Broglia, R.A. (1992) *Ann. Phys.*, **217**, 105.

197 Hovel, H., Fritz, S., Hilger, A., Kreibig, U., and Vollmer, M. (1993) *Phys. Rev. B*, **48**, 18178.

198 Deutsch, B., Hillenbrand, R., and Novotny, L. (2010) *Nano Lett.*, **10**, 652.

199 Ghenuche, P., Cherukulappurath, S., Taminiau, T.H., van Hulst, N.F., and Quidant, R. (2008) *Phys. Rev. Lett.*, **101**, 116805.

200 Sadiq, D., Shirdel, J., Lee, J.S., Selishcheva, E., Park, N., and Lienau, C. (2011) *Nano Lett.*, **11**, 1609.

201 Ropers, C., Solli, D.R., Schulz, C.P., Lienau, C., and Elsaesser, T. (2007) *Phys. Rev. Lett.*, **98**, 043907.

202 Vesseur, E.J.R., de Waele, R., Kuttge, M., and Polman, A. (2007) *Nano Lett.*, **7**, 2843.

203 Nelayah, J. *et al.* (2007) *Nat. Phys*, **3**, 348.

204 Wood, R.W. (1902) *Philos. Mag.*, **3**, 15.

205 Wood, R.W. (1935) *Phys. Rev.*, **48**, 928.

206 Fano, U. (1941) *J. Opt. Soc. Am.*, **31**, 213.

207 Ebbesen, T.W., Lezec, H.J., Ghaemi, H.F., Thio, T., and Wolff, P.A. (1998) *Nature*, **391**, 667.

208 Bethe, H.A. (1944) *Phys. Rev.*, **66**, 163.

209 Bouwkamp, C.J. (1954) *Rep. Prog. Phys.*, **17**, 35.

210 Hohng, S.C. (2002) *et al. Appl. Phys. Lett.*, **81**, 3239.

211 Krishnan, A., Thio, T., Kima, T.J., Lezec, H.J., Ebbesen, T.W., Wolff, P.A., Pendry, J., Martin-Moreno, L., and Garcia-Vidal, F.J. (2001) *Opt. Commun.*, **200**, 1.

212 Ghaemi, H.F., Thio, T., Grupp, D.E., Ebbesen, T.W., and Lezec, H.J. (1998) *Phys. Rev. B*, **58**, 6779.

213 Ropers, C., Park, D.J., Stibenz, G., Steinmeyer, G., Kim, J., Kim, D.S., and Lienau, C. (2005) *Phys. Rev. Lett.*, **94**, 113901.

214 Sarrazin, M., Vigneron, J.P., and Vigoureux, J.M. (2003) *Phys. Rev. B*, **67**, 085415.

215 Genet, C., van Exter, M.P., and Woerdman, J.P. (2003) *Opt. Commun.*, **225**, 331.

216 Cohen-Tannoudji, C., Dupont-Roc, J., and Grynberg, G. (1992) *Atom-Photon Interactions: Basic Processes and Applications*, John Wiley & Sons, Inc., New York.

217 Scully, M.O. and Zubairy, M.S. (1997) *Quantum Optics*, Cambridge University Press, Cambridge.

218 Dicke, R.H. (1954) *Phys. Rev.*, **93**, 99.

219 Garcia-Vidal, F.J., Martin-Moreno, L., Ebbesen, T.W., and Kuipers, L. (2010) *Rev. Mod. Phys.*, **82**, 729.

220 Martin-Moreno, L., Garcia-Vidal, F.J., Lezec, H.J., Pellerin, K.M., Thio, T., Pendry, J.B., and Ebbesen, T.W. (2001) *Phys. Rev. Lett.*, **86**, 1114.

221 Lochbihler, H. (1994) *Phys. Rev. B*, **50**, 4795.

222 Lee, K.G. and Park, Q.H. (2005) *Phys. Rev. Lett.*, **95**, 103902.

223 Muller, R., Malyarchuk, V., and Lienau, C. (2003) *Phys. Rev. B*, **68**, 205415.

224 Muller, R., Ropers, C., and Lienau, C. (2004) *Opt. Express*, **12**, 5067.

225 Lezec, H.J., Degiron, A., Devaux, E., Linke, R.A., Martin-Moreno, L., Garcia-Vidal, F.J., and Ebbesen, T.W. (2002) *Science*, **297**, 820.

226 Tegenfeldt, J.O. *et al.* (2001) *Phys. Rev. Lett.*, **86**, 1378.

227 Neacsu, C.C., Berweger, S., Olmon, R.L., Saraf, L.V., Ropers, C., and Raschke, M.B. (2010) *Nano Lett.*, **10**, 592.

228 Berweger, S., Atkin, J.M., Xu, X.G., Olmon, R.L., and Raschke, M.B. (2011) *Nano Lett.*, **11**, 4309–4313.

229 Ropers, C., Neacsu, C.C., Elsaesser, T., Albrecht, M., Raschke, M.B., and Lienau, C. (2007) *Nano Lett.*, **7**, 2784.

230 Babadjanyan, A.J., Margaryan, N.L., and Nerkararyan, K.V. (2000) *J. Appl. Phys.*, **87**, 3785.

231 Cohen-Tannoudji, C., Diu, B., and Laloë, F. (1977) *Quantum Mechanics*, John Wiley & Sons, Inc., New York.

232 Stockman, M.I., Bergman, D.J., Anceau, C., Brasselet, S., and Zyss, J. (2004) *Phys. Rev. Lett.*, **92**, 057402.

233 Kurihara, K., Yamamoto, K., Takahara, J., and Otomo, A. (2008) *J. Phys. A*, **41**, 295401.

234 Durach, M., Rusina, A., Stockman, M.I., and Nelson, K. (2007) *Nano Lett.*, **7**, 3145.

235 Issa, N.A. and Guckenberger, R. (2007) *Plasmonics*, **2**, 31.

236 Gramotnev, D.K., Vogel, M.W., and Stockman, M.I. (2008) *J. Appl. Phys.*, **104**, 034311.

237 Baida, F.I. and Belkhir, A. (2009) *Plasmonics*, **4**, 51.

238 Gramotnev, D.K. (2005) *J. Appl. Phys.*, **98**, 104302.

239 Verhagen, E., Kuipers, L., and Polman, A. (2007) *Nano Lett.*, **7**, 334.

240 Verhagen, E., Spasenovic, M., Polman, A., and Kuipers, L. (2009) *Phys. Rev. Lett.*, **102**, 203904.

241 Knox, R.S. (1963) *Theory of Excitons*, Academic Press, New York.

242 Yu, P. (2010) *Fundamentals of Semiconductors: Physics and Materials Properties*, Springer, New York.

243 Haug, H. and Koch, S.W. (2009) *Quantum Theory of the Optical and Electronic Properties of Semiconductors*, World Scientific.

244 Davydov, A.S. (1971) *Theory of Molecular Excitons*, Plenum Press, New York.

245 Kobayashi, T. (1996) *J-Aggregates*, World Scientific, Singapore.

246 Olaya-Castro, A. and Scholes, G.D. (2011) *Int. Rev. Phys. Chem.*, **30**, 49.

247 Biolatti, E., Iotti, R.C., Zanardi, P., and Rossi, F. (2000) *Phys. Rev. Lett.*, **85**, 5647.

248 Unold, T., Mueller, K., Lienau, C., Elsaesser, T., and Wieck, A.D. (2005) *Phys. Rev. Lett.*, **94**, 137404.

249 Prodan, E., Radloff, C., Halas, N.J., and Nordlander, P. (2003) *Science*, **302**, 419.

250 Wang, Z.L. and Cowley, J.M. (1987) *Ultramicroscopy*, **23**, 97.

251 Garcia de Abajo, F.J. and Howie, A. (2002) *Phys. Rev. B*, **65**, 115418.

252 Prodan, E. and Nordlander, P. (2004) *J. Chem. Phys.*, **120**, 5444.

253 Nordlander, P., Oubre, C., Prodan, E., Li, K., and Stockman, M.I. (2004) *Nano Lett.*, **4**, 899.

254 Fofang, N.T., Park, T.H., Neumann, O., Mirin, N.A., Nordlander, P., and Halas, N.J. (2008) *Nano Lett.*, **8**, 3481.

255 Lidzey, D.G., Bradley, D.D.C., Armitage, A., Walker, S., and Skolnick, M.S. (2000) *Science*, **288**, 1620.

256 Vasa, P. *et al.* (2010) *ACS Nano*, **4**, 7559.

257 Wiederrecht, G.P., Wurtz, G.A., and Hranisavljevic, J. (2004) *Nano Lett.*, **4**, 2121.

258 Bellessa, J., Bonnand, C., Plenet, J.C., and Mugnier, J. (2004) *Phys. Rev. Lett.*, **93**, 036404.

259 Dintinger, J., Klein, S., Bustos, F., Barnes, W.L., and Ebbesen, T.W. (2005) *Phys. Rev. B*, **71**, 035424.

Index

Dynamics at Solid State Surfaces and Interfaces: Volume 2: Fundamentals, First Edition.
Edited by Uwe Bovensiepen, Hrvoje Petek, and Martin Wolf.